Fundamentals of Creep in Metals and Alloys

Fundamentals of Creep in Metals and Alloys

Michael E. Kassner
Department of Aerospace and Mechanical Engineering
University of Southern California
Los Angeles, USA

María-Teresa Pérez-Prado
Department of Physical Metallurgy
Centro Nacional de Investigaciones Metalúrgicas (CENIM)
Madrid, Spain

2004

ELSEVIER
Amsterdam – Boston – Heidelberg – London – New York – Oxford
Paris – San Diego – San Francisco – Singapore – Sydney – Tokyo

ELSEVIER B.V.
Sara Burgerhartstraat 25
P.O. Box 211
1000 AE Amsterdam
The Netherlands

ELSEVIER Inc.
525 B Street, Suite 1900
San Diego
CA 92101-4495
USA

ELSEVIER Ltd
The Boulevard, Langford Lane
Kidlington,
Oxford OX5 1GB
UK

ELSEVIER Ltd
84 Theobalds Road
London
WC1X 8RR
UK

© 2004 Elsevier Ltd All rights reserved.

This work is protected under copyright by Elsevier Ltd, and the following terms and conditions apply to its use:

Photocopying
Single photocopies of single chapters may be made for personal use as allowed by national copyright laws. Permission of the Publisher and payment of a fee is required for all other photocopying, including multiple or systematic copying, copying for advertising or promotional purposes, resale, and all forms of document delivery. Special rates are available for educational institutions that wish to make photocopies for non-profit educational classroom use.

Permissions may be sought directly from Elsevier's Rights Department in Oxford, UK: phone (+44) 1865 843830, fax (+44) 1865 853333, e-mail: permissions@elsevier.com. Requests may also be completed on-line via the Elsevier homepage (http://www.elsevier.com/locate/permissions).

In the USA, users may clear permissions and make payments through the Copyright Clearance Center, Inc., 222 Rosewood Drive, Danvers, MA 01923, USA; phone: (+1) (978) 7508400, fax: (+1) (978) 7504744, and in the UK through the Copyright Licensing Agency Rapid Clearance Service (CLARCS), 90 Tottenham Court Road, London W1P 0LP, UK; phone: (+44) 20 7631 5555; fax: (+44) 20 7631 5500. Other countries may have a local reprographic rights agency for payments.

Derivative Works
Tables of contents may be reproduced for internal circulation, but permission of the Publisher is required for external resale or distribution of such material. Permission of the Publisher is required for all other derivative works, including compilations and translations.

Electronic Storage or Usage
Permission of the Publisher is required to store or use electronically any material contained in this work, including any chapter or part of a chapter.

Except as outlined above, no part of this work may be reproduced, stored in a retrieval system or transmitted in any form or by any means, electronic, mechanical, photocopying, recording or otherwise, without prior written permission of the Publisher.
Address permissions requests to: Elsevier's Rights Department, at the fax and e-mail addresses noted above.

Notice
No responsibility is assumed by the Publisher for any injury and/or damage to persons or property as a matter of products liability, negligence or otherwise, or from any use or operation of any methods, products, instructions or ideas contained in the material herein. Because of rapid advances in the medical sciences, in particular, independent verification of diagnoses and drug dosages should be made.

First edition 2004

Library of Congress Cataloging in Publication Data
A catalog record is available from the Library of Congress.

British Library Cataloguing in Publication Data
A catalogue record is available from the British Library.

ISBN: 0-08-043637-4

⊗ The paper used in this publication meets the requirements of ANSI/NISO Z39.48-1992 (Permanence of Paper).
Transferred to Digital Printing 2006

Preface

This book on the fundamentals of creep plasticity is both a review and a critical analysis of investigations in a variety of areas relevant to creep plasticity. These areas include five-power-law creep, which is sometimes referred to as dislocation climb-controlled creep, viscous glide or three-power-law creep in alloys, diffusional creep, Harper–Dorn creep, superplasticity, second-phase strengthening, and creep cavitation and fracture. Many quality reviews and books precede this attempt to write an extensive review of creep fundamentals and improvement was a challenge. One advantage with this attempt is the ability to describe the substantial work published subsequent to these earlier reviews. An attempt was made to cover the basic work discussed in these earlier reviews but especially to emphasize more recent developments.

The author wishes to acknowledge support from the U.S. Department of Energy, Basic Energy Sciences under contract DE-FG03-99ER45768. Comments by Profs. F.R.N. Nabarro, W. Blum, T.G. Langdon, J. Weertman, S. Spigarelli, J.H. Schneibel and O. Ruano are appreciated. The assistance with the preparation of the figures by T.A. Hayes and C. Daraio is greatly appreciated. Word processing by Ms. Peggy Blair was vital.

M.E. Kassner
M.T. Pérez-Prado

Contents

Preface	v
List of Symbols and Abbreviations	xi
1. Introduction	1
1.1. Description of Creep	3
1.2. Objectives	7
2. Five-Power-Law Creep	11
2.1. Macroscopic Relationships	13
2.1.1 Activation Energy and Stress Exponents	13
2.1.2 Influence of the Elastic Modulus	20
2.1.3 Stacking Fault Energy and Summary	24
2.1.4 Natural Three-Power-Law	29
2.1.5 Substitutional Solid Solutions	30
2.2. Microstructural Observations	31
2.2.1 Subgrain Size, Frank Network Dislocation Density, Subgrain Misorientation Angle, and the Dislocation Separation within the Subgrain Walls in Steady-State Structures	31
2.2.2 Constant Structure Equations	39
2.2.3 Primary Creep Microstructures	47
2.2.4 Creep Transient Experiments	51
2.2.5 Internal Stress	55
2.3. Rate-Controlling Mechanisms	60
2.3.1 Introduction	60
2.3.2 Dislocation Microstructure and the Rate-Controlling Mechanism	67
2.3.3 In situ and Microstructure-Manipulation Experiments	70
2.3.4 Additional Comments on Network Strengthening	71
2.4. Other Effects on Five-Power-Law Creep	77
2.4.1 Large Strain Creep Deformation and Texture Effects	77
2.4.2 Effect of Grain Size	82
2.4.3 Impurity and Small Quantities of Strengthening Solutes	84
2.4.4 Sigmoidal Creep	87
3. Diffusional-Creep	89
4. Harper–Dorn Creep	97
4.1. The Size Effect	103

	4.2.	The Effect of Impurities	106
5.	Three-Power-Law Viscous Glide Creep		109
6.	Superplasticity		121
	6.1.	Introduction	123
	6.2.	Characteristics of Fine Structure Superplasticity	123
	6.3.	Microstructure of Fine Structure Superplastic Materials	127
		6.3.1 Grain Size and Shape	127
		6.3.2 Presence of a Second Phase	127
		6.3.3 Nature and Properties of Grain Boundaries	127
	6.4.	Texture Studies in Superplasticity	128
	6.5.	High Strain-Rate Superplasticity	128
		6.5.1 High Strain-Rate Superplasticity in Metal–Matrix Composites	129
		6.5.2 High Strain-Rate Superplasticity in Mechanically Alloyed Materials	134
	6.6.	Superplasticity in Nano and Submicrocrystalline Materials	136
7.	Recrystallization		141
	7.1.	Introduction	143
	7.2.	Discontinuous Dynamic Recrystallization (DRX)	145
	7.3.	Geometric Dynamic Recrystallization	146
	7.4.	Particle-Stimulated Nucleation (PSN)	147
	7.5.	Continuous Reactions	147
8.	Creep Behavior of Particle-Strengthened Alloys		149
	8.1.	Introduction	151
	8.2.	Small Volume-Fraction Particles that are Coherent and Incoherent with the Matrix with Small Aspect Ratios	151
		8.2.1 Introduction and Theory	151
		8.2.2 Local and General Climb of Dislocations over Obstacles	155
		8.2.3 Detachment Model	158
		8.2.4 Constitutive Relationships	162
		8.2.5 Microstructural Effects	166
		8.2.6 Coherent Particles	168
9.	Creep of Intermetallics		171
	9.1.	Introduction	173
	9.2.	Titanium Aluminides	175
		9.2.1 Introduction	175
		9.2.2 Rate Controlling Creep Mechanisms in FL TiAl Intermetallics During "Secondary" Creep	178
		9.2.3 Primary Creep in FL Microstructures	186
		9.2.4 Tertiary Creep in FL Microstructures	188

	9.3.	Iron Aluminides	188
		9.3.1 Introduction	188
		9.3.2 Anomalous Yield Point Phenomenon	190
		9.3.3 Creep Mechanisms	194
		9.3.4 Strengthening Mechanisms	197
	9.4.	Nickel Aluminides	198
		9.4.1 Ni_3Al	198
		9.4.2 NiAl	208
10.	Creep Fracture	213
	10.1.	Background	215
	10.2.	Cavity Nucleation	218
		10.2.1 Vacancy Accumulation	218
		10.2.2 Grain-Boundary Sliding	221
		10.2.3 Dislocation Pile-ups	222
		10.2.4 Location	224
	10.3.	Growth	225
		10.3.1 Grain Boundary Diffusion-Controlled Growth	225
		10.3.2 Surface Diffusion-Controlled Growth	228
		10.3.3 Grain-Boundary Sliding	229
		10.3.4 Constrained Diffusional Cavity Growth	229
		10.3.5 Plasticity	234
		10.3.6 Coupled Diffusion and Plastic Growth	234
		10.3.7 Creep Crack Growth	237
	10.4.	Other Considerations	240
References			243
Index			269

List of Symbols and Abbreviations

a	cavity radius
a_0	lattice parameter
$A'-A''''$	constants
A	
A_F	
A_{C-J}	solute dislocation interaction parameters
A_{APB}	
A_{SN}	
A_{CR}	
A_{gb}	grain boundary area
A_{HD}	Harper–Dorn equation constant
A_{PL}	constants
A_v	projected area of void
A_0-A_{12}	constants
b	Burgers vector
B	constant
c	concentration of vacancies
c^*	crack growth rate
c_j	concentration of jogs
c_p	concentration of vacancies in the vicinity of a jog
c_p^*	steady-state vacancy concentration near a jog
c_v	equilibrium vacancy concentration
c_v^D	vacancy concentration near a node or dislocation
c_0	initial crack length
C	concentration of solute atoms
C^*	integral for fracture mechanics of time-dependent plastic materials
C_{1-2}	constant
C_{LM}	Larson–Miller constant
CBED	convergent beam electron diffraction
CGBS	cooperative grain-boundary sliding
CS	crystallographic slip
CSL	coincident site lattice
C_0^*	constant
C_1-C_5	constants
d	average spacing of dislocations that comprise a subgrain boundary
D	general diffusion coefficient
D_c	diffusion coefficient for climb

D_{eff}	effective diffusion coefficient
D_g	diffusion coefficient for glide
D_{gb}	diffusion coefficient along grain boundaries
D_i	interfacial diffusion
D_s	surface diffusion coefficient
D_{sd}	lattice self-diffusion coefficient
D_v	diffusion coefficient for vacancies
D_0	diffusion constant
DRX	discontinuous dynamic recrystallization
$\tilde{D}$	diffusion coefficient for the solute atoms
e	solute-solvent size difference or misfit parameter
E	Young's modulus
E_j	formation energy for a jog
EBSP	electron backscatter patterns
f	fraction
f_m	fraction of mobile dislocations
f_p	chemical dragging force on a jog
f_{sub}	fraction of material occupied by subgrains
F	total force per unit length on a dislocation
g	average grain size (diameter)
g'	constant
G	shear modulus
GBS	grain-boundary sliding
GDX	geometric dynamic recrystallization
GNB	geometrically necessary boundaries
h_r	hardening rate
h_m	average separation between slip planes within a subgrain with gliding dislocations
HAB	high-angle boundary
HVEM	high-voltage transmission electron microscopy
j	jog spacing
J	J integral for fracture mechanics of plastic material
J_{gb}	vacancy flux along a grain boundary
k	Boltzmann constant
$k' - k'''$	constants
k_y	Hall–Petch constant
k_{MG}	Monkman–Grant constant
k_R	relaxation factor
k_1–k_{10}	constants
K	constant

List of Symbols and Abbreviations

K_I	stress intensity factor
$K_0 - K_7$	constants
ℓ	link length of a Frank dislocation network
ℓ_c	critical link length to unstably bow a pinned dislocation
ℓ_m	maximum link length
l	migration distance for a dislocation in Harper–Dorn creep
L	particle separation distance
LAB	low-angle boundary
LM	Larson–Miller parameter
m	strain-rate sensitivity exponent ($= 1/N$)
m′	transient creep time exponent
m″	strain-rate exponent in the Monkman–Grant equation
m_c	constant
$\overline{M}$	average Taylor factor for a polycrystal
M_ρ	dislocation multiplication constant
n	steady-state creep exponent
n*	equilibrium concentration of critical-sized nuclei
n_m	steady-state stress exponent of the matrix in a multi-phase material
N	constant structure stress exponent
$\dot{N}$	nucleation rate
p	steady-state dislocation density stress exponent
p′	inverse grain size stress exponent for superplasticity
PLB	power law breakdown
POM	polarized light optical microscopy
PSB	persistent slip band
q	dislocation spacing, d, stress exponent
Q_c	activation energy for creep (with E or G compensation)
Q'_c	apparent activation energy for creep (no E or G compensation)
Q_p	activation energy for dislocation pipe diffusion
Q_{sd}	activation energy for lattice self-diffusion
Q_v	formation energy for a vacancy
$Q*$	effective activation energies in composites where load transfer occurs
r_r	recovery rate
R_0	diffusion distance
R_s	radius of solvent atoms
s	structure
SAED	selected area electron diffraction
t	time
t_c	time for cavity coalescence on a grain-boundary facet
t_f	time to fracture (rupture)

t_s	time to the onset of steady-state
T	temperature
T_d	dislocation line tension
T_m	melting temperature
T_p	temperature of the peak yield strength
TEM	transmission electron microscopy
v	dislocation glide velocity
v_c	dislocation climb velocity
v_{cr}	critical dislocation velocity at breakaway
v_D	Debye frequency
v_p	jog climb velocity
$\bar{v}$	average dislocation velocity
$\bar{v}_\ell$	climb velocity of dislocation links of a Frank network
w	width of a grain boundary
$\bar{x}_c$	average dislocation climb distance
$\bar{x}_g$	average dislocation slip length due to glide
XRD	x-ray diffraction
α	Taylor equation constant
α'	climb resistance parameter
α_{1-3}	constants
β_{1-3}	constants
γ	shear strain
γ_A	anelastic unbowing strain
γ_{gb}	interfacial energy of a grain boundary
γ_m	surface energy of a metal
$\dot{\gamma}$	shear creep-rate
$\dot{\gamma}_{ss}$	steady-state shear creep-rate
δ	grain boundary thickness
Δa	activation area
ΔG	Gibbs free energy
ΔV_C	activation volume for creep
ΔV_L	activation volume for lattice self-diffusion
ε	uniaxial strain
ε_0	instantaneous strain
$\dot{\varepsilon}$	strain-rate
$\dot{\varepsilon}_{min}$	minimum creep-rate
$\dot{\varepsilon}_{ss}$	steady-state uniaxial strain-rate
$\bar{\varepsilon}$	effective uniaxial or von Mises strain
θ	misorientation angle across high-angle grain boundaries
θ_λ	misorientation angle across (low-angle) subgrain boundaries

List of Symbols and Abbreviations

$\theta_{\lambda\mathrm{ave}}$	average misorientation angle across (low-angle) subgrain boundaries
λ	average subgrain size (usually measured as an average intercept)
λ_s	cavity spacing
λ_{ss}	average steady-state subgrain size
ν	Poisson's ratio
ρ	density of dislocations not associated with subgrain boundaries
ρ_m	mobile dislocation density
ρ_{ms}	mobile screw dislocation density
ρ_{ss}	steady-state dislocation density not associated with subgrain walls
σ	applied uniaxial stress
σ_i	internal stress
σ_0	single crystal yield strength
σ'_0	annealed polycrystal yield strength
σ''_0	sintering stress for a cavity
σ_p	Peierls stress
σ_{ss}	uniaxial steady-state stress
σ_T	transition stress between five-power-law and Harper–Dorn creep
σ_{TH_s}	threshold stress for superplastic deformation
$\sigma_y\vert_{T,\dot{\varepsilon}}$	yield or flow stress at a reference temperature and strain-rate
$\bar{\sigma}$	effective uniaxial, or von Mises, stress
τ	shear stress
τ_b	breakaway stress of the dislocations from solute atmospheres
τ_c	critical stress for climb over a second-phase particle
τ_d	detachment stress from a second-phase particle
τ_j	stress to move screw dislocations with jogs
τ_{or}	Orowan bowing stress
τ_B	shear stress necessary to eject dislocation from a subgrain boundary
τ_{BD}	maximum stress from a simple tilt boundary
τ_L	stress to move a dislocation through a boundary resulting from jog creation
τ_N	shear strength of a Frank network
$(\tau/G)_t$	normalized transition stress
$\phi(P)$	Frank network frequency distribution function
χ	stacking fault energy
χ'	primary creep constant
ψ	angle between cavity surface and projected grain-boundary surface
ω_m	maximum interaction energy between a solute atom and an edge dislocation
Ω	atomic volume
ω	fraction of grain-boundary area cavitated

Chapter 1
Introduction

1.1. Description of Creep   3
1.2. Objectives   7

Chapter 1
Introduction

1.1 DESCRIPTION OF CREEP

Creep of materials is classically associated with time-dependent plasticity under a fixed stress at an elevated temperature, often greater than roughly 0.5 T_m, where T_m is the absolute melting temperature. The plasticity under these conditions is described in Figure 1 for constant stress (a) and constant strain-rate (b) conditions. Several aspects of the curve in Figure 1 require explanation. First, three regions are delineated: Stage I, or primary creep, which denotes that portion where [in (a)] the creep-rate (plastic strain-rate), $\dot{\varepsilon} = d\varepsilon/dt$ is changing with increasing plastic strain or time. In Figure 1(a) the primary-creep-rate decreases with increasing strain, but with some types of creep, such as solute drag with "3-power creep," an "inverted" primary occurs where the strain-rate increases with strain. Analogously, in (b), under constant strain-rate conditions, the metal hardens, resulting in increasing flow stresses. Often, in pure metals, the strain-rate decreases or the stress increases to a value that is constant over a range of strain. The phenomenon is termed Stage II, secondary, or steady-state creep. Eventually, cavitation and/or cracking increase the apparent strain-rate or decrease the flow stress. This regime is termed Stage III, or tertiary creep, and leads to fracture. Sometimes, Stage I leads directly to Stage III and an "inflection" is observed. Thus, care must sometimes be exercised in concluding a mechanical steady-state (ss).

The term "creep" as applied to plasticity of materials likely arose from the observation that at modest and constant stress, at or even below the macroscopic yield stress of the metal (at a "conventional" strain-rate), plastic deformation occurs over time as described in Figure 1(a). This is in contrast with the *general* observation, such as at ambient temperature, where, a material deformed at, for example, 0.1–0.3 T_m, shows very little plasticity under constant stress at or below the yield stress, again, at "conventional" or typical tensile testing strain-rates (e.g., 10^{-4}–10^{-3}s^{-1}). {The latter observation is not always true as it has been observed that some primary creep is observed (e.g., a few percent strain, or so) over relatively short periods of time at stresses less than the yield stress in some "rate-sensitive" and relatively low strain-hardening alloys such as titanium [1] and steels [2].}

We observe in Figure 2 that at the "typical" testing strain rate of about 10^{-4}s^{-1}, the yield stress is σ_{y_1}. However, if we decrease the testing strain-rate to, for example, 10^{-7}s^{-1}, the yield stress decreases significantly, as will be shown is common for metals and alloys at high temperatures. To a "first approximation," we might

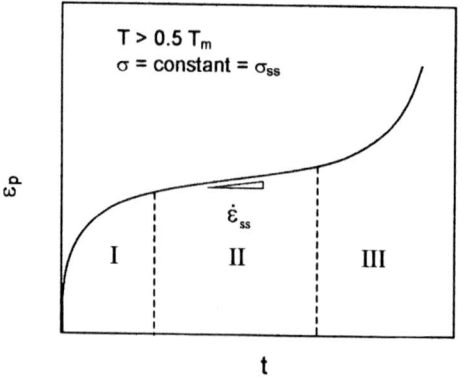

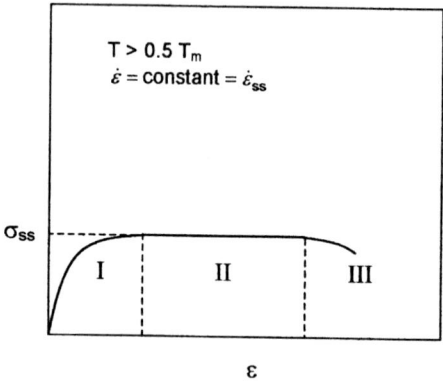

Figure 1. Constant true stress and constant strain-rate creep behavior in pure and Class M (or Class I) metals.

consider the microstructure (created by dislocation microstructure evolution with plasticity) at just 0.002 plastic strain to be independent of $\dot{\varepsilon}$. In this case, we might describe the change in yield stress to be the sole result of the $\dot{\varepsilon}$ change and predicted by the "constant structure" stress-sensitivity exponent, N, defined by

$$N = [\partial \ln \dot{\varepsilon} / \partial \ln \sigma]_{T,s} \tag{1}$$

where T and s refer to temperature and the substructural features, respectively. Sometimes, the sensitivity of the creep rate to changes in stress is described by a constant structure strain-rate sensitivity exponent, $m = 1/N$. Generally, N is relatively high at lower temperatures [3] which implies that significant changes in the strain-rate do not dramatically affect the flow stress. In pure fcc metals,

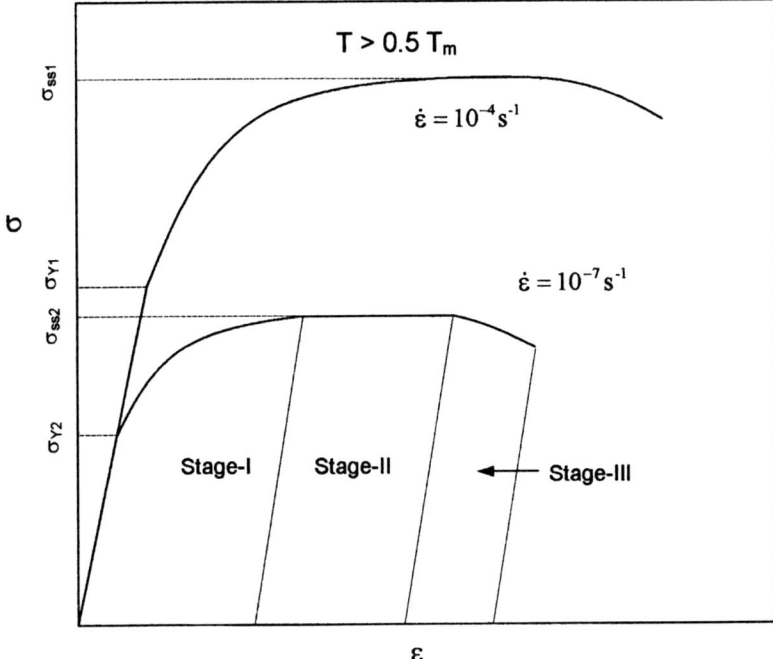

Figure 2. Creep behavior at two different constant strain-rates.

N is roughly between 50 and 250 [3]. At higher temperatures, the values may approach 10, or so [3–10]. N is graphically described in Figure 3. The trends of N versus temperature for nickel are illustrated in Figure 4.

Another feature of the hypothetical behaviors in Figure 2 is that (at the identical temperature) not only is the yield stress at a strain rate of $10^{-7}\,\text{s}^{-1}$ lower than at $10^{-4}\,\text{s}^{-1}$, but also the peak stress or, perhaps, steady-state stress, which is maintained over a substantial strain range, is *less* than the yield stress at a strain rate of $10^{-4}\,\text{s}^{-1}$. (Whether steady-state occurs at, for example, ambient temperature has not been fully settled, as large strains are not easily achievable. Stage IV and/or recrystallization may preclude this steady-state [11–13]). Thus, if a constant stress σ_{ss_2} is applied to the material then a substantial strain may be easily achieved at a low strain-rate despite the stress being substantially below the "conventional" yield stress at the higher rate of $10^{-4}\,\text{s}^{-1}$. Thus, creep is, basically, a result of significant strain-rate sensitivity together with low strain hardening. We observe in Figure 4 that N decreases to relatively small values above about $0.5\,T_m$, while N is relatively high below about this temperature. This implies that we would expect that "creep" would

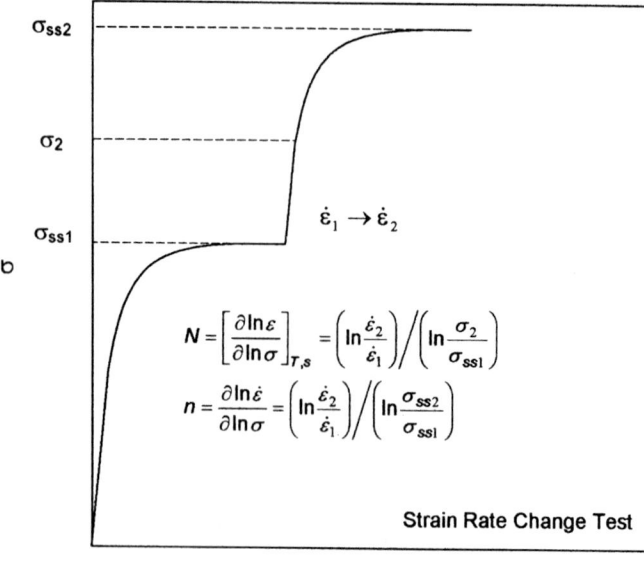

Figure 3. A graphical description of the constant-structure strain-rate sensitivity experiment, $N\ (=1/m)$ and the steady-state stress exponent, n.

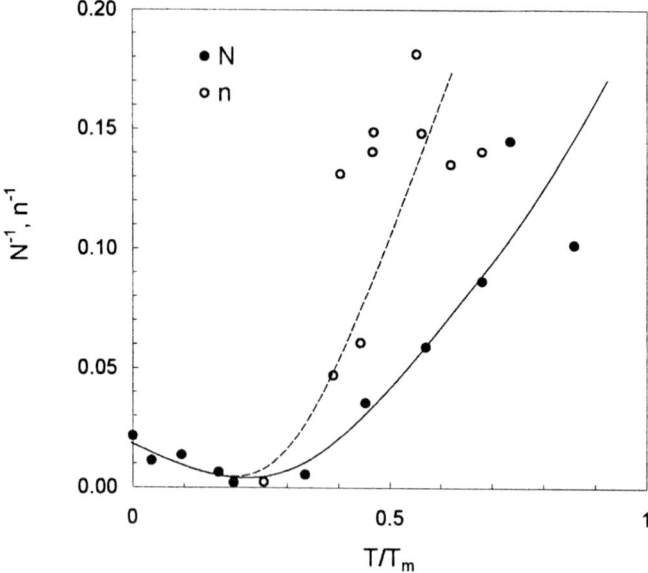

Figure 4. The values of n and N as a function of temperature for nickel. Data from Ref. [7].

be more pronounced at higher temperatures, and less obvious at lower temperatures, since, as will be shown subsequently, work-hardening generally diminishes with increasing temperature and N also decreases (more strain-rate sensitive). The above description/explanation for creep is consistent with earlier descriptions [14]. Again, it should be emphasized that the maximum stress, σ_{ss_2}, in a constant strain rate ($\dot{\varepsilon}$) test, is often referred to as a steady-state stress (when it is the result of a balance of hardening and dynamic recovery processes, which will be discussed later). The creep rate of $10^{-7} s^{-1}$ that leads to the steady-state stress (σ_{ss_2}) is the *same* creep-rate that would be achieved in a constant stress test at σ_{ss_2}. Hence, at σ_{ss_2}, $10^{-7} s^{-1}$ is the steady-state creep rate. The variation in the steady-state creep rate with the applied stress is often described by the steady-state stress exponent, n, defined by

$$n = [\delta \ln \dot{\varepsilon}_{ss} / \delta \ln \sigma_{ss}]_T \qquad (2)$$

This exponent is described in Figure 3. Of course, with hardening, n is expected to be less than N. This is illustrated in Figures 3 and 4. As just mentioned, generally, the lower the strain-rate, or higher the temperature, the less pronounced the strain hardening. This is illustrated in Figure 5, reproduced from [15], where the stress versus strain behavior of high-purity aluminum is illustrated over a wide range of temperatures and strain-rates. All these tests utilize a constant strain-rate. The figure shows that with increasing temperature, the yield stress decreases, as expected. Also, for a given temperature, increases in strain rate are associated with increases in the yield stress of the annealed aluminum. That is, increases in temperature and strain-rate tend to oppose each other with respect to flow stress. This can be rationalized by considering plasticity to be a thermally activated process. Figure 5 also illustrates that hardening is more dramatic at lower temperatures (and higher strain-rates). The general trend that the strain increases with increasing stress to achieve steady-state (decreasing temperature and/or increasing strain-rate) is also illustrated.

1.2 OBJECTIVES

There have been other, often short, reviews of creep, notably, Sherby and Burke [16], Takeuchi and Argon [17], Argon [18], Orlova and Cadek [19], Cadek [20], Mukherjee, [21], Blum [22], Nabarro and de Villiers [23], Weertman [24,25], Nix and Ilschner [26], Nix and Gibeling [27], Evans and Wilshire [28], Kassner and Pérez-Prado [29] and others [30–32]. These, however, often do not include some important recent work, and have sometimes been relatively brief (and, as a result, are

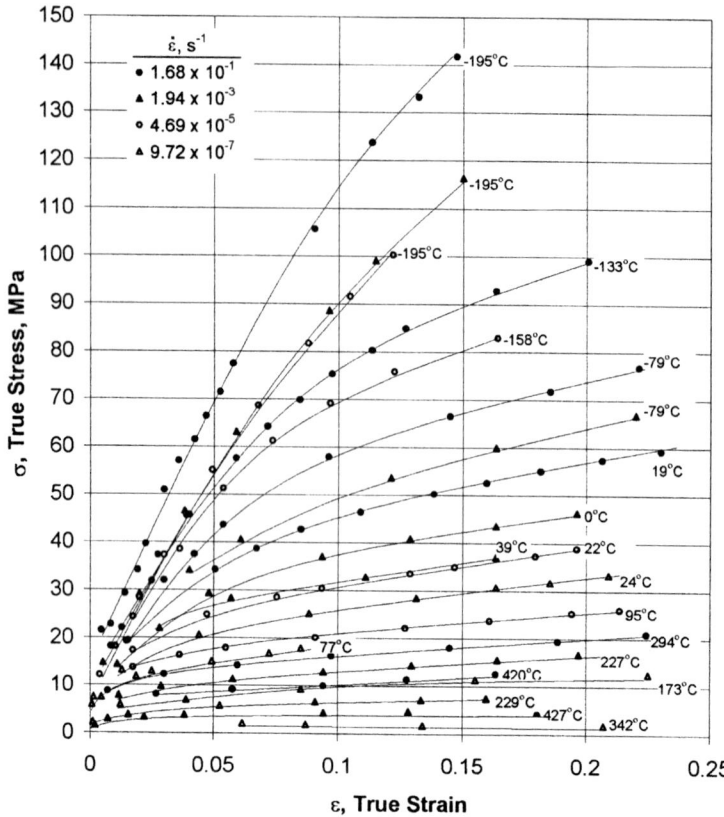

Figure 5. The stress versus strain behavior of high-purity aluminum. Data from Ref. [15].

not always very comprehensive). Thus, it was believed important to provide a new description of creep that is both extensive, current and balanced. Creep is discussed in the context of traditional Five-Power-Law Creep, Nabarro-Herring, Coble, diffusional creep, Harper-Dorn, low-temperature creep (power-law-breakdown or PLB) as well as with 3-power Viscous Glide Creep. Each will be discussed separately. Figure 6 shows a deformation map of silver [33]. Here, several deformation regimes are illustrated as a function of temperature and grain size. Five-Power-Law Creep is indicated by the "dislocation creep" regime bounded by diffusional creep (Coble and Nabarro-Herring) and "dislocation glide" at low temperatures and high stress. Deformation maps have been formulated for a variety of metals [33]. Additionally, Superplasticity particle-strengthening in creep and Creep Fracture will be discussed.

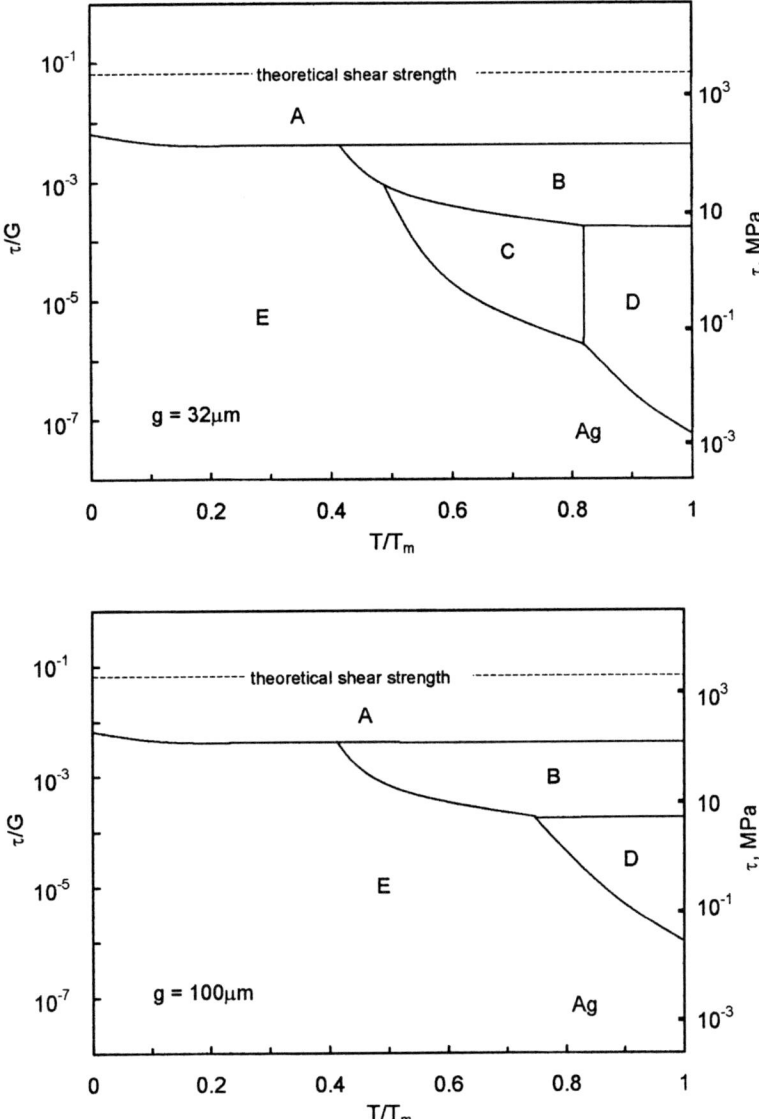

Figure 6. Ashby deformation map of silver from [33]. grain sizes 32 and 100 μm, $\dot{\varepsilon} = 10^{-8}\,\mathrm{s}^{-1}$, A – dislocation glide, B – Five-Power-Law Creep, C – Coble creep, D – Nabarro-Herring creep, E – elastic deformation.

Chapter 2
Five-Power-Law-Creep

2.1.	Macroscopic Relationships	13
	2.1.1. Activation Energy and Stress Exponents	13
	2.1.2. Influence of the Elastic Modulus	20
	2.1.3. Stacking Fault Energy and Summary	24
	2.1.4. Natural Three-Power-Law	29
	2.1.5. Substitutional Solid Solutions	30
2.2.	Microstructural Observations	31
	2.2.1. Subgrain Size, Frank Network Dislocation Density, Subgrain Misorientation Angle, and the Dislocation Separation within the Subgrain Walls in Steady-State Structures	31
	2.2.2. Constant Structure Equations	39
	2.2.3. Primary Creep Microstructures	47
	2.2.4. Creep Transient Experiments	51
	2.2.5. Internal Stress	55
2.3.	Rate-Controlling Mechanisms	60
	2.3.1. Introduction	60
	2.3.2. Dislocation Microstructure and the Rate-Controlling Mechanism	67
	2.3.3. In situ and Microstructure-Manipulation Experiments	70
	2.3.4. Additional Comments on Network Strengthening	71
2.4.	Other Effects on Five-Power-Law-Creep	77
	2.4.1. Large Strain Creep Deformation and Texture Effects	77
	2.4.2. Effect of Grain Size	82
	2.4.3. Impurity and Small Quantities of Strengthening Solutes	84
	2.4.4. Sigmoidal Creep	87

Chapter 2
Five-Power-Law-Creep

2.1 MACROSCOPIC RELATIONSHIPS

2.1.1 Activation Energy and Stress Exponents

In pure metals and Class M alloys (similar creep behavior similar to pure metals), there is an established, largely phenomenological, relationship between the steady-state strain-rate, $\dot{\varepsilon}_{ss}$, (or creep rate) and stress, σ_{ss}, for steady-state 5-power-law (PL) creep:

$$\dot{\varepsilon}_{ss} = A_0 \exp[-Q_c/kT](\sigma_{ss}/E)^n \tag{3}$$

where A_0 is a constant, k is Boltzmann's constant, and E is Young's modulus (although, as will be discussed subsequently, the shear modulus, G, can also be used). This is consistent with Norton's Law [34]. The activation energy for creep, Q_c, has been found to often be about that of lattice self-diffusion, Q_{sd}. The exponent n is constant and is about 5 over a relatively wide range of temperatures and strain-rates (hence "five-power-law" behavior) until the temperature decreases below roughly 0.5–0.6 T_m, where power-law-breakdown (PLB) occurs, and n increases and Q_c generally decreases. steady-state creep is often emphasized over primary or tertiary creep due to the relatively large fraction of creep life within this regime. The importance of steady-state is evidenced by the empirical Monkman–Grant relationship [35]:

$$\dot{\varepsilon}_{ss}^{m''} t_f = k_{MG} \tag{4}$$

where t_f is the time to rupture and k_{MG} is a constant.

A hyperbolic sine (sinh) function is often used to describe the transition from PL to PLB.

$$\dot{\varepsilon}_{ss} = A_1 \exp[-Q_c/kT][\sinh \alpha_1(\sigma_{ss}/E)]^5 \tag{5}$$

(although some have suggested that there is a transition from 5 to 7-power-law behavior prior to PLB [25,36], and this will be discussed more later). Equations (3) and (5) will be discussed in detail subsequently. The discussion of 5-power-law creep will be accompanied by a significant discussion of the lower temperature companion, PLB.

As discussed earlier, time-dependent plasticity or creep is traditionally described as a permanent or plastic extension of the material under fixed applied stress. This is usually illustrated for pure metals or Class M alloys (again, similar quantitative behavior to pure metals) by the constant-stress curve of Figure 1, which also illustrates, of course, that creep plasticity can occur under constant strain-rate conditions as well. Stage I, or primary creep, occurs when the material experiences hardening through changes in the dislocation substructure. Eventually Stage II, or secondary, or steady-state creep, is observed. In this region, hardening is balanced by dynamic recovery (e.g., dislocation annihilation). The consequence of this is that the creep-rate or plastic strain-rate is constant under constant true von Mises stress (tension, compression or torsion). In a constant strain-rate test, the flow stress is independent of plastic strain except for changes in texture (e.g., changes in the average Taylor factor of a polycrystal), often evident in larger strain experiments (such as $\varepsilon > 1$) [37–39]. It will be illustrated that a genuine mechanical steady-state is achievable. As mentioned earlier, this stage is particularly important as large strains can accumulate during steady-state at low, constant, stresses leading to failure.

Since Stage II or steady-state creep is important, the creep behavior of a material is often described by the early plots such as in Figure 7 for high purity aluminum [16]. The tests were conducted over a range of temperatures from near the melting temperature to as low as 0.57 T_m. Data has been considered unreliable below about 0.3 T_m, as it has recently been shown that dynamic recovery is not the exclusive

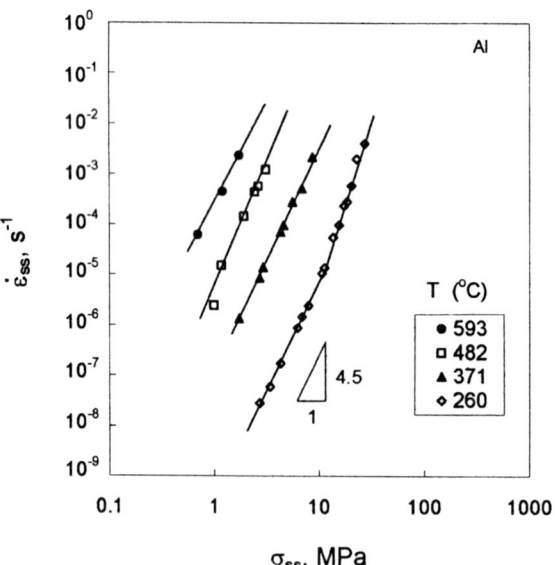

Figure 7. The steady-state stress versus strain-rate for high-purity aluminum at four temperatures, from Ref. [136].

restoration mechanism [11], since dynamic recrystallization in 99.999% pure Al has been confirmed. Dynamic recrystallization becomes an additional restoration mechanism that can preclude a constant flow stress (for a constant strain-rate) or a "genuine" mechanical steady-state, defined here as a balance between dynamic recovery and hardening. The plots in Figure 7 are important for several reasons. First, the steady-state data are at fixed temperatures and it is not necessary for the stress to be modulus-compensated to illustrate stress dependence [e.g., equation (3)]. Thus, the power-law behavior is clearly evident for each of the four temperature sets of high-purity aluminum data without any ambiguity (from modulus compensation). The stress exponent is about 4.5 for aluminum. Although this is not precisely five, it is constant over a range of temperature, stress, and strain-rate, and falls within the range of 4–7 observed in pure metals and class M alloys. This range has been conveniently termed "five power". "Some have referred to five-power-law-creep as "dislocation climb controlled creep", but this term may be misleading as climb control appears to occur in other regimes such as Harper–Dorn, Superplasticity, PLB, etc. We note from Figure 7 that slope increases with increasing stress and the slope is no longer constant with changes in the stress at higher stresses (often associated with lower temperatures). Again, this is power-law breakdown (PLB) and will be discussed more later. The activation energy for steady-state creep, Q_c, calculations have been based on plots similar to Figure 7. The activation energy, here, simply describes the change in (steady-state) creep-rate for a given substructure (strength), at a fixed applied "stress" with changes in temperature. It will be discussed in detail later; for at least steady-state, the microstructures of specimens tested at different temperatures appears approximately identical, by common microstructural measures, for a fixed *modulus-compensated* stress, σ_{ss}/E or σ_{ss}/G. [Modulus compensation (modest correction) will be discussed more later]. For a given substructure, s, and relevant "stress," σ_{ss}/E, (again, it is often assumed that a constant σ_{ss}/E or σ_{ss}/G implies constant structure, s) the activation energy for creep, Q_c, can be defined by

$$Q_c = -k[\delta(\ln \dot{\varepsilon}_{ss})/\delta(1/T)]_{\sigma_{ss}/E,s} \quad (6)$$

It has been very frequently observed that Q_c seems to be essentially equal to the activation energy for lattice self-diffusion Q_{sd} for a large class of materials. This is illustrated in Figure 8, where over 20 (bcc, fcc, hcp, and other crystal structures) metals show excellent correlation between Q_c and Q_{sd} (although it is not certain that this figure includes the (small) modulus compensation). Another aspect of Figure 8 which is strongly supportive of the activation energy for five-power-law-creep being equal to Q_{sd} is based on activation volume (ΔV) analysis by Sherby and Weertman [5].

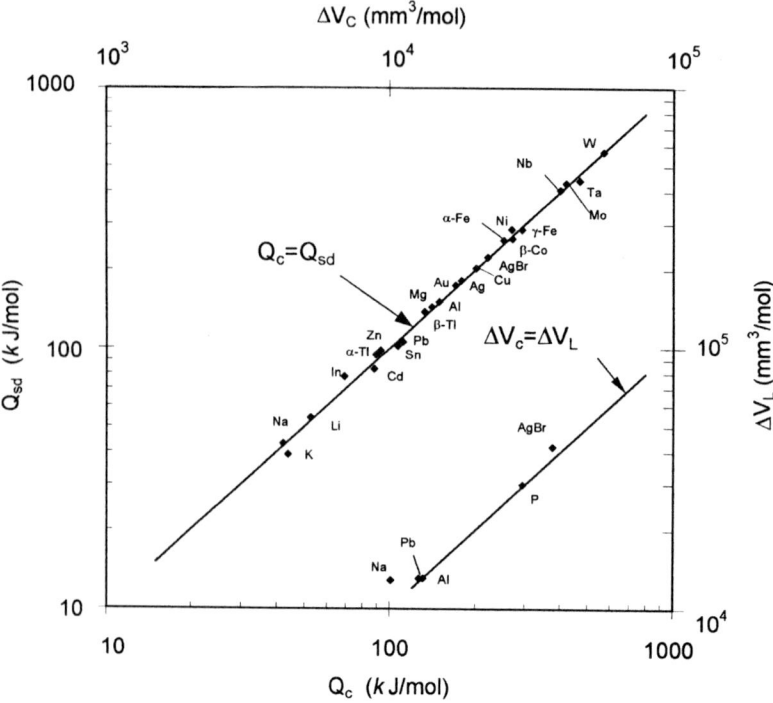

Figure 8. The activation energy and volume for lattice self-diffusion versus the activation energy and volume for creep. Data from Ref. [26].

That is, the effect of (high) pressure on the creep rate $(\partial \dot{\varepsilon}_{ss}/\partial P)_{T,\sigma_{ss}/E(\text{or } G)} = \Delta V_c$ is the same as the known dependence of self-diffusion on the pressure $(\partial D_{sd}/\partial P)_{T,\sigma_{ss}/E(\text{or } G)}$. Other more recent experiments by Campbell, Tao, and Turnbull on lead have shown that additions of solute that affect self-diffusion also appear to identically affect the creep-rate [40].

Figure 9 describes the data of Figure 7 on what appears as a nearly single line by compensating the steady-state creep-rates ($\dot{\varepsilon}_{ss}$) at several temperatures by the lattice self-diffusion coefficient, D_{sd}. At higher stresses PLB is evident, where n continually increases. The above suggests for power-law creep, typically above 0.6 T_m (depending on the creep rate)

$$\dot{\varepsilon}_{ss} = A_2 \exp[-Q_{sd}/kT](\sigma_{ss})^{n(\cong 5)} \tag{7}$$

where A_2 is a constant, and varies significantly among metals. For aluminum, as mentioned earlier, $n = 4.5$ although for most metals and class M alloys $n \cong 5$, hence "five-power" (steady-state) creep. Figure 9 also shows that, phenomenologically,

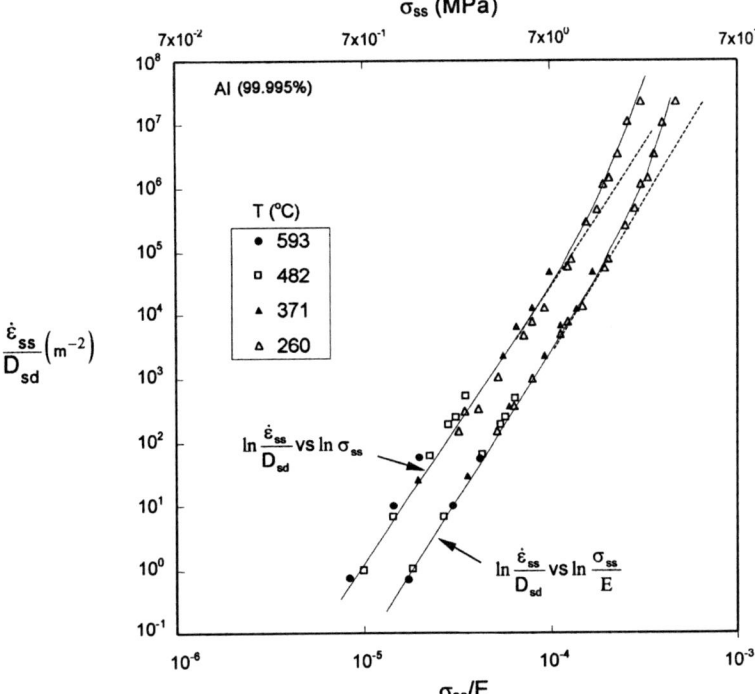

Figure 9. The lattice self-diffusion coefficient compensated steady-state strain-rate versus the Young's modulus compensated steady-state stress, from Ref. [16].

the description of the data may be improved by normalizing the steady-state stress by the elastic (Young's in this case) modulus. This will be discussed more later. {The correlation between Q_c and Q_{sd} [the former calculated from equation (6)] utilized modulus compensation of the stress. Hence, equation (7) actually implies modulus compensation.}

It is now widely accepted that the activation energy for five-power-law-creep closely corresponds to that of lattice self-diffusion, D_{sd}, or $Q_c \cong Q_{sd}$, although this is not a consensus judgment [41–43]. Thus, most have suggested that the mechanism of five-power-law-creep is associated with dislocation climb.

Although within PLB, Q_c generally decreases as n increases, some still suggest that creep is dislocation climb controlled, but Q_c corresponds to the activation energy for dislocation-pipe diffusion [5,44,45]. Vacancy supersaturation resulting from deformation, associated with moving dislocations with jogs, could explain this decrease with decreasing temperature (increasing stress) and still be consistent with dislocation climb control [4]. Dislocation glide mechanisms may be important [26] and the rate-controlling mechanism for plasticity in PLB is still speculative. It will be discussed more later, but recent studies observe very well-defined subgrain boundaries that form from dislocation reaction (perhaps as a consequence of the dynamic recovery process), suggesting that substantial dislocation climb is at least occurring [11,12,46,47] in PLB. Equation (7) can be extended to additionally phenomenologically describe PLB including changes in Q_c with temperature and stress by the hyperbolic sine function in equation (5) [44,48].

Figure 10, taken from the reviews by Sherby and Burke [16], Nix and Ilschner [26], and Mukherjee [21], describes the steady-state creep behavior of hcp, bcc, and fcc metals (solid solutions will be presented later). The metals all show approximate five-power-law behavior over the specified temperature and stress regimes. These plots confirm a range of steady-state stress exponent values in a variety of metals from 4 to 7 with 5 being a typical value [49].

Normalization of the stress by the shear modulus G (rather than E) and the inclusion of additional normalizing terms (k, G, b, T) for the strain rate will be discussed in the next section. It can be noted from these plots that for a fixed

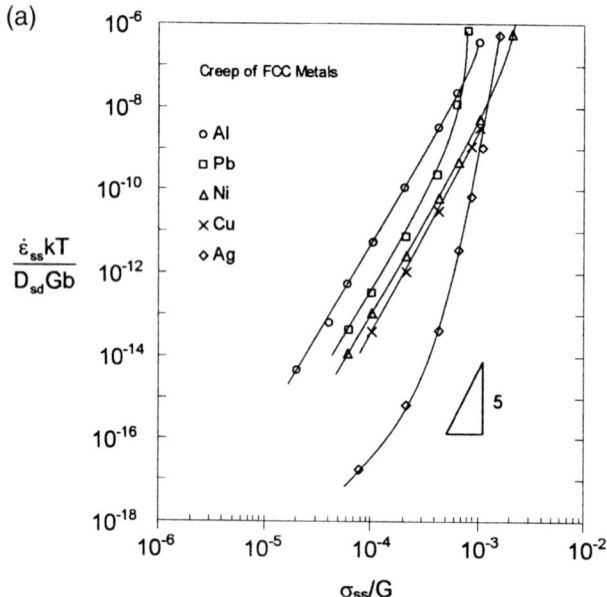

Figure 10. (a) The compensated steady-state strain-rate versus modulus-compensated steady-state stress, based on Ref. [26] for selected FCC metals. (b) The compensated steady-state strain-rate versus modulus-compensated steady-state stress, based on Ref. [26] for selected BCC metals. (c) The compensated steady-state strain-rate versus modulus-compensated steady-state stress, based on Ref. [21] for selected HCP metals.

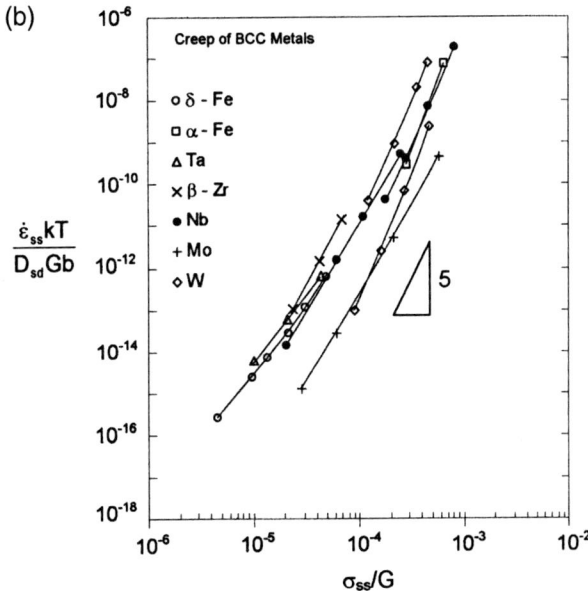

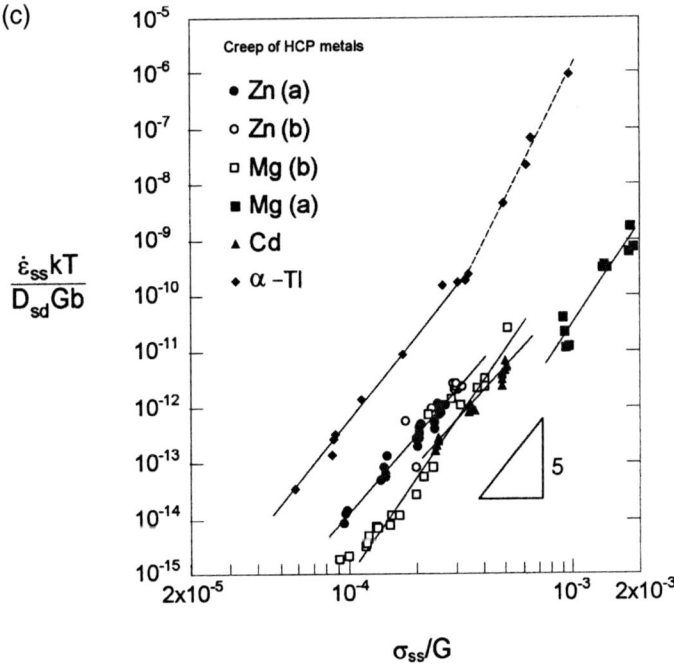

Figure 10. Continued.

steady-state creep-rate, the steady-state flow-stress of metals may vary by over two orders of magnitude for a given crystal structure. The reasons for this will be discussed later. A decreasing slope (exponent) at lower stresses may be due to diffusional creep or Harper–Dorn Creep [50]. Diffusional creep includes Nabarro–Herring [51] and Coble [52] creep. These will be discussed more later, but briefly, Nabarro–Herring consists of volume diffusion induced strains in polycrystals while Coble consists of mass transport under a stress by vacancy migration via short circuit diffusion along grain boundaries. Harper–Dorn is not fully understood [53–55] and appears to involve dislocations within the grain interiors. There has been some recent controversy as to the existence of diffusional creep [56–61], as well as Harper–Dorn [55].

2.1.2 Influence of the Elastic Modulus

Figure 11 plots the steady-state stress versus the Young's modulus at a *fixed* lattice self-diffusion-coefficient compensated steady-state creep-rate. Clearly, there is an associated increase in creep strength with Young's modulus and the flow stress can be described by,

$$\sigma_{ss}|_{\dot{\varepsilon}_{ss}/D_{sd}} = K_0 G \qquad (8)$$

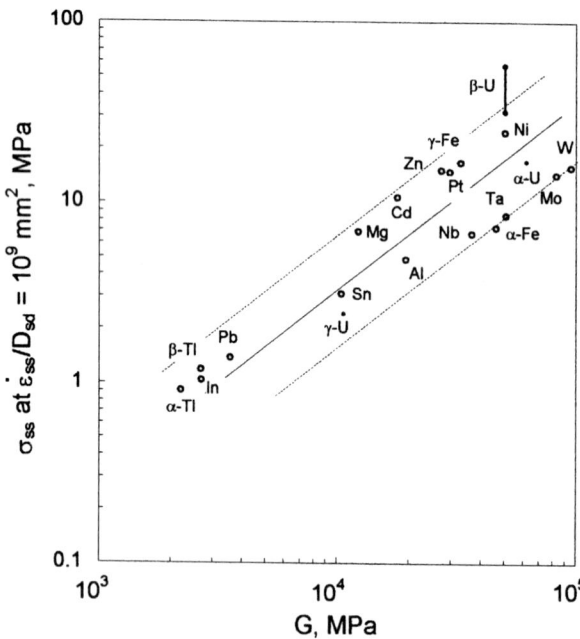

Figure 11. The influence of the shear modulus on the steady-state flow stress for a fixed self-diffusion-coefficient compensated steady-state strain-rate, for selected metals, based on Ref. [26].

where K_0 is a constant. This, together with equation (7), can be shown to imply that five-power-law-creep is described by the equation utilizing modulus-compensation of the stress, such as with equation (3),

$$\dot{\varepsilon}_{ss} = A_3 \exp[-Q_{sd}/kT](\sigma_{ss}/G)^5 \qquad (9)$$

where A_3 is a constant. Utilizing modulus compensation produces less variability of the constant A_3 among metals, as compared to A_2, in equation (7). It was shown earlier that the aluminum data of Figure 9 could, in fact, be more accurately described by a simple power-law if the stress is modulus compensated. The modulus compensation of equation (9) may also be sensible for a given material as the dislocation substructure is better related to the modulus-compensated stress rather than just the applied stress. The constant A_3 will be discussed more later. Sherby and coworkers compensated the stress using the Young's modulus, E, while most others use the shear modulus, G. The choice of E versus G is probably not critical in terms of improving the ability of the phenomenological equation to describe the data. The preference by some for use of the shear modulus may be based on a theoretical "palatability", and is also used in this review for consistency.

Thus, the "apparent" activation energy for creep, Q'_c, calculated from plots such as Figure 7 *without* modulus compensation, is not exactly equal to Q_{sd} even if dislocation climb is the rate-controlling mechanism for Five-power-law-creep. This is due to the temperature dependence of the Elastic Modulus. That is,

$$Q'_c = Q_{sd} + 5k[d(\ln G)/d(1/T)] \qquad (10)$$

Thus, $Q'_c > Q_{sd} \cong Q_c$ the magnitude being material and temperature dependent. The differences are relatively small near $0.5\,T_m$ but become more significant near the melting temperature.

As mentioned in advance, dislocation features in creep-deformed metals and alloys can be related to the *modulus*-compensated stress. Thus, the "s" in equation (6), denoting constant structure, can be omitted if constant modulus-compensated stress is indicated, since for steady-state structures in the power-law regime, a constant σ_{ss}/E (or σ_{ss}/G) will imply, at least approximately, a fixed structure. Figure 12 [6] illustrates some of the Figure 7 data, as well as additional (PLB) data on a strain-rate versus modulus-compensated stress plot. This allows a direct determination of the activation energy for creep, Q_c, since changes in $\dot{\varepsilon}_{ss}$ can be associated with changes in T for a fixed structure (or σ_{ss}/G). Of course, Konig and Blum [62] showed that with a change in temperature at a constant applied stress, the substructure changes, due, at least largely, to a change in σ/G in association with a change in temperature. We observe in Figure 12 that activation energies are comparable to that

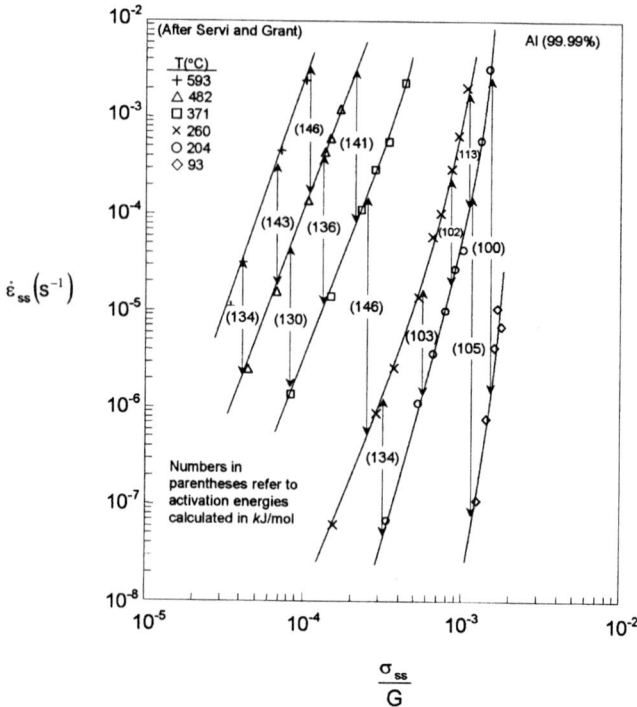

Figure 12. The steady-state strain-rate versus the modulus-compensated stress for six temperatures. This plot illustrates the effect of temperature on the strain-rate for a fixed modulus-compensated steady-state stress (constant structure) leading to the calculation for activation energies for creep, Q_c. Based on [6].

of lattice self-diffusion of aluminum (e.g., 123 kJ/mol [63]). Again, below about 0.6 T_m, or so, depending on the strain-rate, the activation energy for creep Q_c begins to significantly decrease *below* Q_{sd}. This occurs at about PLB where $n > 5$ (> 4.5 for Al). Figure 13 plots steady-state aluminum data along with steady-state silver activation energies [12,44]. Other descriptions of Q_c vs T/T_m for Al [64] and Ni [65] are available that utilize temperature-change tests in which constant structure is assumed (but not assured) and σ/G is not constant. The trends observed are nonetheless consistent with those of Figure 13. The question as to whether the activation energy for steady-state and primary (or transient, i.e., from one steady-state to another) creep are identical does not appear established and this is an important question. However, some [66,67] have suggested that the activation energy from primary to steady-state does not change substantially. Luthy *et al.* [44] and Sherby and Miller [68] present a Q_c versus T/T_m plot for steady-state deformation of W, NaCl, Sn and Cu that is frequently referenced. This plot suggests

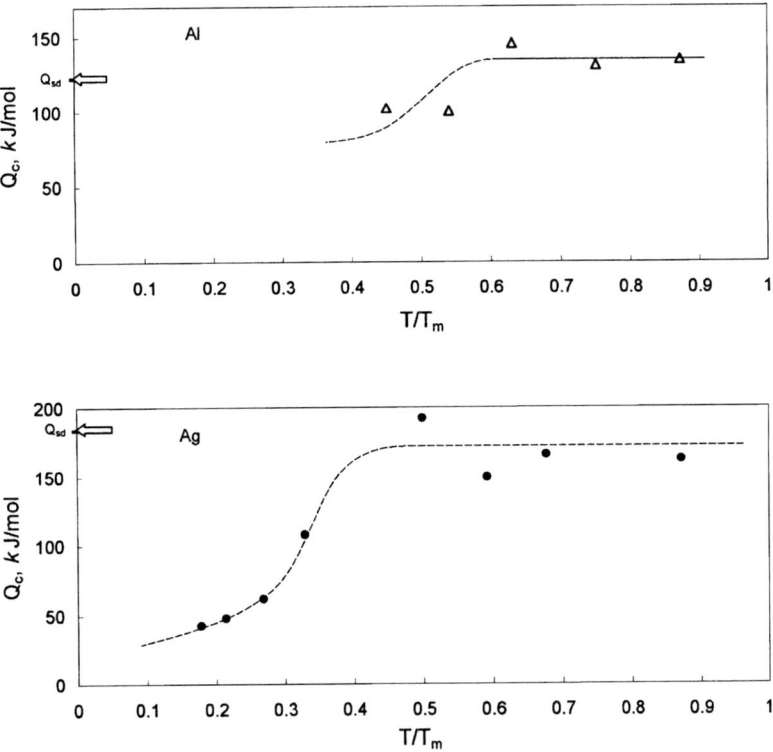

Figure 13. The variation of the activation energy for creep versus fraction of the melting temperature for Al (based on Ref. [44]) and Ag (based on Ref. [12]).

two activation energies, one regime where $Q_c \cong Q_{sd}$ (as Figure 13 shows for Ag and Al) from 0.60 to 1.0 T/T_m. They additionally suggest that, with PLB, Q_c is approximately equal to that of vacancy diffusion through dislocation pipes, Q_P. That is, it was suggested that the rate-controlling mechanism for steady-state creep in PLB is still dislocation climb, but facilitated by short circuit diffusion of vacancies via the elevated density of dislocations associated with increased stress between 0.3 and about 0.6 T/T_m. (The interpretation of the NaCl results are ambiguous and may actually be more consistent with Figure 13 if the original NaCl data is reviewed [16].) Spingarn et al. [69] also suggest such a transition and use this transition to rationalize five-power-law exponents. Luthy et al. suggested that aluminum Q_c trends are possible, but caution should be exercised in the PLB interpretations. The situation for Cu is ambiguous. Raj and Langdon [70] reviewed activation energy data and it appears that Q_c may decrease continuously below at least 0.7 T_m from Q_{sd}, in contrast to earlier work on Cu that suggested $Q_c = Q_{sd}$ above about 0.7 T_m

and "suddenly" decreases to Q_P. As mentioned earlier, the steady-state torsion creep data of Luthy et al., on which the lower temperature activation energy calculations were based, are probably unreliable. Dynamic recrystallization is certainly occurring in their high purity aluminum along with probable (perhaps 20%) textural softening (decrease in the average Taylor factor, $\overline{M}$) along with adiabatic heating. Use of solid specimens also complicates the interpretation of steady-state as outer portions may soften while inner portions are hardening. Lower purity specimens could be used to avoid dynamic recrystallization, but Stage IV hardening [11,13] may occur and may preclude (although sometimes just postpone) a mechanical steady-state. Thus, steady-state, as defined here, as a balance between dislocation hardening and, exclusively, dynamic recovery, is not relevant. Thus, their activation energies as well as their steady-state stress values below about 0.4 T/T_m are not used here. Weertman [25] suggested that the Sn results may show an activation energy transition to a value of Q_p over a range of elevated temperatures. This transition occurs already at about 0.8 T_m and Q_c values at temperatures less than 0.6 T_m do not appear available. Quality activation energy measurements over a wide range of temperatures both for steady-state and primary creep for a variety of pure metals are surprisingly unavailable. Thus, the values of activation energy, between 0.3 and 0.6 T_m (PLB), and the question as to whether these can be related to the activation energy of dislocation pipe diffusion, are probably unsettled.

Sherby and Burke have suggested that vacancy supersaturation may occur at lower temperatures where PLB occurs (as have others [71]). Thus, vacancy diffusion may still be associated with the rate-controlling process despite a low, non-constant, activation energy. Also, as suggested by others [9,26,41–43], cross-slip or the cutting of forest dislocations (glide) may be the rate-controlling dislocation mechanisms rather than dislocation climb.

2.1.3 Stacking Fault Energy and Summary

In the above, the steady-state creep rate for five-power-law-creep was described by,

$$\dot{\varepsilon}_{ss} = A_4 D_{sd} (\sigma_{ss}/G)^5 \qquad (11)$$

where

$$D_{sd} = D_o \exp(-Q_{sd}/kT). \qquad (12)$$

Many investigators [4,21,26,72] have attempted to decompose A_4 into easily identified constants. Mukherjee et al. [72] proposed that

$$\dot{\varepsilon}_{ss} = A_5 (D_{sd} G b / kT)(\sigma_{ss}/G)^5 \qquad (13)$$

This review will utilize the form of equation (13) since this form has been more widely accepted than equation (11). Equation (13) allows the expression of the power law on a logarithmic plot more conveniently than equation (11), due to dimensional considerations.

The constants A_0 through A_5 depend on stacking fault energy, in at least fcc metals, as illustrated in Figure 14. The way by which the stacking fault energy affects the creep rate is unclear. For example, it does not appear known whether the climb-rate of dislocations is affected (independent of the microstructure) and/or whether the dislocation substructure, which affects creep rate, is affected. In any case, for fcc metals, Mohamed and Langdon [73] suggested:

$$\dot{\varepsilon}_{ss} = A_6(\chi/Gb)^3 (D_{sd}Gb/kT)(\sigma_{ss}/G)^5 \qquad (14)$$

where χ is the stacking fault energy.

Thus, in summary it appears that, over five-power creep, the activation energy for steady-state creep is approximately that of lattice self-diffusion. (Exceptions with

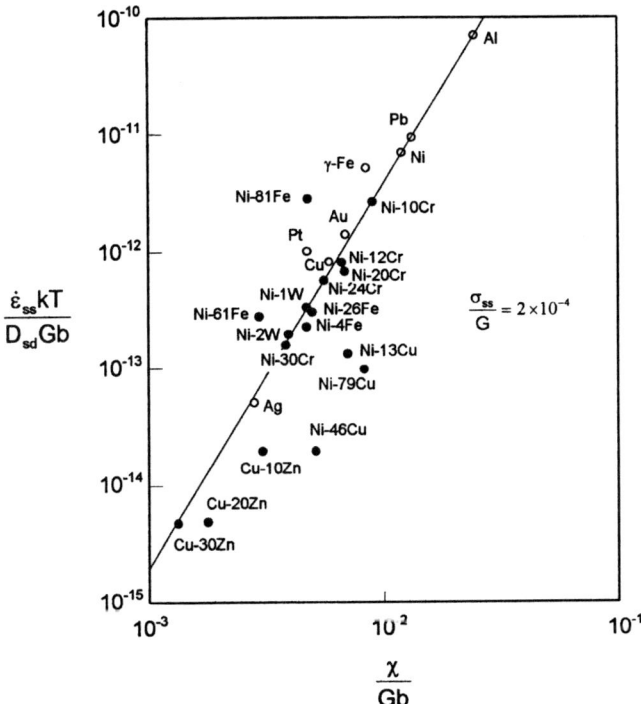

Figure 14. The effect of stacking fault energy on the (compensated) steady-state strain-rate for a variety of metals and Class M alloys based on Ref. [73].

pure metals above 0.5 T/T_m have been suggested. One example is Zr, where a glide control mechanism [74] has been suggested to be rate controlling, but self-diffusion may still be viable [75], just obscured by impurity effects.) This suggests that dislocation climb is associated with the rate-controlling process for Five-power-law-creep. The activation energy decreases below about 0.5 T_m, depending, of course, on the strain-rate. There is a paucity of reliable steady-state activation energies for creep at these temperatures and it is difficult to associate these energies with specific mechanisms. The classic plot of effective diffusion coefficient D_{eff} compensated strain rate versus modulus-compensated stress for aluminum by Luthy et al. may be the most expansive in terms of the ranges of stress and temperature. It appears in other creep reviews [23,24] and may have some critical flaws. They modified equation (11) to a Garofalo (hyperbolic sine) [48] equation to include PLB:

$$\dot{\varepsilon}_{ss} = BD_{eff}[\sinh \alpha_1 (\sigma_{ss}/E)]^5 \tag{15}$$

where α_1 and B are constants. Here, again, D_{eff} reflects the increased contribution of dislocation pipe diffusion with decreasing temperature. D_{eff} compensated strain-rate utilizes a "composite" strain-rate controlled by lattice and dislocation pipe-diffusion. The contributions of each of these to D_{eff} depend on both the temperature and the dislocation density (which at steady-state is non-homogeneous, as will be discussed). Equation (15), above, was later modified by Wu and Sherby [53] for aluminum to account for Internal stresses although a dramatic improvement in the modeling of the data of 5-power-law and PLB was not obvious. The subject of Internal stresses will be discussed later. Diffusion is not a clearly established mechanism for plastic flow in PLB and D_{eff} is not precisely known. For this reason, this text will avoid the use of D_{eff} in compensating strain-rate.

Just as PLB bounds the high-stress regime of five-power-law-creep, a diffusional creep mechanism or Harper–Dorn creep may bound the low-stress portion of five-power-law-creep (for alloys, superplasticity [2-power] or viscous glide [3-power] may also be observed as will be discussed later). For pure aluminum, Harper–Dorn creep is generally considered to describe the low-stress regime and is illustrated in Figure 15. The stress exponent for Harper–Dorn is 1 with an activation energy often of Q_{sd}, and Harper–Dorn is grain size independent. The precise mechanism for Harper–Dorn Creep is not understood [53,54] and some have suggested that it may not exist [55]. Figure 15, from Blum and Straub [76,77], is a compilation of high quality steady-state creep in pure aluminum and describes the temperature range from about 0.5 T_m to near the melting temperature, apparently showing three separate creep regimes. It appears that the range of steady-state data for Al may be more complete than any other metal. The aluminum data presented in earlier figures are consistent with the data plotted by Blum and Straub. It is intended that this data

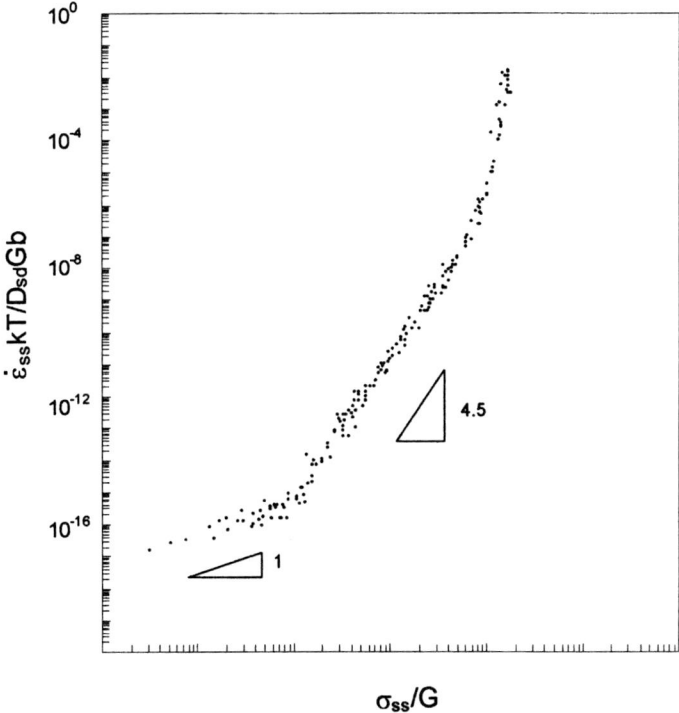

Figure 15. The compensated steady-state strain-rate versus the modulus-compensated steady-state stress for 99.999 pure Al, based on Refs. [76,77].

does not include the temperature/stress regime where Stage IV and recrystallization may obfuscate recovery-controlled steady-state. This plot also (probably not critical to the PLB transition) uses the same activation energy, Q_{sd}, (142 kJ mol^{-1}) [76] over the entire stress/strain-rate/temperature regime. As discussed earlier, Q_c seems to decrease with decreasing temperature (increasing strain-rate) within PLB. The aluminum data shows a curious undulation at $\sigma_{ss}/G = 2 \times 10^{-5}$, that is not understood, although impurities were a proposed explanation [76].

There are other metallic systems for which a relatively large amount of data from several investigators can be summarized. One is copper, which is illustrated in Figure 16. The summary was reported recently in Ref. [78]. Again, a well-defined "five" power law regime is evident. Again, this data is consistent with the (smaller quantity of reported) data of Figure 10. Greater scatter is evident here as compared with Figure 15, as the results of numerous investigators were used and the purity varied. Copper is a challenging experimental metal as oxygen absorption and

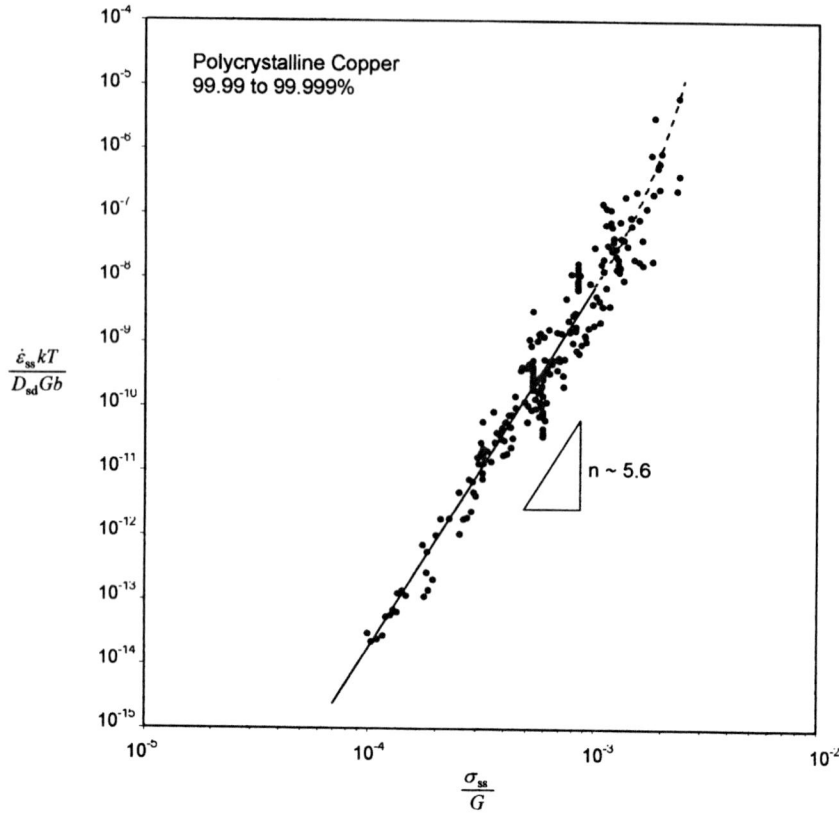

Figure 16. Summary of the diffusion-coefficient compensated steady-state strain-rate versus the modulus-compensated steady-state stress for copper of various high purities from various investigations. From Ref. [78].

discontinuous dynamic recrystallization can obfuscate steady-state behavior in five-power-law-creep, which is a balance between dislocation hardening and dynamic recovery.

Also, at this point, it should be mentioned that it has been suggested that some metals and class M alloys may be deformed by glide-control mechanisms (e.g., jogged-screw, as will be discussed later in this chapter) [79] as mentioned earlier. Ardell and Sherby and others [74] suggested a glide mechanism for zirconium. Recent analysis of zirconium, however, suggests that this HCP metal behaves as a classic five-power-law metal [80]. Figure 17, just as Figure 16, is a compilation of numerous investigations on zirconium of various purity. Here, with zirconium, as with copper, oxygen absorption and DRX can be complicating

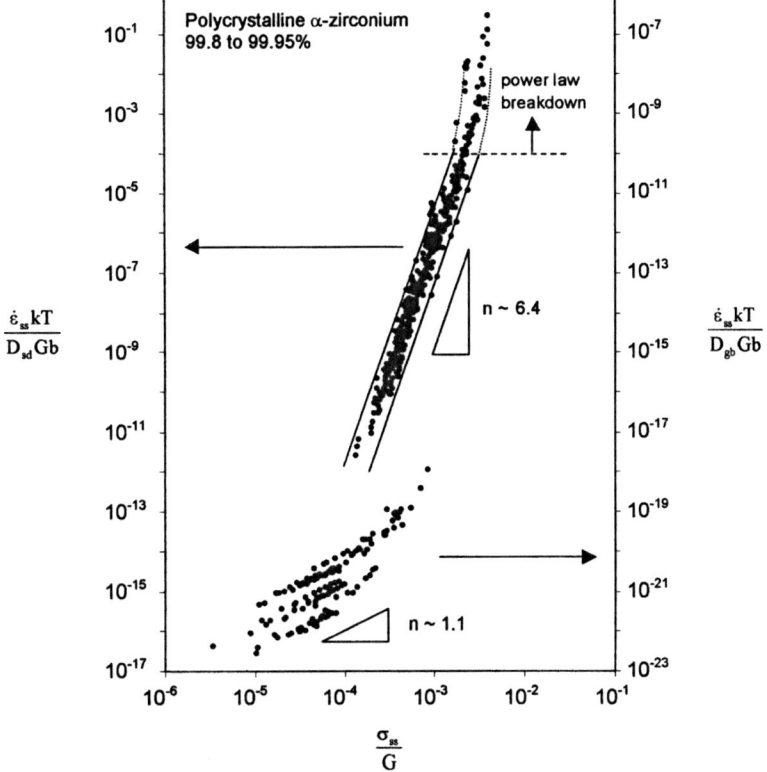

Figure 17. The diffusion-coefficient compensated steady-state strain-rate versus modulus-compensated steady-state stress for polycrystalline zirconium of various purities from various investigations. From Ref. [80].

factors. The lower portion of the figure illustrates lower stress exponent creep (as in Figure 15), of uncertain origin. Harper–Dorn creep, grain boundary sliding, and diffusional creep have all been suggested.

2.1.4 Natural Three-Power-Law

It is probably important to note here that Blum [22] suggests the possibility of some curvature in the 5-power-law regime in Figure 15. Blum cautioned that the effects of impurities in even relatively high-purity aluminum could obscure the actual power-law relationships and that the value of "strain rate-compensated creep" at $\sigma/G \cong 10^{-5}$ was consistent with Three-Power-Law creep theory [81]. Curvature was also suggested by Nix and Ilschner [26] in bcc metals (in Figure 10) at lower stresses, and suggested a possible approach to a lower slope of 3 or, "natural" power law exponent consistent with some early arguments by Weertman [25] (although Springarn, Barnett and Nix [69] earlier suggest that dislocation core diffusion may rationalize five-power-law behavior). Both groups interpreted five-power behavior as a disguised "transition" from Three-Power-Law to PLB. Weertman suggested that five-power-law behavior is unexpected. The Three-Power-Law exponent, or so-called natural law, has been suggested to be a consequence of

$$\dot{\varepsilon} = (1/2)\bar{v}b\rho_m \qquad (16)$$

where $\bar{v}$ is the average dislocation velocity and ρ_m is the mobile dislocation density. As will be discussed later in a theory section, the dislocation climb-rate, which controls $\bar{v}$, is proportional to σ. It is assumed that $\sigma^2 \propto \rho_m$ (although dislocation hardening is *not* assumed) which leads to 3-power behavior in equation (16). This text will later attempt to illustrate that this latter equation has no mandate.

Wilshire [82] and Evans and Wilshire [83] have described and predicted steady-state creep rates phenomenologically over wide temperature regimes without assumptions of transitions from one rate-controlling process to another across the range of temperature/strain-rates/stresses in earlier plots (which suggested to include, for example, Harper–Dorn creep, five-power-law-creep, PLB, etc). Although this is not a widely accepted interpretation of the data, it deserves mention, particularly as some investigators, just referenced, have questioned the validity of five-power-law. A review confirms that nearly all investigators recognize that power-law behavior in pure metals and Class M alloys appears to be generally fairly well defined over a considerable range of modulus-compensated steady-state stress (or diffusion coefficient-compensated steady-state creep rate). Although this value varies, a typical value is 5. Thus, for this review, the designation of "five-power-law-creep" is judged meaningful.

2.1.5 Substitutional Solid Solutions

Two types of substitutional solutions can be considered; cases where a relatively large fraction of solute alloying elements can be dissolved, and those cases with small amounts of either intentional or impurity (sometimes interstitial) elements. The addition of solute can lead to basically two phenomena within the five-power-law regime of the solvent. Hardening or softening while maintaining five-power-law behavior can be observed, or three-power, viscous glide behavior, the latter being discussed in a separate section, may be evident. Figure 18 shows the effects of substitutional solid-solution additions for a few alloy systems for which five-power-law behavior is maintained. This plot was adapted from Mukherjee [21].

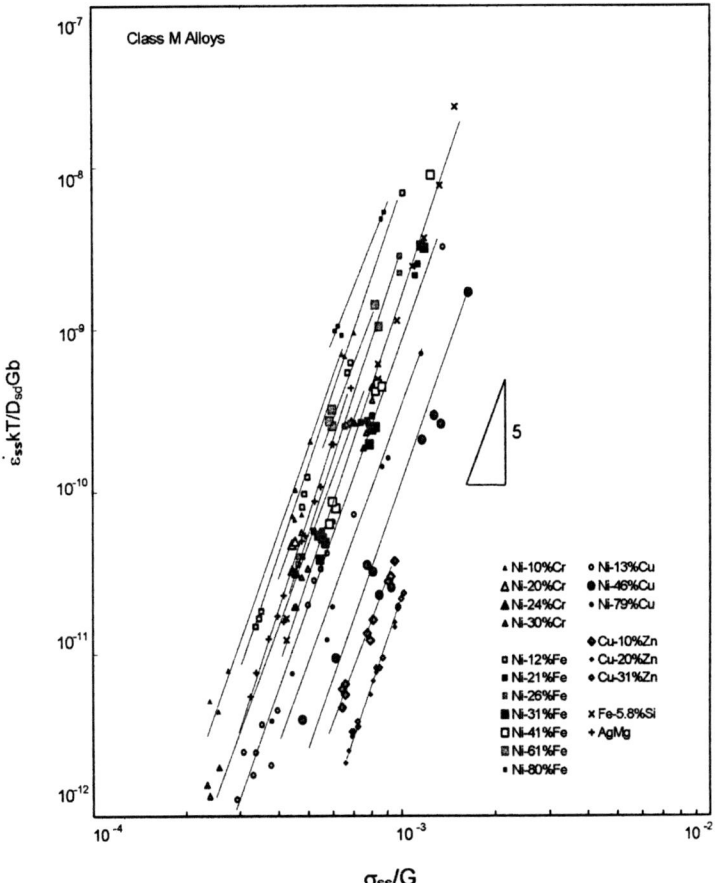

Figure 18. The compensated steady-state strain-rate versus modulus-compensated steady-state stress for a variety of Class M (Class I) alloys. Based on Ref. [21].

2.2 MICROSTRUCTURAL OBSERVATIONS

2.2.1 Subgrain Size, Frank Network Dislocation Density, Subgrain Misorientation Angle, and the Dislocation Separation within the Subgrain Walls in Steady-State Structures

Certain trends in the dislocation substructure evolution have been fairly well established when an annealed metal is deformed at elevated temperature (e.g., under constant stress or strain-rate) within the five-power-law regime. Basically, on commencement of plastic deformation, the total dislocation density increases and this is associated with the eventual formation of low-misorientation subgrain walls.

That is, in a polycrystalline aggregate, where misorientations, θ (defined here as the minimum rotation required to bring two lattices, separated by a boundary, into coincidence), between grains are typically (e.g., cubic metals) 10–62°, the individual grains become filled with subgrains. The subgrain boundaries are low-energy configurations of the dislocations generated from creep plasticity.

The misorientations, θ_λ, of these subgrains are generally believed to be low at elevated temperatures, often roughly 1°. The dislocations within the subgrain interior are generally believed to be in the form of a Frank network [84–87]. A Frank network is illustrated in Figure 19 [88]. The dislocation density can be reported as dislocation line length per unit volume (usually reported as mm/mm^3) or intersections per unit area (usually reported as mm^{-2}). The former is expected to be about a factor of two larger than the latter. Sometimes the method by which ρ is

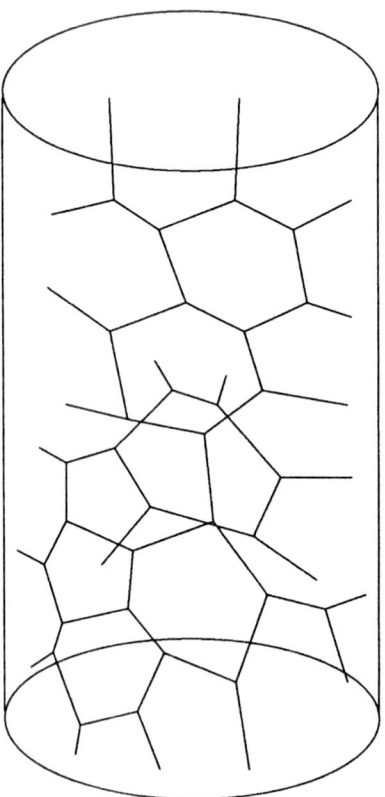

Figure 19. A three-dimensional Frank network of dislocations within a subgrain or grain. Based on Ref. [88].

determined is not reported. Thus, microstructurally, in addition to the average grain size of the polycrystalline aggregate, g, the substructure of materials deformed within the five-power-law regime is more frequently characterized by the average Subgrain size λ (often measured as an average intercept), the average misorientation across subgrain boundaries, $\theta_{\lambda_{ave}}$, and the density of dislocations not associated with subgrain boundaries, ρ. Early reviews (e.g., [16]) did not focus on the dislocation substructure as these microstructural features are best investigated by transmission electron microscopy (TEM). A substantial amount of early dislocation substructure characterization was performed using metallographic techniques such as polarized light optical microscopy (POM) and dislocation etch-pit analysis. As TEM techniques became refined and widely used, it became clear that the optical techniques are frequently unreliable, often, for example, overestimating Subgrain size [89] partly due to a lack of ability (particularly POM) to detect lower-misorientation-angle subgrain boundaries. Etch-pit-based dislocation density values may be unreliable, particularly in materials with relatively high ρ values [90]. Unambiguous TEM characterization of the dislocation substructure is not trivial. Although dislocation density measurements in metals may be most reliably performed by TEM, several short-comings of the technique must be overcome. Although the TEM can detect subgrain boundaries of all misorientations, larger areas of thin (transparent to the electron beam) foil must be examined to statistically ensure a meaningful average intercept, λ. The dislocation density within a subgrain may vary substantially within a given specimen. This variation appears independent of the specimen preparation procedures. The number of visible dislocations can be underestimated by a factor of two by simply altering the TEM imaging conditions [91]. Furthermore, it has been suggested that in high stacking fault energy materials, such as pure aluminum, dislocations may "recover" *from* the thin foil, leading, again, to an underestimation of the original density [92,93]. As mentioned earlier, there are at least two different means by which the dislocation density is reported; one is the surface intersection technique where the density is reported as the number of dislocation-line intersections per unit surface area of foil and another is the dislocation line length per unit volume [91]. The latter is typically a factor of two larger than the former. Misorientation angles of subgrain boundaries, θ_λ, have generally been measured by X-rays, selected area electron diffraction (SAED, including Kikuchi lines) and, more recently, by electron backscattered patterns (EBSP) [94] although this latter technique, to date, cannot easily detect lower (e.g. $<2^{-4\circ}$) misorientation boundaries. Sometimes the character of the subgrain boundary is alternatively described by the average spacing of dislocations, d, that constitute the boundary. A reliable determination of d is complicated by several considerations. First, with conventional bright-field or weak-beam TEM, there are limitations as to the minimum discernable separation of dislocations. This appears to be within the range of d frequently possessed by subgrain boundaries formed at elevated temperatures. Second, boundaries in at least some fcc metals [95–97] may have 2–5 separate sets of Burgers vectors, perhaps with different separations, and can be of tilt, twist, or mixed character. Often, the separation of the most closely separated set is reported. Determination of θ_λ by SAED requires significant effort since for a *single* orientation in the TEM, only the tilt angle by Kikuchi shift is accurately determined. The rotation component cannot be accurately measured for small total misorientations. θ_λ can only be determined using Kikuchi lines with some effort, involving examination of the crystals separated by a boundary using several orientations [42,43]. Again, EBSP cannot detect lower θ_λ boundaries which may comprise a large fraction of subgrain boundaries.

Figure 20 is a TEM micrograph of 304 austenitic stainless steel, a class M alloy, deformed to steady-state within the five-power-law regime. A well-defined subgrain

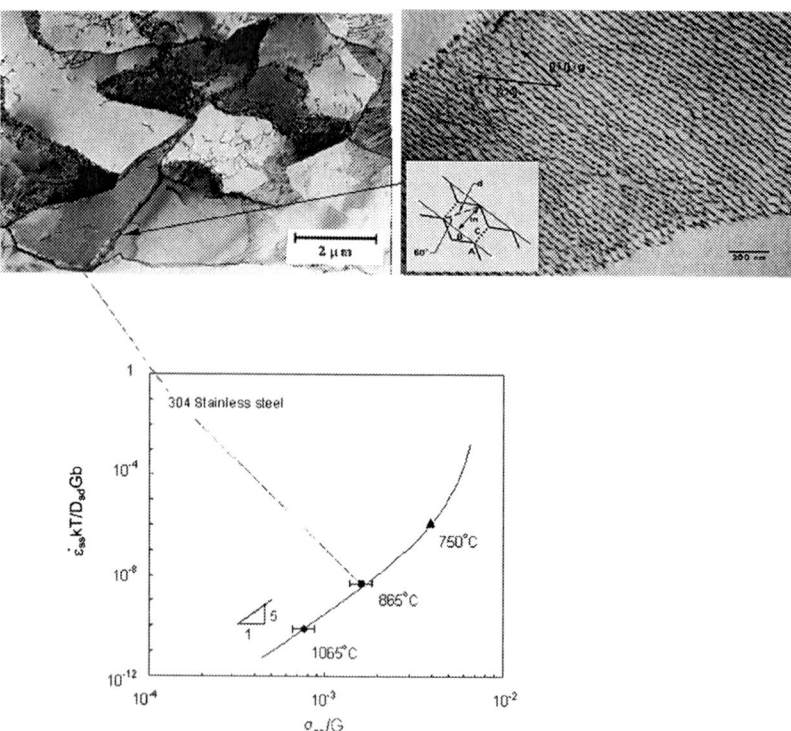

Figure 20. TEM micrographs illustrating the dislocation microstructure of 304 stainless steel deformed at the indicated conditions on the compensated steady-state strain-rate versus modulus-compensated steady-state stress plot. The micrograph on the right is a high magnification image of the subgrain boundaries such as illustrated on the left. Based on Refs. [98,99]. (Modulus based on 316 values.)

substructure is evident and the subgrain walls are "tilted" to expose the sets of dislocations that comprise the walls. The dislocations not associated with subgrain walls are also evident. Because of the finite thickness of the foil ($\cong 100$ nm), any Frank network has been disrupted. The heterogeneous nature of the dislocation substructure is evident. A high magnification TEM micrograph of a hexagonal array of screw dislocations comprising a subgrain boundary with one set (of Burgers vectors) satisfying invisibility [98,99] is also in the figure.

It has long been observed, phenomenologically, that there is an approximate relationship between the subgrain size and the steady-state flow stress;

$$\sigma_{ss}/G = C_1(\lambda_{ss})^{-1} \quad (17)$$

It should be emphasized that this relationship between the steady-state stress and the subgrain size is not for a fixed temperature and strain rate. Hence, it is not of a same type of equation as, for example, the Hall–Petch relationship, which relates the strength to the grain size, g, at a reference T and $\dot{\varepsilon}$. That is, equation (17) only predicts the subgrain size for a given steady-state stress which varies with the temperature and strain rate. Some (e.g., [89,100]) have normalized the subgrain size in equation (17) by the Burgers vector, although this is not a common practice.

The subgrains contain a dislocation density in excess of the annealed values, and are often believed to form a three-dimensional, Frank, network. The conclusion of a Frank network is not firmly established for five-power creep, but indirect evidence of a large number of nodes in thin foils [54,85,101–103] supports this common contention [84,104–109]. Analogous to equation (17), there appears to be a relationship between the density of dislocations not associated with subgrain boundaries [17] and the steady-state stress:

$$\sigma_{ss}/G = C_2(\rho_{ss})^p \quad (18)$$

where C_2 is a constant and ρ_{ss} is the density of dislocations not associated with subgrain boundaries. The dislocation density is not normalized by the Burgers vector as suggested by some [18]. The exponent "p" is generally considered to be about 0.5 and the equation, with some confusion, reduces to:

$$\sigma_{ss}/G = C_3\sqrt{\rho_{ss}} \quad (19)$$

As will be discussed in detail later in this text, this relationship between the steady-state stress and the dislocation density is *not* independent of temperature and strain-rate. Hence, it is not of a same type of microstructure-strength equation as the classic Taylor relationship which *is* generally independent of strain-rate and temperature.

That is, this equation tells us the dislocation density not associated with subgrain walls that can be expected for a given steady-state stress which varies with the temperature and strain-rate, analogous to equation (17). However, the flow stress associated with this density depends on $\dot{\varepsilon}$ and T.

As mentioned in the discussion of the "natural stress exponent," ρ_{ss} is sometimes presumed equal, or at least proportional, to the mobile dislocation density, ρ_m. This appears unlikely for a Frank network and the fraction mobile over some time interval, dt, is unknown, and may vary with stress. In Al, at steady-state, ρ_m may only be about $1/3\,\rho$ or less [107] although this fraction was not firmly established.

Proponents of dislocation hardening also see a resemblance of equation (19) to

$$\tau = \frac{Gb}{\ell_c} \tag{20}$$

leading to

$$\tau \approx Gb\sqrt{\rho_{ss}}$$

where τ is the stress necessary to activate a Frank Read source of critical link length, ℓ_c. Again, the above equation must be regarded as "athermal" or valid only at a specific temperature and strain-rate. It does not consider the substantial solute *and* impurity strengthening evident (even in 99.999% pure) in metals and alloys, as will be discussed in Figures 29 and 30, which is very temperature dependent. Weertman [25] appears to justify equation (19) by the dislocation (all mobile) density necessarily scaling with the stress in this manner, although he does not regard these dislocations as the basis for the strength.

The careful TEM data for aluminum by Blum and coworkers [22] for Al alloys and 304 and 316 stainless steels by others [66,98,103,110,111] are plotted in Figures 21 and 22. Blum and coworkers appear to have been reluctant to measure ρ in pure aluminum due to possible recovery effects during unloading and sample preparation and examination of thin foils. Consequently, Al-5at%Zn alloy, which has basically identical creep behavior, but possible precipitation pinning of dislocations during cooling, was used to determine the ρ, vs. σ_{ss} trends in Al instead. These data (Al and stainless steel) were used for Figures 21 and 22 as particular reliability is assigned to these investigations. They are reflective of the general observations of the community, and are also supportive of equation (17) with an exponent of -1. Both sets of data are consistent with equation (18) with $p \cong 0.5$. It should, however, be mentioned that there appears to be some variability in the observed exponent, p, in equation (18). For example, some TEM work on αFe [112] suggests 1 rather than 2. Hofmann and Blum [113] more recently suggested 0.63 for Al modeling.

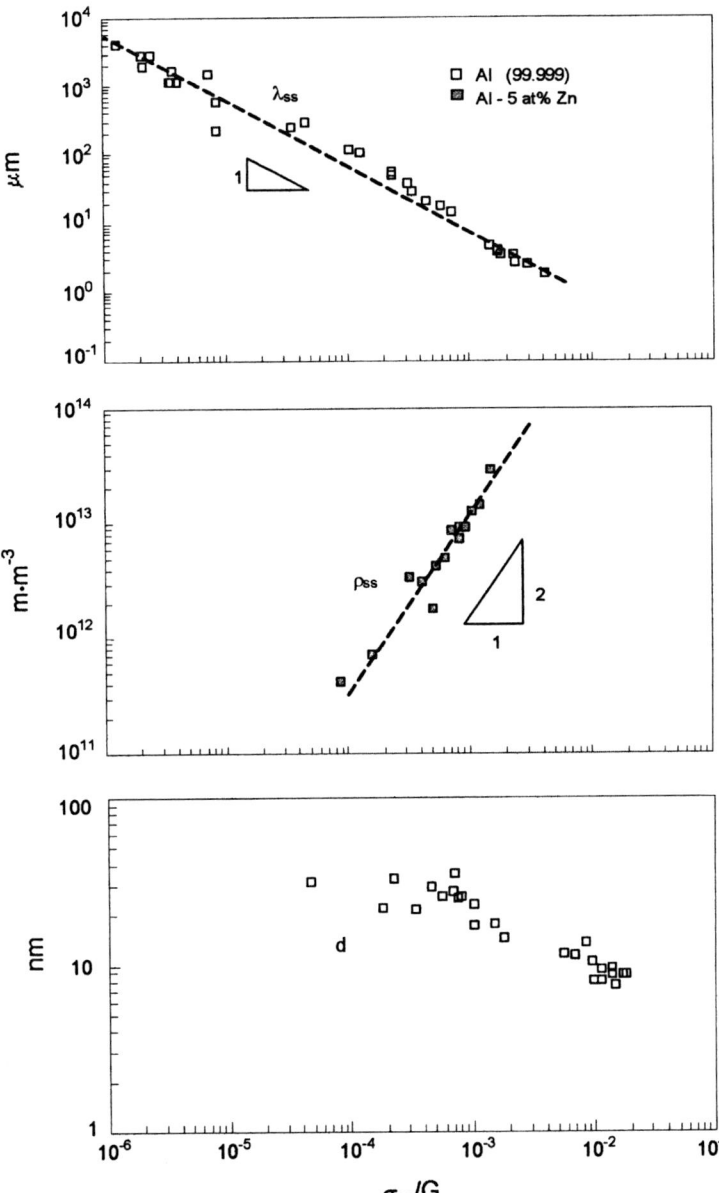

Figure 21. The average steady-state subgrain intercept, λ, density of dislocations not associated with subgrain walls, ρ, and the average separation of dislocations that comprise the subgrain boundaries for Al [and Al-5 at%Zn that behaves, mechanically, essentially identical to Al, but is suggested to allow for a more accurate determination of ρ by TEM]. Based on Refs. [22,90].

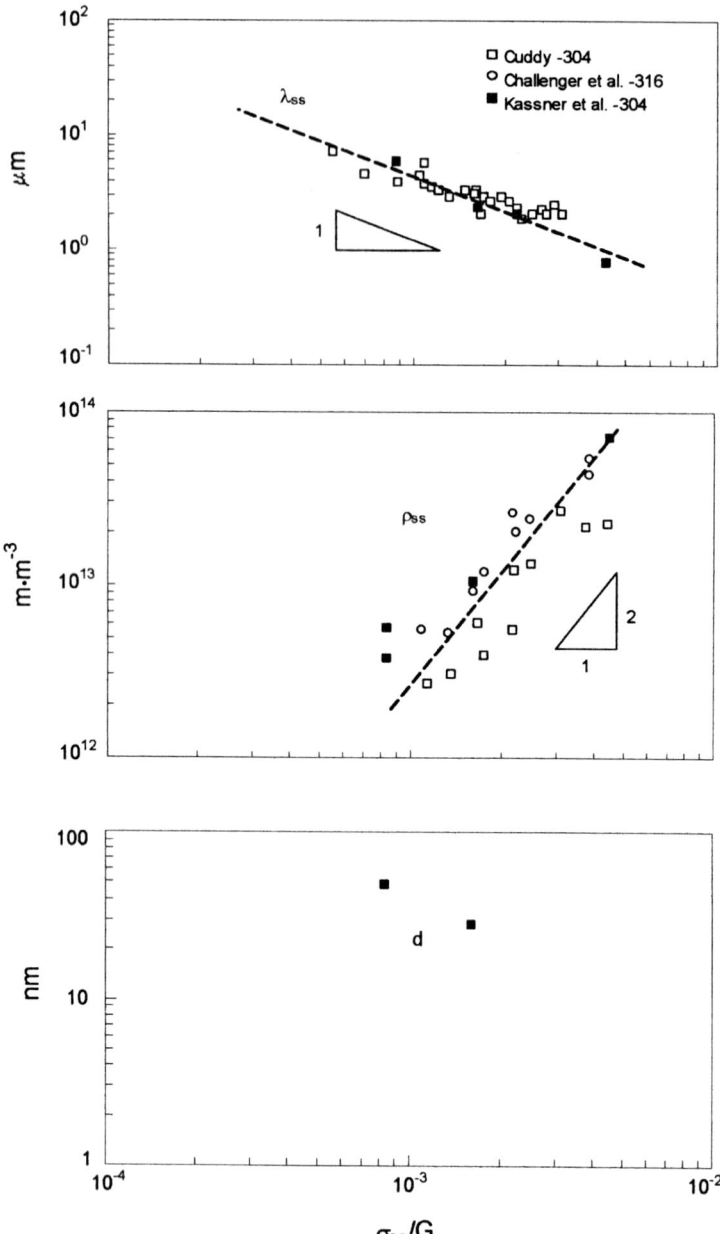

Figure 22. The average steady-state subgrain intercept, λ, density of dislocations not associated with subgrain boundaries, ρ, and average separation of dislocations that comprise the subgrain boundaries, d, for 304 stainless steel. Data from Refs. [66,98,103,110,111].

This latter value depicts the general procedure to derive a "natural three power law." The above two equations, of course, mandate a relationship between the steady-state Subgrain size and the density of dislocations not associated with subgrain boundaries,

$$\lambda_{ss} = C_4(\rho_{ss})^{-p} \qquad (21)$$

rendering it difficult, simply by microstructural inspection of steady-state substructures, to determine which feature is associated with the rate-controlling process for steady-state creep or elevated temperature strength.

Figures 21 and 22 additionally report the spacing, d, of dislocations that constitute the subgrain walls. The relationship between d and σ_{ss} is not firmly established, but Figures 21 and 22 suggest that

$$\frac{\sigma_{ss}}{G} \cong d^{-q} \qquad (22)$$

where q may be between 2 and 4. Other [90] work on Fe- and Ni-based alloys suggests that q may be closer to values near 4. One possible reason for the variability may be that d (and $\theta_{\lambda_{ave}}$) may vary during steady-state as will be discussed in a later section. Figure 22 relies on d data well into steady-state using torsion tests ($\bar{\varepsilon} \cong 1.0$). Had d been selected based on the onset of steady-state, a q value of about 4 would also have been obtained. It should, of course, be mentioned that there is probably a relationship between d and θ_λ. Straub and Blum [90] suggested that,

$$\theta_\lambda \cong 2\arcsin(b/2d) \qquad (23)$$

2.2.2 Constant Structure Equations
a. Strain-Rate Change Tests. A discussion of constant structure equations necessarily begins with strain-rate change tests. The constant-structure strain-rate sensitivity, N, can be determined by, perhaps, two methods. The first is the strain-rate increase test, as illustrated in Figure 3, where the change in flow stress with a "sudden" change in the imposed strain-rate (cross-head rate) is measured. The "new" flow stress at a fixed substructure is that at which plasticity is initially discerned. There are, of course, some complications. One is that the stress at which plastic deformation proceeds at a "constant dislocation" substructure can be ambiguous. Also, the plastic strain-rate at this stress is not always the new cross-head rate divided by the specimen gage length due to substantial machine compliances. These complications notwithstanding, there still is value in the concept of equation (1) and strain-rate increase tests.

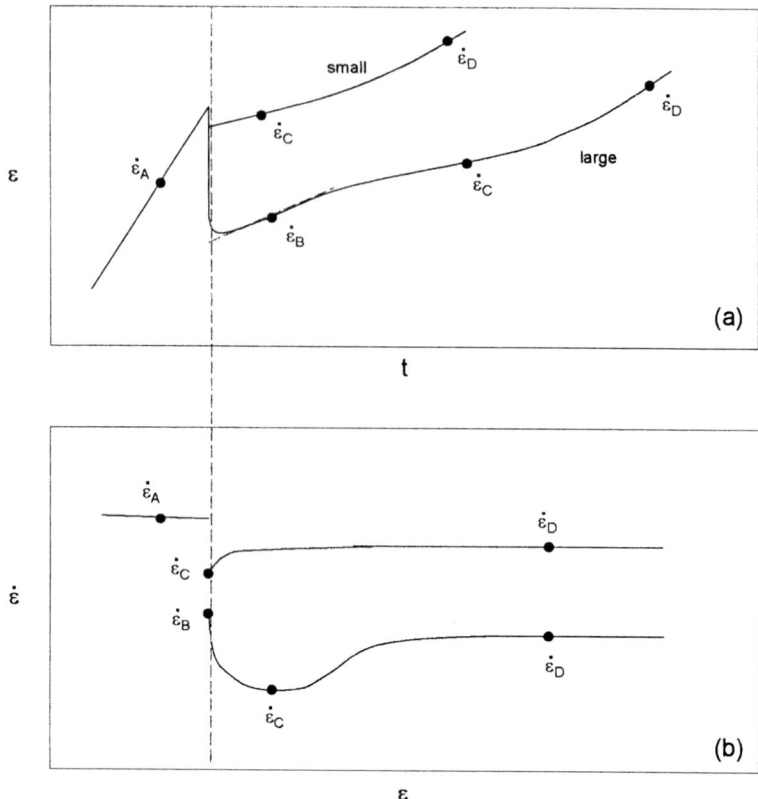

Figure 23. Description of the strain (a) and strain-rate (b) versus time and strain for stress-dip (drop) tests associated with relatively small and large decreases in the applied stress. From Ref. [10].

Another method to determine N (or m) is with stress-drop (or dip) tests illustrated in Figure 23 (based on Ref. [10]). In principle, N could be determined by noting the "new" creep-rate or strain-rate at the lower stress for the same structure as just prior to the stress dip. These stress dip tests originated over 30 years ago by Gibbs [114] and adopted by Nix and coworkers [115,116] and their interpretation is still ambiguous [10]. Biberger and Gibeling [10] published an overview of creep transients in pure metals following stress drops, emphasizing work by Gibeling and coworkers [10,87,117–120], Nix and coworkers [27,115–119], and Blum and coworkers [22,77, 92,121–125] all of whom have long studied this area as well as several others [66,104,114,126–131]. The following discussion on the stress dip test relies on this overview.

With relatively large stress drops, there are "quick" contractions that may occur as a result of an initial, rapid, anelastic component, in addition, of course, to elastic contractions, followed by slower anelastic backflow [119]. Researchers in this area tend to report a "first maximum" creep rate which occurs at "B" in Figure 23. It has been argued that the plastic strain preceding "B" is small and that the dislocation microstructure at "B" is essentially "identical" to that just prior to the stress drop. Some investigators have shown, however, that the interior or network density, ρ, may be different [131]. Also, since the creep-rate *decreases* further to a minimum value at "C" (in Figure 23), the creep at "B" has been, occasionally, termed "anomalous." $\dot{\varepsilon}_C$ (point "C") has also been referred to as "constant structure," probably, primarily, as a consequence that the subgrain size, λ, at "C" is probably very close to the same value as just prior to the stress drop, despite the observation that the interior dislocation density, ρ, appears to change [4,6,120,125,132]. Thus, with the stress dip test, the constant structure stress exponent, N, or related constant-structure descriptors, such as the activation area, Δa, or volume, ΔV_0, [132,133] may be ambiguous as the strain-rate immediately on unloading is negative, and the material does not have a fixed substructure at definable stages such as at "B" and "C." Eventually, the material "softens" to "D" as a consequence of deformation at the lower stress, to the strain-rate that most investigators have concluded corresponds to that which would have been obtained on loading the annealed metal at same temperature and stress although this is not, necessarily, a consensus view, e.g. [128,134]. Parker and Wilshire [134], for example, find that at lower temperatures, Cu, with a stress drop, did not return to the creep-rate for the uninterrupted test. Of course, it is unclear whether the rate would have eventually increased to the uninterrupted rate with larger strains that can be precluded by fracture in tensile tests.

The anelastic strains are very small for small stress-reductions and may not be observed. The creep-rate cannot be easily defined until $\dot{\varepsilon}_C$ and an "anomalous" creep is generally not observed as with large stress reductions. Again, the material eventually softens to a steady-state at "D."

The stress-dip test appears to at least be partially responsible for the introduction of the concept of an internal "backstress." That is, the backflow associated with the stress dip, observed in polycrystals and single crystals alike, has been widely presumed to be the result of an internal stress. At certain stress reductions, a zero initial creep-rate can result, which would, presumably, be at an applied stress about equal to the backstress [81,115,126,135]. Blum and coworkers [22,77,122,125, 135,136], Nix and coworkers [26,115–117], Argon and coworkers [18,137], Morris and Martin [43,44], and many others [20,53,138] (to reference a few) have suggested that the backflow or backstress is a result of high local internal stresses that are associated with the heterogeneous dislocation substructure, or subgrain walls. Recent justification for high internal stresses beyond the stress-dip test has included X-ray diffraction (XRD) and convergent beam electron diffraction (CBED) [136].

Gibeling and Nix [117] have performed stress relaxation experiments on aluminum single crystals and found large anelastic "backstrains", which they believed were substantially in excess of that which would be expected from a homogeneous stress state [26,117,139]. This experiment is illustrated in Figure 24 for a stress drop from 4 to 0.4 MPa at 400°C. If a sample of cubic structure is assumed with only one active slip system, and an orthogonal arrangement of dislocations, with a density ρ, and all segments are bowed to a critical radius, then the anelastic unbowing strain is about

$$\gamma_A = \cong \frac{\pi b \sqrt{\rho}}{8\sqrt{3}} \qquad (24)$$

Figure 18 suggests that γ_A from the network dislocations would be about 10^{-4}, about the same as suggested by Nix et al. [26,117,139]. The above equation only assumes one-third of ρ are bowing as two-thirds are on planes without a resolved shear stress. In reality, slip on {111} and {110} [130] may lead to a higher fraction of bowed dislocations. Furthermore, it is known that the subgrain boundaries are mobile. The motion may well involve (conservative) glide, leading to line tension and the potential for substantial backstrain. It is known that substantial elastic incompatibilities are associated with grain boundaries [140]. Although Nix et al. suggest that backstrain from grain boundary sliding (GBS) is not a consideration for single crystals, it has been later demonstrated that for single crystals of Al, in creep, such as in Figure 24, high angle boundaries are readily formed in the absence of classic discontinuous dynamic recrystallization (DRX) [141,142]. These incompatibility stresses may relax during (forward) creep, but "reactivate" on

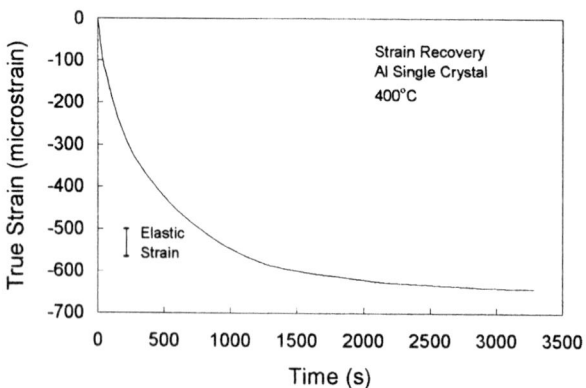

Figure 24. The backstrain associated with unloading an aluminum single crystal from 4 to 0.4 MPa at 400°C. From Ref. [117].

unloading, leading to strains that may be a fraction of the elastic strain. This may explain large (over 500 microstrain) backflow observed in near α Ti alloys after just 0.002 strain creep, where subgrains may not form [143], but fine grains are present. The subject of internal stress will be discussed more later. Substructural changes (that almost surely occur during unloading) may lead to the backstrains, not a result of long-range internal stress, such as in Figure 24, although it is not clear how these strains would develop. More recently, Muller et al. [125] suggested that subgrain boundary motion after a stress-dip may be associated with "back" strain. Hence, the "back" strains of Figure 24 do not deny the possibility for a homogeneous stress state. The arguments by Nix et al. still seem curious in view of the fact that such anelastic backstrains, if a result only of internal stresses from subgrains, as they suggest, would seem to imply internal stress value well over an order of magnitude greater than the applied stress rather than the factor of that 3 Nix et al. suggest. It is not known to the author of this review as to whether the results of Figure 21 have been reproduced.

b. Creep Equations. Equations such as (14)

$$\dot{\varepsilon}_{ss} = A_6(\chi/Gb)^3(D_{sd}Gb/kT)(\sigma_{ss}/G)^5$$

are capable of relating, at a fixed temperature, the creep rate to the steady-state flow stress. However, in associating different steady-state creep rates with (steady-state) flow stresses, it must be remembered that the dislocation structures are different. This equation does not relate different stresses and substructures at a fixed temperature *and* strain rate as, for example, the Hall–Petch equation.

Sherby and coworkers reasoned that relating the flow stress to the (e.g., steady-state) substructure at a fixed strain-rate and temperature may be performed with knowledge of N (or m) in equation (1)

$$N = \left(\frac{\partial \ln \dot{\varepsilon}}{\partial \ln \sigma}\right)_{T,s}$$

Sherby and coworkers suggested that the flow stress at a fixed elevated temperature and strain-rate is predictable through [4,6]:

$$\dot{\varepsilon} = A_7(\lambda^3)\exp[-Q_{sd}/kT](\sigma/E)^N \qquad (25)$$

for substructures resulting from steady-state creep deformation in the five-power regime. It was suggested that $N \cong 8$. Steady-state (ss) subscripts are notably absent in this equation. This equation is important in that the flow stress can be directly related to the microstructure at any fixed (e.g., reference) elevated temperature and strain-rate. Sherby and coworkers, at least at the time that this equation was formulated, believed that subgrain boundaries were responsible for elevated-temperature strength. Sherby and coworkers believed that particular value was inherent in this equation since if equation (17),

$$\frac{\sigma_{ss}}{G} = C_1 \lambda_{ss}^{-1}$$

were substituted into equation (25), then the well-established five-power-law equation [equation (9)] results,

$$\dot{\varepsilon}_{ss} = A_3 \exp[-Q_{sd}/kT]\left(\frac{\sigma_{ss}}{G}\right)^5$$

Equation (25) suggests that at a fixed temperature and strain-rate:

$$\left.\frac{\sigma}{E}\right|_{\dot{\varepsilon},T} = C_5(\lambda)^{-3/8} \quad (26)$$

(Sherby normalized stress with the Young's modulus although the shear modulus could have been used.) This equation, of course, does not preclude the importance of the interior dislocation network over the heterogeneous dislocation substructure (or subgrain walls) for steady-state substructures (on which equation (25) was based). This is because there is a fixed relationship between the steady-state subgrain size and the steady-state interior dislocation density. Equation (26) could be reformulated, without a loss in accuracy, as

$$\left.\frac{\sigma}{G}\right|_{\dot{\varepsilon},T} = k_2(\rho)^{-3p/8 \simeq -3/16} \quad (27)$$

c. Dislocation Density and Subgrain-Based Constant-Structure Equations.

Equation (26) for *sub*grain strengthening does not have a strong resemblance to the well-established Hall–Petch equation for *high*-angle grain-boundary strengthening:

$$\left.\sigma_y\right|_{\dot{\varepsilon},T} = \sigma_o + k_y g^{-1/2} \quad (28)$$

where $\sigma_y|_{\dot{\varepsilon},T}$ is the yield or flow stress (at a reference or fixed temperature and strain-rate), and k_y is a constant, g is the average grain diameter, and σ_o is the single crystal strength and can include solute strengthening as well as dislocation hardening. [Of course, subgrain boundaries may be the microstructural feature associated with elevated temperature strength and the rate-controlling process for creep, without obedience to equation (28).] Nor does equation (27) resemble the classic dislocation hardening equation [144]:

$$\sigma_y|_{\dot{\varepsilon},T} = \sigma_0' + \alpha M G b (\rho)^{1/2} \tag{29}$$

where $\sigma_y|_{\dot{\varepsilon},T}$ is the yield or flow stress (at a reference or fixed temperature and strain-rate), σ_0' is the near-zero dislocation density strength and can include solute strengthening as well as grain-size strengthening, and M is the Taylor factor, 1–3.7 and α is a constant, often about 0.3 at ambient temperature. (This constant will be dependent upon the units of ρ, as line-length per unit volume, or intersections per unit area, the latter being a factor of 2 lower for identical structure.) Both equations (28) and (29) assume that these hardening features can be simply summed to obtain their combined effect. Although this is reasonable, there are other possibilities [145]. Equation (29) can be derived on a variety of bases (e.g., bowing stress, passing stress in a "forest" of dislocations, etc.), all essentially athermal, and may not always include a σ_0' term.

Even with high-purity aluminum experiments (99.999% pure), it is evident in constant strain-rate mechanical tests, that annealed polycrystal has a yield strength (0.002 plastic strain offset) that is about one-half the steady-state flow stress [4,146] that cannot be explained by subgrain (or dislocation) hardening; yet this is not explicitly accounted in the phenomenological equations (e.g., equations 26 and 27). When accounted, by assuming that σ_0 (or σ_0') = $\sigma_y|_{T,\dot{\varepsilon}}$ for annealed metals, Sherby and coworkers showed that the resulting subgrain-strengthening equation that best describes the data form would not resemble equation (28), the classic Hall–Petch equation; the best-fit $(1/\lambda)$ exponent is somewhat high at about 0.7. Kassner and Li [147] also showed that there would be problems with assuming that the creep strength could be related to the subgrain size by a Hall–Petch equation. The constants in equation (28), the Hall–Petch equation, were experimentally determined for high-purity annealed aluminum with various (HAB) grain sizes. The predicted (extrapolated) strength (at a fixed elevated temperature and strain-rate) of aluminum with grain sizes comparable to those of steady-state subgrain sizes was substantially *lower* than the observed value. Thus, even if low misorientation subgrain walls strengthen in a manner analogous to HABS, then an "extra strength" in steady-state, subgrain containing, structures appears from sources other than that provided by boundaries. Kassner and Li suggested that this extra strength may be due to the steady-state dislocation density not associated with the subgrains, and dislocation hardening was observed. Additional discussion of grain-size effects on the creep properties will be presented later.

The hypothesis of dislocation strengthening was tested using data of high-purity aluminum as well as a Class M alloy, AISI 304 austenitic stainless steel (19Cr-10Ni)

[107,148]. It was discovered that the classic dislocation hardening (e.g., Taylor) equation is reasonably obeyed if σ'_0 is approximately equal to the annealed yield strength. Furthermore, the constant α in equations (29), at 0.29, is comparable to the observed values from ambient temperature studies of dislocation hardening [144,149–151] as will be discussed more later. Figure 25 illustrates the polycrystalline stainless steel results. The λ and ρ values were manipulated by combinations of creep and cold work. Note that the flow stress at a reference temperature and strain-rate [that corresponds to nearly within 5-power-law-creep (750°C in Figure 20)] is independent of λ for a fixed ρ. The dislocation strengthening conclusions are consistent with the experiments and analysis of Ajaja and Ardell [152,153] and Shi and Northwood [154,155] also on austenitic stainless steels.

Henshall et al. [156] also performed experiments on Al-5.8at% Mg in the three-power regime where subgrain boundaries only sluggishly form. Again, the flow stress was completely independent of the subgrain size (although these tests were relevant to three-power creep). The Al-Mg results are consistent with other

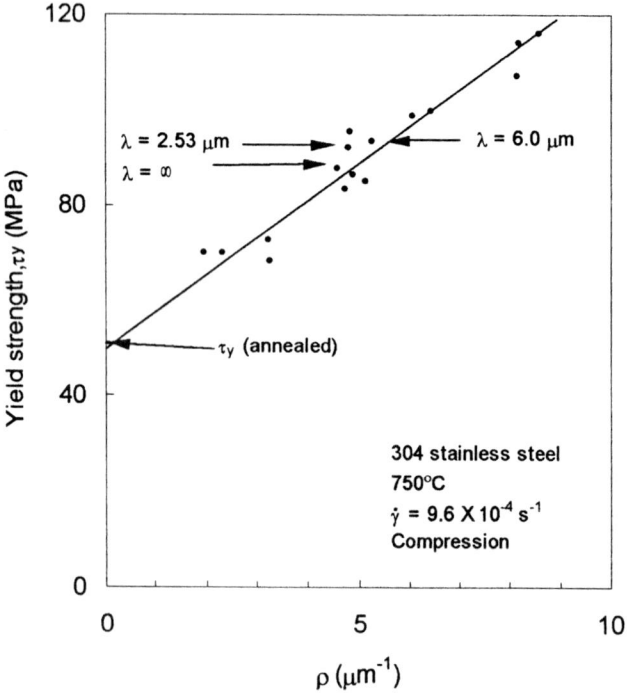

Figure 25. The elevated temperature yield strength of 304 stainless steel as a function of the square root of the dislocation density (not associated with subgrain boundaries) for specimens of a variety of subgrain sizes. (Approximately five-power-law temperature/strain-rate combination.) Based on Ref. [148].

experiments by Weckert and Blum [121] and the elevated temperature In situ TEM experiments by Mills [157]. The latter experiments did not appear to show interaction between subgrain walls and gliding dislocations. The experiments of this paragraph will be discussed in greater detail later.

2.2.3 Primary Creep Microstructures

Previous microstructural trends in this review emphasized steady-state substructures. This section discusses the development of the steady-state substructure during primary creep where hardening is experienced. A good discussion of the phenomenological relationships that describe primary creep was presented by Evans and Wilshire [28]. Primary creep is often described by the phenomenological equation,

$$\ell = \ell_o(1 + \beta t^{1/3})e^{\chi' t} \tag{30}$$

This is the classic Andrade [158] equation. Here, P is the instantaneous gage length of a specimen and P_0 is the gage length on loading (apparently including elastic deflection) and β and χ' are constants. This equation leads to equations of the form,

$$\varepsilon = at^{1/3} + ct + dt^{4/3} \tag{31}$$

which is the common phenomenological equation used to describe primary creep. Modifications to this equation include [159]

$$\varepsilon = at^{1/3} + ct \tag{32}$$

and [160]

$$\varepsilon = at^{1/3}bt^{2/3} + ct \tag{33}$$

or

$$\varepsilon = at^b + c^t \tag{34}$$

where [161]

$$0 < b < 1.$$

These equations cannot be easily justified, fundamentally [23].

For a given steady-state stress and strain-rate, the steady-state microstructure appears to be independent as to whether the deformation occurs under constant stress or constant strain-rate conditions. However, there are some differences between the substructural development during a constant stress as compared to constant strain-rate primary-creep. Figure 26 shows Al-5at%Zn at 250°C at a

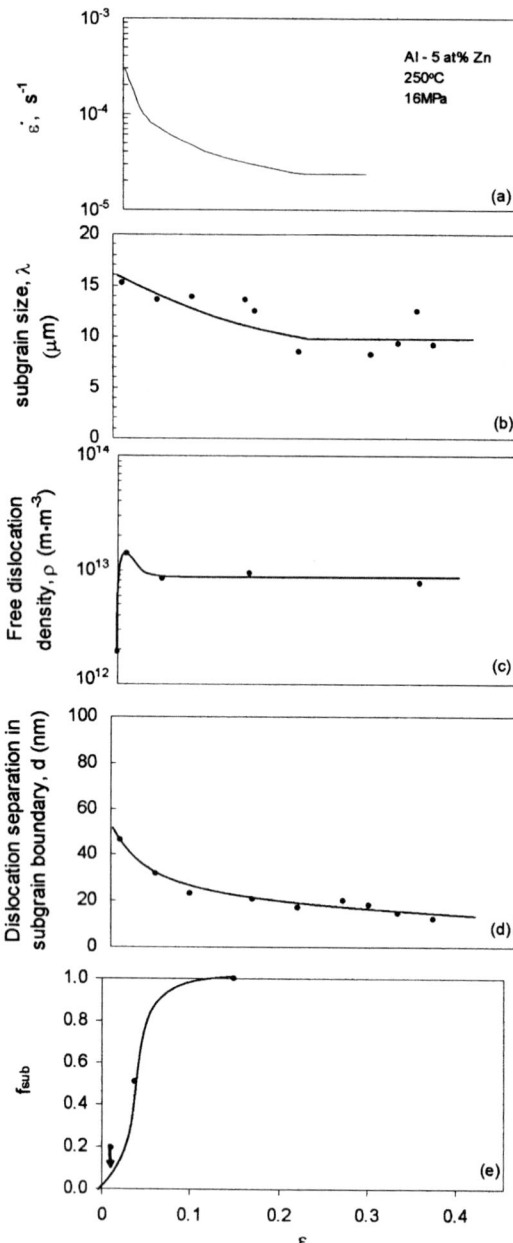

Figure 26. The constant-stress primary Creep transient in Al-5at%Zn (essentially identical behavior to pure Al) illustrating the variation of the average subgrain intercept, λ, density of dislocations not associated with subgrain walls, ρ, and the spacing, d, of dislocations that comprise the boundaries. The fraction of material occupied by subgrains is indicated by f_{sub}. Based on Ref. [77].

constant stress of 16 MPa [77]. Again, this is a class M alloy which, mechanically, behaves essentially identical to pure Al. The strain-rate continually decreases to a strain of about 0.2, where mechanical steady-state is achieved. The density of dislocations not associated with subgrain boundaries is decreasing from a small strain after loading (< 0.01) to steady-state. This constant-stress trend with the "free" dislocation density is consistent with early etch pit analysis of Fe-3% Si [162], and the TEM analysis 304 stainless steel [163], α-Fe [164] and Al [165,166]. Some have suggested that the decrease in dislocation density in association with hardening is evidence that hardening cannot be associated with dislocations and is undisputed proof that subgrains influence the rate of plastic deformation [81]. However, as will be discussed later, this may not be accurate. Basically, Kassner [107,149] suggested that for constant-stress transients, the network dislocations cause hardening but the fraction of mobile dislocations may decrease, leading to strain-rate decreases not necessarily associated with subgrain formation. Figure 26 plots the average subgrain size only in areas of grains that contain subgrains. The volume of Al 5at.% Zn is not completely filled with subgrains until steady-state, at $\varepsilon \cong 0.2$. Thus, the Subgrain size averaged over the entire volume would show a more substantial decrease during primary creep. The average spacing, d, of dislocations that comprises subgrain walls decreases *both* during primary, and, at least, during early steady-state. This trend in d and/or θ_λ was also observed by Suh *et al.* in Sn [167], Morris and Martin [42,43] and Petry *et al.* [168] in Al-5at%Zn, Orlova *et al.* [166] in Al, Karashima *et al.* [112] in αFe, and Kassner *et al.* in Al [146] and 304 stainless steel [98]. These data are illustrated in Figure 27.

Work-hardening, for constant strain-rate creep, microstructural trends were examined in detail by Kassner and coworkers [98,107,146,169] and are illustrated in Figures 28 and 29 for 304 stainless steel and Figure 30 for pure Al. Figure 28 illustrates the dislocation substructure, quantitatively described in Figure 29. Figure 30(a) illustrates the small strain region and that steady-state is achieved by $\bar{\varepsilon} = 0.2$. Figure 30(b) considers larger strains achieved using torsion of solid aluminum specimens. Figure 31 illustrates a subgrain boundary in a specimen deformed in Figure 30(b) to an equivalent uniaxial strain (torsion) of 14.3 (a) with all dislocations in contrast in the TEM under multiple beam conditions and (b) one set out of contrast under two beam conditions (as in Figure 20). The fact that, at these large strains, the misorientations of subgrains that form from dislocation reactions remain relatively low ($\theta_{\lambda_{ave}} < 2°$) and subgrains remain equiaxed suggests boundaries migrate and annihilate. Here, with constant strain-rate, we observe similar subgrain trends to the constant stress trends of Blum in Figure 26 at a similar fraction of the absolute melting temperature. Of course, 304 has a relatively low stacking fault energy while aluminum is relatively high. In both cases, the average subgrain size (considering the *entire* volume) decreases over primary creep. The lower stacking fault energy 304 austenitic

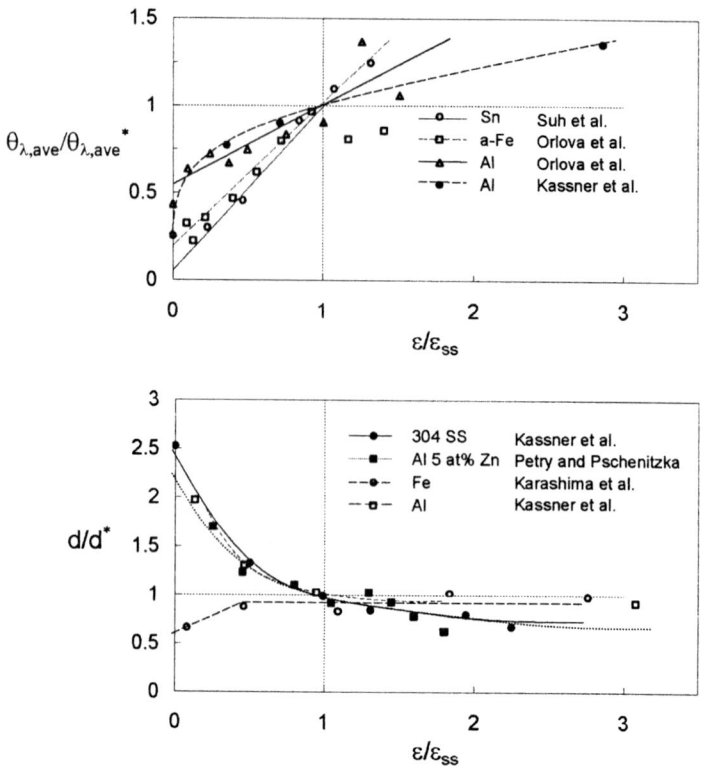

Figure 27. The variation of the average misorientation angle across subgrain walls, $\theta_{\lambda,ave}$, and separation of dislocations comprising subgrain walls with fraction of strain required to achieve steady-state, $\varepsilon/\varepsilon_{ss}$ for various metals and alloys. $\theta^*_{\lambda,ave}$ and d^* are values at the onset of steady-state.

stainless steel, however, requires substantially more primary creep strain (0.4 vs. 0.2) to achieve steady-state at a comparable fraction of the melting temperature. It is possible that the subgrain size in 304 stainless steel continues to decrease during steady-state. Under constant strain-rate conditions, the density of dislocations not associated with subgrain boundaries *monotonically increases* with increased flow stress for both austenitic stainless steel and high-purity aluminum. This is opposite to constant-stress trends. Similar to the constant-stress trends, both pure Al and 304 stainless steel show decreasing d (increasing θ) during primary and "early" steady-state creep. Measurements of "d" were considered unreliable at strains beyond 0.6 in Al and only misorientation angles are reported in Figure 30(b). [It should be mentioned that HABs form by elongation of the starting grains through geometric dynamic recrystallization, but these are not included in Figure 30(b). This mechanism is discussed in greater detail in a later section.]

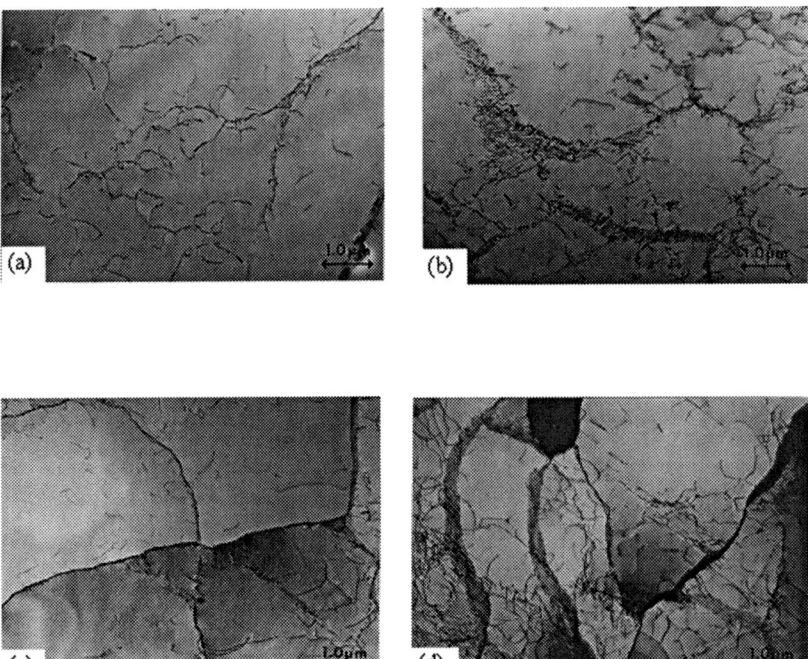

Figure 28. TEM micrographs illustrating the evolution of the dislocation substructure during primary creep of AISI 304 stainless steel torsionally deformed at 865°C at $\bar{\dot{\varepsilon}} = 3.2 \times 10^{-5}\,\text{s}^{-1}$, to strains of 0.027 (a), 0.15 (b), 0.30 (c), and 0.38 (d).

2.2.4 Creep Transient Experiments

As mentioned earlier, Creep transient experiments have been performed by several investigators [127,129,170] on high- and commercial-purity aluminum, where a steady-state is achieved at a fixed stress/strain-rate followed by a change in the stress/strain-rate. The strain-rate/stress change is followed by a creep "transient," which leads to a new steady-state with, presumably, the characteristic dislocation substructure associated with an uninterrupted test at the (new) stress/strain-rate. These investigators measured the subgrain size during the transient and subsequent mechanical steady-state, particularly following a drop in stress/strain-rate. Although Ferriera and Stang [127] found, using less reliable polarized light optical metallography (POM), that changes in λ in Al correlate with changes in $\dot{\varepsilon}$ following a stress-drop, Huang and Humphreys [129] and Langdon et al. [170] found the opposite using TEM; the λ continued to change even once a new mechanical steady-state was reached. Huang and Humphreys [129] and Langdon et al. [170] showed that the dislocation microstructure changes with a stress drop, but the dislocation

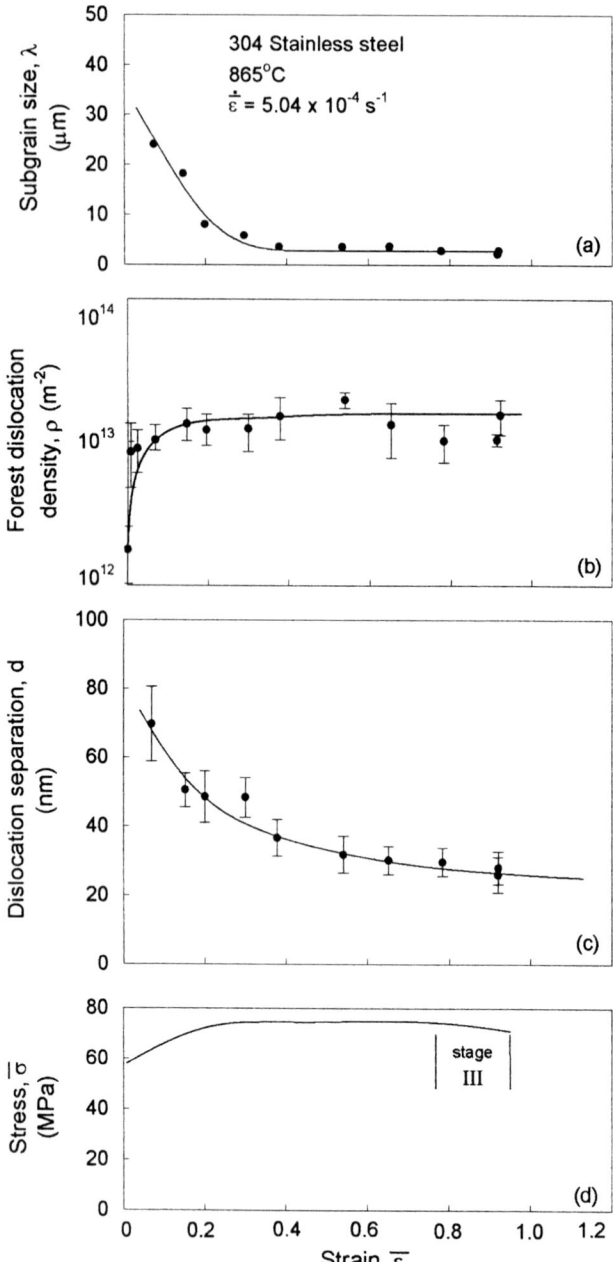

Figure 29. Work-hardening at a constant strain-rate primary creep transient in AISI 304 stainless steel, illustrating the changes in λ, ρ, and d with strain. Based on Ref. [98].

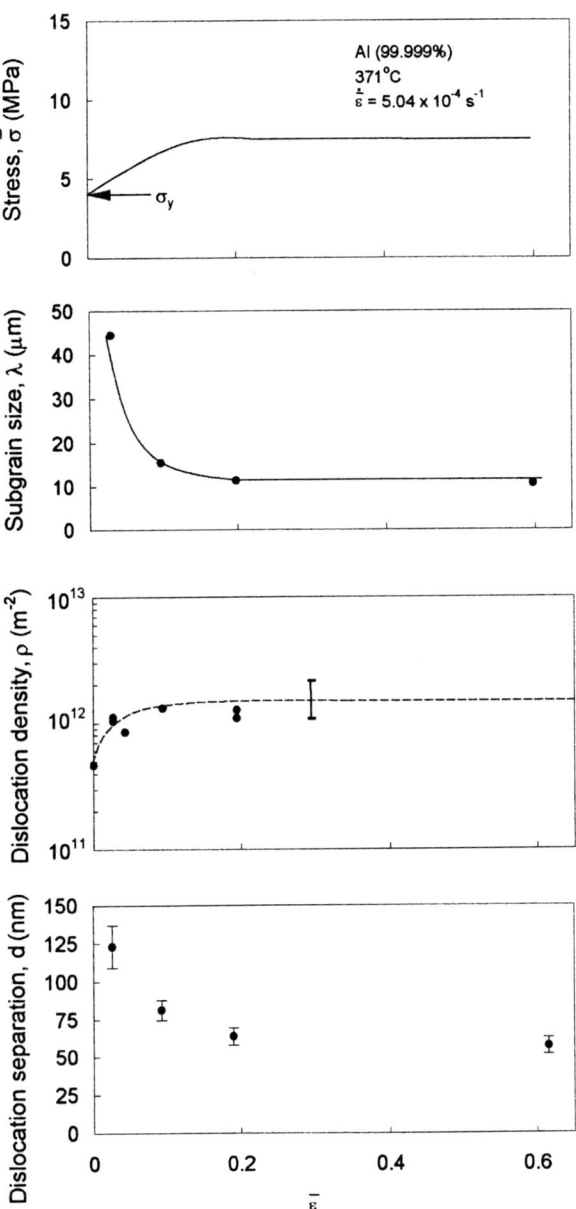

Figure 30. The work-hardening during a constant strain-rate creep transient for Al, illustrating the variation of λ, ρ, d, and $\theta_{\lambda_{ave}}$ over primary and secondary creep. The bracket refers to the range of steady-state dislocation density values observed at larger strains [e.g., see (b)]. From Ref. [146].

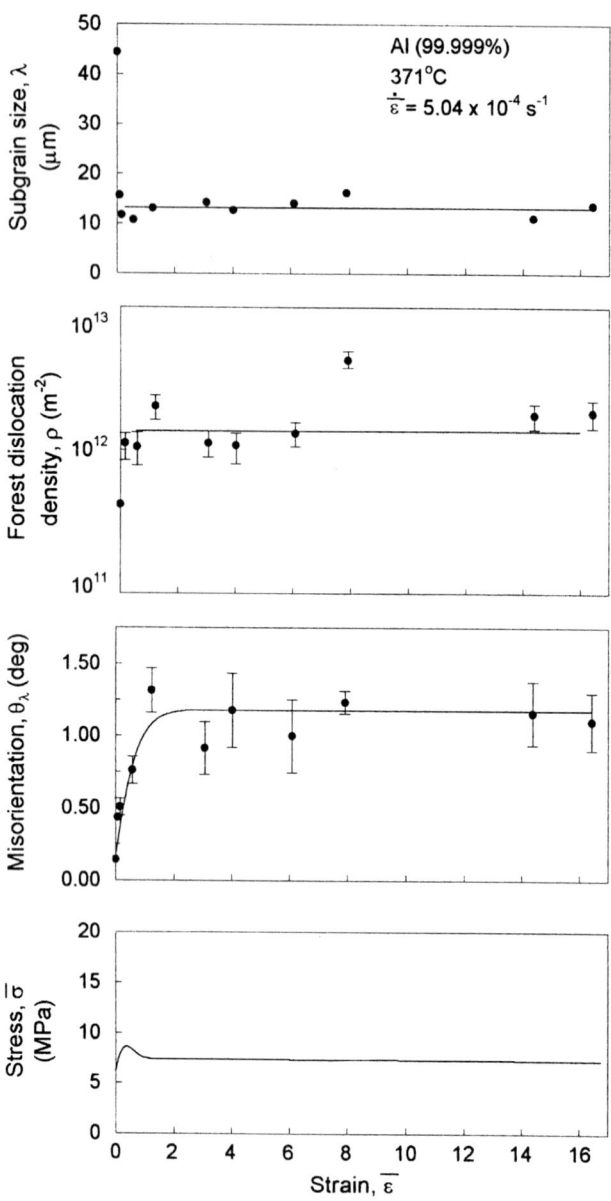

Figure 30. Continued.

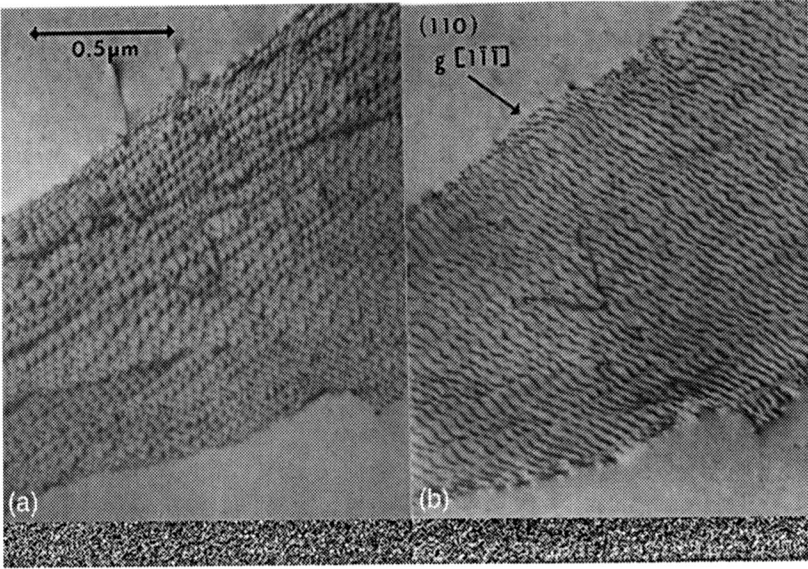

Figure 31. TEM micrographs of a subgrain boundary in Al deformed at 371°C at $\dot{\bar{\varepsilon}} = 5.04 \times 10^{-4}\,\text{s}^{-1}$, to steady-state under (a) multiple and (b) two-beam diffraction conditions. Three sets of dislocations, of, apparently, nearly screw character. From Ref. [146].

density follows the changes in creep-rate more closely than the subgrain size in high-purity aluminum. This led Huang and Humphreys to conclude as did Evans et al. [171] the "free" dislocation density to be critical in determining the flow properties of high-purity aluminum. Parker and Wilshire [134] made similar conclusions for Cu in the five-power-law regime. Blum [22] and Biberger and Gibeling [10] suggest that interior dislocations can be obstacles to gliding dislocations, based on stress drop experiments leading to aluminum activation area calculations.

2.2.5 Internal Stress

One of the important suggestions within the creep community is that of the internal (or back) stress which, of course, has been suggested for plastic deformation, in general. The concept of internal or backstress stresses in materials may have first been discussed in connection with the Bauschinger effect, which is observed both at high and low temperatures and is illustrated in Figure 32 for Al single crystal oriented for single slip at $-196°C$, from Ref. [172]. The figure illustrates that the metal strain hardens after some plastic straining. On reversal of the direction of straining, the metal plastically flows at a stress less in magnitude than in the forward direction, in contrast to what would be expected based on isotropic hardening.

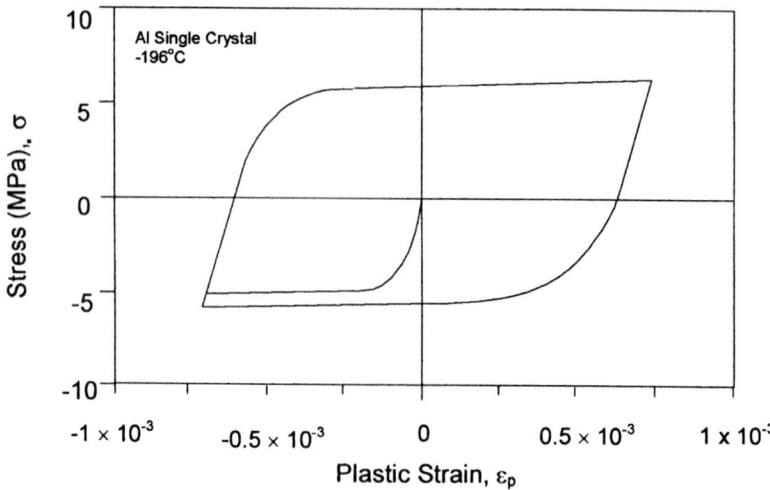

Figure 32. The hysteresis loop indicating the Bauschinger effect in an Al single crystal deformed at 77 K. From Ref. [172].

Figure 32 shows that not only is the flow stress lower on reversal, but that the hardening features are different as well. Sleeswyk et al. [173] analyzed the hardening features in several materials at ambient temperature and found that the hardening behavior on reversal can be modeled by that of the monotonic case provided a small (e.g., 0.01) "reversible" strain is subtracted from the early plastic strain associated with each reversal. This led Sleeswyk and coworkers to conclude the Orowan-type mechanism (no internal or backstress) [174] with dislocations easily reversing their motion across cells. Sleeswyk suggested that gliding dislocations, during work hardening, encounter increasingly effective obstacles and the stress necessary to activate further dislocation motion as plasticity continually increases. On reversal of the direction of straining, however, the dislocations will need to only move past those obstacles they have already surmounted. Thus, the flow stress is initially relatively low. High temperature work by Hasegawa et al. [175] suggested that dissolution of the cell/subgrains occurred with a reversal of the strain, indicating an "unraveling" of the substructure in Cu-16 at% Al, perhaps consistent with the ideas of Sleeswyk and coworkers. Others, as mentioned earlier, have suggested that a nonhomogeneous state of (back) stress may assist plasticity on reversal. There are two broad categories for back or Internal stresses.

In a fairly influential development, Mughrabi [138,176] advanced the concept of relatively high internal stresses in subgrain walls and cell structures. He advocated the simple case where "hard" (high dislocation density walls or cells) and soft (low dislocation density) elastic-perfectly-plastic regions are compatibly sheared

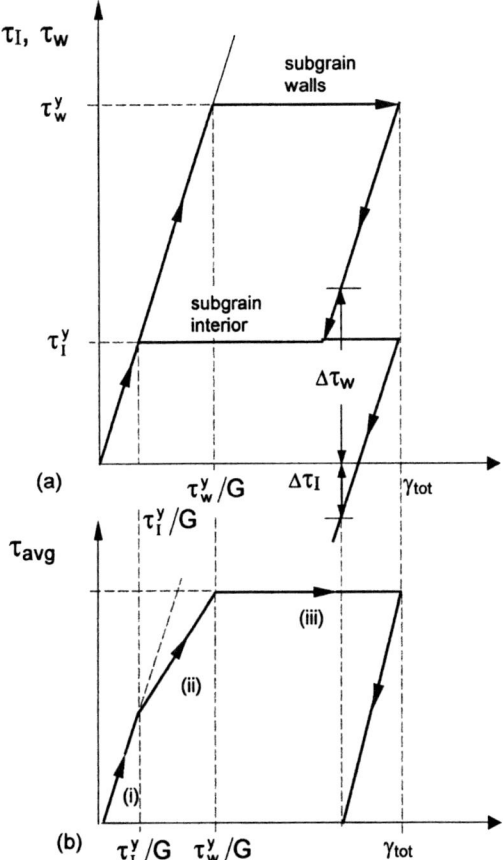

Figure 33. The composite model illustrating the Bauschinger effect. The different stress versus strain behaviors of the (hard) subgrain walls and the (soft) subgrain interiors are illustrated in (a), while the stress versus strain behavior of the composite is illustrated in (b). When the composite is completely unloaded, the subgrain interior is under compressive stress. This leads to a yielding of the softer component in compression at a "macroscopic" stress less than τ_I^y under initial loading. Hence, a Bauschinger effect due to inhomogeneous (or internal) stresses is observed. Note that the individual components are elastic-perfectly-plastic.

in parallel. The "composite" model is illustrated in Figure 33. Basically, the figure shows that each component yields at different stresses and, hence, the composite is under a heterogeneous stress-state with the cell walls (subgrains at high temperatures) having the higher stress. This composite may also rationalize the Bauschinger effect. As the "hard" and "soft" regions are unloaded in parallel, the hard region eventually, while its stress is still positive, places the soft region in compression. Figure 30 indicates that when the "total" or "average" stress is zero, the stress in the hard region is positive, while negative in the soft region. Thus, a Bauschinger effect can be observed, where plasticity occurs on reversal at a lower "average" magnitude of stress than that on initial unloading. This leads to an interpretation of an inhomogeneous stress-state and "back-stresses." As will be discussed subsequently, this "composite" model appears to have been embraced by Derby and Ashby [177], Blum and coworkers [122,136,178], as well as others for Five-power-law-creep.

In situ deformation experiments by Lepinoux and Kubin [179] and the well-known neutron irradiation experiments by Mughrabi [138] find evidence for such internal stresses that are, roughly, a factor of 3 higher than the applied stress. These are based on dislocation curvature measurements that assume fairly precise measurement of elastic strain energy of a dislocation and that dislocation configurations are in equilibrium, neither of which are satisfied.

More recent *in situ* reversed deformation experiments by Kassner and coworkers (as well as other experiments), however, suggest that this conclusion may not be firm [180].

Morris and Martin [42,43] concluded that dislocations are ejected from sources at the subgrain boundary by high local stresses. High local stresses, perhaps a factor of 20 higher than the applied stress, were concluded by observing the radius of curvature of "ejecting dislocations" "frozen in place" (amazingly) by a precipitation reaction in Al-5at%Zn on cooling from the creep temperature. As mentioned earlier, stress-dip tests have often been interpreted to suggest internal stresses [112,115,116,135].

Another concept of backstress is related to dislocation configurations. This suggestion was proposed by Argon and Takeuchi [137], and subsequently adopted by Gibeling and Nix [117] and Nix and Ilschner [26]. With this model, the subgrain boundaries that form from dislocation reaction, bow under action of the shear stress and this creates relatively high local stresses. The high stresses in the vicinity of the boundary are suggested to be roughly a factor of 3 larger than the applied stress. On unloading, a negative stress in the subgrain interior causes reverse plasticity (or anelasticity). There is a modest anelastic back-strain that is associated with this backstress that is illustrated in Figure 24.

One of the most recent developments in this area of internal stresses in creep-deformed metals was presented by Straub and coworkers [136] and Borbély and coworkers [181]. This work consisted of X-ray diffraction (XRD) and convergent beam electron diffraction (CBED) of specimens creep tested to steady-state. Some X-ray creep-experiments were performed *in situ* or under stress. The CBED was performed at $-196°C$ on unstressed thin (TEM) foils. Basically, both sets of experiments are interpreted by the investigators to suggest that the lattice parameters within the specimens are larger near subgrain (and cell) walls than in the subgrain interior. These are important experiments.

Basically, X-ray peaks broaden with plastic deformation as illustrated in Figure 34. A "deconvolution" is performed that results in two nearly symmetric peaks. One peak is suggested to represent the small amount of metal in the vicinity of subgrains where high local stresses are presumed to increase the lattice parameter, although there does not appear to be direct evidence of this. The resolution limit of X-rays is about 10^{-4} [182], rendering it sensitive to internal stresses that are of the order or larger than the applied stress. Aside from resolution difficulties associated with small changes in the lattice parameter with small changes in stress with X-rays, there may be inaccuracies in the "deconvolution" or decomposition exercise as suggested by Levine [183] and others [184]. These observations appear to be principally relevant to

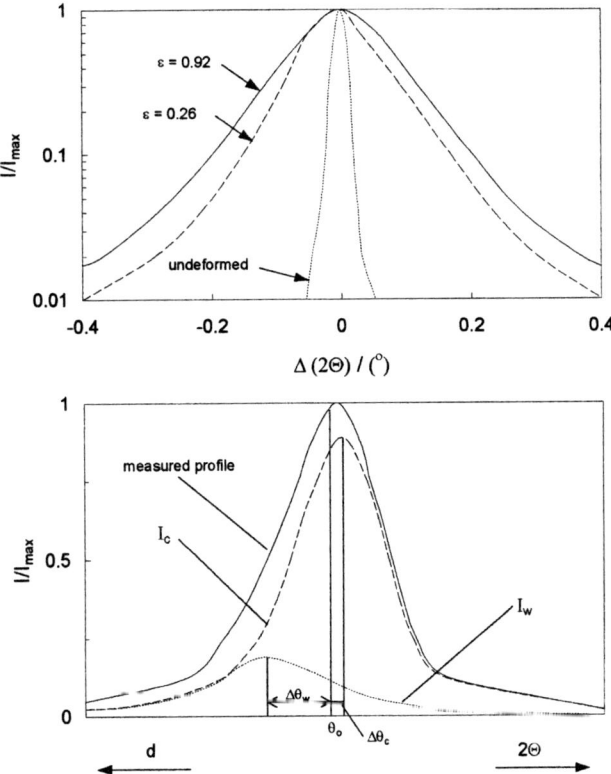

Figure 34. The X-ray diffraction peak in Cu deformed to various strains, showing broadening. A "deconvolution" is performed that leads to two symmetric peaks that may suggest a heterogeneous stress state. Based on Ref. [181].

Cu, with other materials, such as Al, not generally evincing such asymmetry. CBED can probe smaller areas with a 20 nm beam size rather than the entire sample as with X-rays and is potentially more accurate. However, the results by Borbély et al. that suggest high local stresses in the vicinity of subgrain boundaries in copper, based on CBED, may be speculative. The results in Figure 3 of Ref. [181] are ambiguous. Of course, there is always the limitation with performing diffraction experiments in unloaded specimens, as any stress heterogeneities may relax before the diffraction experiments are completed. Performing a larger number CBED on foils under load would, although difficult, have been preferable. Recent CBED experiments by Kassner et al. [185,186] on cyclically deformed Cu and unloaded Al single crystals deformed at a temperature/strain-rate regime similar to Figure 30 did not detect the presence of any residual stresses in the vicinity of dislocation heterogeneities.

The importance of the backstress concept also appears within the phenomenological equations where the applied stress is decomposed. Nix and Ilschner [26] attempted to rationalize power-law breakdown by suggesting that the backstress is a decreasing fraction of the applied stress with decreasing temperature, particularly below about 0.5 T_m. The decreasing fraction of backstress was attributed to less-defined subgrains (cell walls) at decreasing stress leading to a dominating contribution by a glide-controlled mechanism. However, it is now well established that well-defined subgrains form with sufficient strain even at very low temperatures [11,12,47].

2.3 RATE-CONTROLLING MECHANISMS

2.3.1 Introduction

The mechanism for plastic flow for five-power-law-creep is generally accepted to be diffusion controlled. Evidence in addition to the activation energy being essentially equal to that of lattice self diffusion includes Sherby and Weertman's analysis showing that the activation volume for creep is also equal to that of self-diffusion [5]. More recent elegant experiments by Campbell et al. [40] showed that impurity additions that alter the self diffusivity also correspondingly affect the creep-rate. However, an established theory for five-power-law-creep is not available although there have been numerous attempts to develop a fundamental mathematical description based on dislocation-climb control. This section discusses some selected attempts.

a. Weertman [25,187–189]. This was one of the early attempts to fundamentally describe creep by dislocation climb. Here, the creep process consists of glide of dislocations across relatively large distances, $\bar{x}_g$, followed by climb at the rate-controlling velocity, v_c, over a distance, $\bar{x}_c$. The dislocations climb and annihilate at a rate predictable by a concentration gradient established between the equilibrium vacancy concentration,

$$c_v = c_0 \exp(-Q_v/kT) \tag{35}$$

and the concentration near the climbing dislocation.

The formation energy for a vacancy, Q_v, is altered in the vicinity of a dislocation in a solid under an applied stress, σ, due to work resulting from climb,

$$c_v^d = c_0 \exp(-Q_v/kT)\exp(\pm\sigma\Omega/kT) \tag{36}$$

where Ω is the atomic volume. Again, the (steady-state) flux of vacancies determines the climb velocity,

$$v_c \cong 2\pi\left(\frac{D}{b}\right)(\sigma\Omega/kT)\ell n(R_0/b) \tag{37}$$

where R_0 is the diffusion distance, related to the spacing of dislocations [Weertman suggests $\rho\, nR_0/b \cong 3\, \rho n\, 10$].

Weertman approximates the average dislocation velocity, $\bar{v}$,

$$\bar{v} \cong v_c \bar{x}_g/\bar{x}_c \tag{38}$$

$$\dot{\varepsilon}_{ss} = \rho_m b\bar{v} \tag{39}$$

Weertman assumes

$$\rho_m \cong \left(\frac{\sigma_{ss}}{Gb}\right)^2 \tag{40}$$

Weertman appears to suggest that the density of dislocations $\rho \cong \rho_m$ and dislocation interaction suggests that ρ_m should scale with σ_{ss} by equation (40). This is also analogous to the phenomenological equation (19), leading to

$$\dot{\varepsilon}_{ss} = K_6 \frac{D_{sd}}{b^2}(G\Omega/kT)\left(\frac{\bar{x}_g}{\bar{x}_c}\right)\left(\frac{\sigma}{G}\right)^3 \tag{41}$$

the classic "natural" or Three-Power-Law equation.

b. Barrett and Nix [190]. Several investigators considered five-power-law-creep as controlled by the non-conservative (climb) motion of (edge) jogs of screw dislocations [112,190,191]. The models appear similar to the earlier description by

Weertman, in that climb motion controls the average dislocation velocity and that the velocity is dictated by a vacancy flux. The flux is determined by the diffusivity and the concentration gradient established by climbing jogs.

The model by Barrett and Nix [190] is reviewed as representative of these models. Here, similar to the previous equation (39),

$$\dot{\gamma}_{ss} = \rho_{ms} \bar{v} b \qquad (42)$$

where $\dot{\gamma}_{ss}$ is the steady-state (shear) creep-rate from screws with dragging jogs and ρ_{ms} is the density of mobile screw dislocations.

For a vacancy producing jog in a screw segment of length, j, the chemical dragging force on the jog is:

$$f_p = \frac{kT}{b} \ln \frac{c_p}{c_v} \qquad (43)$$

where c_p is the concentration of vacancies in the vicinity of the jogs. The jog is considered a moving point source for vacancies, and it is possible to express c_p as a function of D_v, and the velocity of the jog, v_p,

$$c_p^* - c_v = \frac{v_p}{4\pi D_v b^2} \qquad (44)$$

where c_p^* is the steady-state vacancy concentration near the jog. Substitution of (44) into (43) leads to

$$f_p = \frac{kT}{b} \ln \left[1 + \frac{v_p}{4\pi D_v b^2 c_v} \right] \qquad (45)$$

and with $\tau bj = f_p$,

$$v_p = 4\pi D_v b^2 c_v \left[\exp\left(\frac{\tau b^2 j}{kT}\right) - 1 \right] \qquad (46)$$

Of course, both vacancy producing and vacancy absorbing jogs are present but, for convenience, the former is considered and substituting (46) into (42) yields

$$\dot{\gamma}_{ss} = 4\pi D \beta_2 \left(\frac{b}{a_0}\right) \rho_{ms} \left[\exp\left(\frac{\tau b^2 j}{kT}\right) - 1 \right] \qquad (47)$$

where a_0 is the lattice parameter.

Barrett and Nix suggest that $\rho_{ms} = A_8\sigma^3$ (rather than $\rho \propto \sigma^2$) and $\dot{\gamma}_{ss} \cong A_9\tau^4$ is obtained. One difficulty with the theory [equation (47)] is that all (at least screw) dislocations are considered mobile. With stress drops, the strain rate is predicted to decrease, as observed. However, ρ_{ms} is also expected to drop (with decreasing ρ_{ss}) with time and a *further* decrease in $\dot{\gamma}_{ss}$ is predicted, despite the observation that $\dot{\gamma}$ (and $\dot{\gamma}_{ss}$) *increases*.

c. Ivanov and Yanushkevich [192]. These investigators were among the first to explicitly incorporate subgrain boundaries into a fundamental climb-control theory. This model is widely referenced. However, the described model (after translation) is less than very lucid, and other reviews of this theory [187] do not provide substantial help in clarifying all the details of the theory.

Basically, the investigators suggest that there are dislocation sources within the subgrain and that the emitted dislocations are obstructed by subgrain walls. The emitted dislocations experience the stress fields of boundary and other emitted dislocations. Subsequent slip or emission of dislocations requires annihilation of the emitted dislocations at the subgrain wall, which is climb controlled. The annihilating dislocations are separated by a mean height,

$$\bar{h}_m = k_3 \frac{Gb}{\tau} \qquad (48)$$

where the height is determined by equating a calculated "backstress" to the applied stress, τ. The creep-rate

$$\dot{\varepsilon}_{ss} = \frac{\lambda^2 b \bar{v} \rho'_m}{\bar{h}_m} \qquad (49)$$

where $\rho'_m = 1/\bar{h}_m\lambda^2$; where ρ'_m is the number of dislocation loops per unit volume. The average dislocation velocity

$$\bar{v} = k_4 D_v b^2 \exp(\tau b^3/kT - 1) \qquad (50)$$

similar to Weertman's previous analysis. This yields,

$$\dot{\varepsilon}_{ss} = \frac{k_5 D_{sd} bG}{kT}\left(\frac{\tau}{G}\right)^3 \qquad (51)$$

In this case, the third power is a result of the inverse dependence of the climb distance on the applied stress. Modifications to the model have been presented by Nix and Ilschner, Blum and Weertman [26,193–195].

d. Network Models (Evans and Knowles [105]). Several investigators have developed models for five-power-creep based on climb control utilizing dislocation networks, including early work by McLean and coworkers [84,196], Lagneborg and coworkers [85,197], Evans and Knowles [105], Wilshire and coworkers [104,126], Ardell and coworkers [54,86,101,102,152,153], and Mott and others [106,109, 198–200]. There are substantial similarities between the models. A common feature is that the dislocations interior to the subgrains are in the form of a Frank network [87]. This is a three-dimensional mesh with an average link length between nodes of ℓ illustrated in Figure 19. The network coarsens through dislocation climb and eventually links of a critical length ℓ_c are activated dislocation sources (e.g., Frank–Read). Slip is caused by activated dislocations that have a presumed slip length, $\bar{x}_g \cong \ell$. The emitted dislocations are absorbed by the network and this leads to mesh refinement. There is a distribution of link lengths inferred from TEM of dislocation node distributions, such as illustrated in Figure 35 for Al in the five-power regime. Here $\phi(\ell)$ is the observed frequency function for link-length distribution.

Evans and Knowles [105] developed a creep equation based on networks, basically representative of the other models, and this relatively early model is chosen for illustrative purposes. Analogous to Weertman, discussed earlier, Evans and Knowles suggest that the vacancy concentration near a climbing

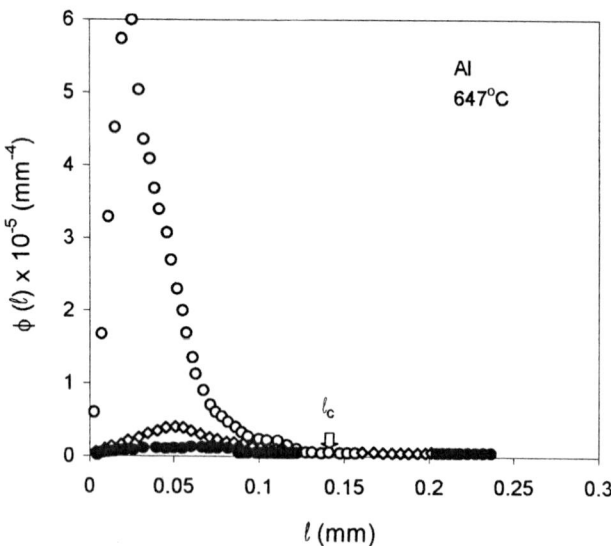

Figure 35. The observed distribution of links in a Frank network [101] in Al. A calculated critical length, P_c, for 0.15 MPa is indicated. The stresses are 0.08, 0.1, and 0.15 MPa.

node or dislocation is

$$c_v^D = c_v \exp\left[\frac{F\Omega}{bkT}\right] \tag{52}$$

They show that the climb rate of nodes is faster than that of the links and the links are, therefore, controlling. A flux of vacancies leads to a climb velocity

$$v_\ell = \frac{2\pi DFb}{kT \ln(\ell/2b)} \tag{53}$$

very similar to that of Weertman's analysis where, here, F is the total force per unit length of dislocation favoring climb. Three forces are suggested to influence the total force F; the climb force due to the applied stress ($\sigma b/2$), elastic interaction forces from other links ($Gb^2/2\pi(1-\nu)\ell$), and a line tension ($\cong Gb^2/\ell$) due to the fact that with coarsening, the total elastic strain-energy decreases. This leads to the equation for creep-rate utilizing the usual equation ($\ell = \alpha Gb/\sigma$) and assuming that the contribution of dislocation pipe diffusion is not important,

$$\dot{\varepsilon}_{ss} = \frac{4.2\sqrt{3}\sigma_{ss}^3 b}{\alpha_2^2 G^2 kT}\left[\frac{D_{sd}}{\ln(\alpha_2 G/2\sigma)}\right]\left[1 + \frac{2}{\alpha_2}\left(1 + \frac{1}{2\pi(1-\nu)}\right)\right] \tag{54}$$

This suggests about 3-power behavior. The dislocation line is treated as an ideal vacancy source rather than each jog.

e. Recovery-Based Models. One shortcoming of the previously discussed models is that a recovery aspect is not included, in detail. It has been argued by many (e.g., [81,201]) that steady-state, for example, reflects a balance between dislocation hardening processes, suggested to include strain-driven network refinements, subgrain-size refinement or subgrain-boundary mesh-size refinement, and thermally activated softening processes that result in coarsening of the latter features.

Maruyama et al. [202] attempted to determine the microstructural feature associated with the rate-controlling (climb) process for creep by examining the hardening and recovery rates during transients in connection with the Bailey–Orowan [203,204] equation,

$$\dot{\varepsilon}_{ss} = \frac{r_r}{h_r} \tag{55}$$

where $r_r = d\sigma/dt$ is the recovery rate and $h_r = d\sigma/d\varepsilon$ is the hardening rate.

The recovery rates in several single phase metals and alloys were estimated by stress reduction tests, while work hardening rates were calculated based on the observed network dislocation densities within the subgrains and the average dislocation separation within the subgrain walls, d. Determinations of $\dot{\varepsilon}_{ss}$ were made as a function of σ_{ss}. The predictions of equation (35) were inconclusive in determining whether a subgrain wall or network hardening basis was more reasonable, although a somewhat better description was evident with the former.

More recently, Daehn et al. [205,206] attempted to formulate a more basic objective of rationalizing the most general phenomenology such as five-power-law behavior, which has not been successfully explained. Hardening rates (changes in the [network] dislocation density) are based on experimentally determined changes in ρ with strain at low temperatures,

$$\rho_{t+dt} = \rho_t + M_\rho \left(\frac{\rho}{\rho_o}\right)^c \dot{\gamma} dt \tag{56}$$

where M_ρ is the dislocation breeding constant, and c and ρ_o are constants. Refinement is described by the changes in a substructural length-scale ℓ' (ρ, d, or λ) by,

$$\frac{d\ell'}{dt} = -\left(\frac{M_\rho \cdot (\ell'_o)^{2c}(\ell)^{3-2c}}{2g'^2}\right)\dot{\gamma} \tag{57}$$

where ℓ'_o is presumably a reference length scale, and g' is a constant.

The flow stress is vaguely related to the substructure by

$$\tau = \frac{\hat{k}}{b\ell'} \tag{58}$$

Daehn et al. note that if network strengthening is relevant, the above equation should reduce to the Taylor equation although it is unclear that this group is really including equation (29), or an "athermal" dislocation relationship without a σ_0 term.

Coarsening is assumed to be independent of concurrent plastic flow and diffusion controlled

$$d(\ell')^{m_c} = KD dt \tag{59}$$

where m_c and K are constants and D is the diffusivity. Constants are based on microstructural coarsening observations at steady-state, refinement and coarsening are equal and the authors suggest that

$$\dot{\gamma} = BD\left(\frac{\tau}{G}\right)^n \tag{60}$$

results with $n = 4\text{--}6$.

This approach to understanding five-power behavior does seem particularly attractive and can, potentially, allow descriptions of primary and transient creep.

2.3.2 Dislocation Microstructure and the Rate-Controlling Mechanism

Consistent with the earlier discussion, the details by which the dislocation climb-control which is, of course, diffusion controlled, is specifically related to the creep-rate, are not clear. The existing theories (some prominent models discussed earlier) basically fall within two broad categories: (a) those that rely on the heterogeneous dislocation substructure (i.e., the subgrain boundaries) and, (b) those that rely on the more uniform Frank dislocation network (not associated with dislocation heterogeneities such as cells or subgrain walls).

a. Subgrains. The way by which investigators rely upon the former approach varies but, basically, theories that rely on the dislocation heterogeneities believe that one or more of the following are relevant:

(a) The subgrain boundaries are obstacles for gliding dislocations, perhaps analogous to suggestions for high angle grain boundaries in a (e.g., annealed) polycrystal described by the Hall–Petch relation. In this case, the misorientation across the subgrain boundaries, which is related to the spacing of the dislocations that constitute the boundaries, has been suggested to determine the effectiveness of the boundary as an obstacle [207]. [One complication with this line of reasoning is that it now appears well established that, although these features may be obstacles, the mechanical behavior of metals and alloys during five-power-law-creep appears independent of the details of the dislocation spacing, d, or misorientation across subgrain boundaries $\theta_{\lambda,\mathrm{ave}}$ as shown in Figure 27.]

(b) It has been suggested that the boundary is a source for Internal stresses, as mentioned earlier. Argon *et al.* [137], Gibeling and Nix [117], Morris and Martin [42,43], and Derby and Ashby [177] suggested that subgrain boundaries may bow and give rise to an internal back stress that is the stress relevant to the rate-controlling mechanism. Morris and Martin claim to have measured high local stresses that were 10–20 times larger than the applied stress near Al-5at%Zn subgrains formed within the five-power-law regime. Their stress calculations were based on dislocation loop radii measurements. Many have suggested that subgrain boundaries are important as they may be "hard" regions, such as, according to Mughrabi [138], discussed earlier, extended to the case of creep by Blum and coworkers in a series of articles (e.g., [136]). Basically, here, the subgrain wall is considered three-dimensional with a high-yield stress compared to the subgrain or cell interior. Mughrabi originally suggested that there is elastic compatibility between the subgrain wall and the matrix with parallel straining. This gives rise to a high internal backstress or internal stress. These investigators appear to suggest that these elevated stresses are the relevant stress for the rate-controlling process (often involving dislocation climb) for creep, usually presumed to be located in the vicinity of subgrain walls.

(c) Others have suggested that the ejection of dislocations from the boundaries is the critical step [42,43,202]. The parameter that is important here is basically the spacing between the dislocations that comprise the boundary which is generally related to the misorientation angle across the boundary. Some additionally suggest that the relevant stress is not the applied stress, but the stress at the boundary which may be high, as just discussed in (b) above.

(d) Similar to (a), it has been suggested that boundaries are important in that they are obstacles for gliding dislocations (perhaps from a source within the subgrain) and that, with accumulation at the boundary (e.g., a pile-up at the boundary), a backstress is created that "shuts off" the source which is only reactivated once the number of dislocations within a given pile-up is diminished. It has been suggested that this can be accomplished by climb and annihilation of dislocations at the same subgrain boundary [26,192–195]. This is similar to the model discussed in Section 4.1.3. Some suggest that the local stress may be elevated as discussed in (b) above.

b. Dislocations. Others have suggested that the rate-controlling process for creep plasticity is associated with the Frank dislocation network within the subgrains such as discussed in 2(C)1(d). That is, the strength associated with creep is related to the details (often, the density) of dislocations in the subgrain interior [54,84–86, 98,101,102,104–106,108–110,129,146,152,153,155,196–200,208]. One commonly proposed mechanism by which the dislocation network is important is that dislocation sources are the individual links of the network. As these bow, they can become unstable, leading to Frank–Read sources, and plasticity ensues. The density of links that can be activated sources depends on the link length distribution and, thus, related to the density of dislocation line length within the subgrains. The generated dislocation loops are absorbed by the network, leading to refinement or decreasing P. The network also "naturally" coarsens at elevated temperature and plasticity is activated as links reach the critically "long" segment length, P_c. Hence, climb (self-diffusion) control is justified. Some of the proponents of the importance of the interior dislocation-density have based their judgments on experimental evidence that shows that creep-strength (resistance) is associated with higher dislocation density and appears independent of the subgrain size [110,129,141,153].

c. Theoretical Strength Equations. In view of the different microstructural features (e.g., λ, d, $\theta_{\lambda_{ave}}$, ρ, ℓ_c) that have been suggested to be associated with the strength or rate-controlling process for five-power-law-creep, it is probably worthwhile to assess strength associated with different obstacles. These are calculable from simple (perhaps simplistic) equations. The various models for the rate-controlling, or strength-determining process, are listed below. Numerical calculations are based on pure Al creep deforming as described in Figure 30.

(i) The network stress τ_N. Assuming a Frank network, the average link length, ℓ (assumed here to be uniform $\cong \ell_c$)

$$\tau_N \cong \frac{Gb}{\ell} \tag{61}$$

Using typical aluminum values for five-power-law-creep (e.g., $P1/\sqrt{\rho_{ss}}$) $\tau_N \cong 5$ MPa (fairly close to $\tau_{ss} \cong 7$ MPa for Al at the relevant ρ_{ss}).

(ii) If the critical step is regarded as ejection of dislocations from the subgrain boundary,

$$\tau_B \cong \frac{Gb}{d} \tag{62}$$

$\tau_B \cong 80$ MPa, much higher (by an order of magnitude or so) than the applied stress.

(iii) If subgrain boundaries are assumed to be simple tilt boundaries with a single Burgers vector, an attractive or repulsive force will be exerted on a slip dislocation approaching the boundary. The maximum stress

$$\tau_{bd} \cong \frac{0.44 Gb}{2(1-\nu)d} \tag{63}$$

from Ref. [88] based on Ref. [209]. This predicts a stress of about $\tau_{bd} = 50$ MPa, again, much larger than the observed applied stress.

(iv) For dragging jogs resulting from passing through a subgrain boundary, assuming a spacing $\cong j \cong d$,

$$\tau_j \cong \frac{E_j}{b^2 j} \tag{64}$$

from Ref. [88]. For Al, E_j, the formation energy for a jog, $\cong 1$ eV [88] and $\tau_j \cong 45$ MPa, a factor 6–7 higher than the applied stress.

(v) The stress associated with the increase in dislocation line length (jog or kinks) to pass a dislocation through a subgrain wall (assuming a wall dislocation spacing, d)

from above, is expected to be

$$\tau_L \cong \frac{Gb^3}{b^2 d} \cong \frac{Gb}{d} \cong 80 \text{ MPa} \tag{65}$$

over an order of magnitude larger than the applied stress.

Thus, it appears that stresses associated with ejecting dislocations from, or passing dislocations through subgrain walls, are typically 30–80 MPa for Al within the five-power-law regime. This is roughly an order of magnitude larger than the applied stress. Based on the simplified assumptions, this disparity may not be considered excessive and does not eliminate subgrain walls as important, despite the favorable agreement between the network-based (using the average link length, P) strength and the applied stress (internal stresses not considered). These calculations indicate why some subgrain-based strengthening models utilize elevated Internal stresses. It must be mentioned that care must be exercised in utilizing the above, athermal, equations for time-dependent plasticity. These equations do not consider other hardening variables (solute, etc.) that may be a substantial fraction of the applied stress, even in relatively pure metals. Thus, these very simple "theoretical" calculations do not provide obvious insight into the microstructural feature associated with the rate controlling process, although a slight preference for network-based models might be argued as the applied stress best matches network predictions for dislocation activation.

2.3.3 In situ and Microstructure-Manipulation Experiments

a. In Situ Experiments. *In situ* straining experiments, particularly those of Calliard and Martin [95], are often referenced by the proponents of subgrain (or heterogeneous dislocation arrangements) strengthening. Here thin foils (probably less than 1 μm thick) were strained at ambient temperature (about 0.32 T_m). It was concluded that the interior dislocations were not a significant obstacle for gliding dislocations, rather, the subgrain boundaries were effective obstacles. This is an important experiment, but is limited in two ways: first, it is of low temperature (i.e., $\cong 0.32\ T_m$) and may not be relevant for the five-power-law regime and second, in thin foils, such as these, as McLean mentioned [84] long ago, a Frank network is disrupted as the foil thickness approaches ℓ. Henderson–Brown and Hale [210] performed *in situ* high-voltage transmission electron microscope (HVEM) creep experiments on Al–1Mg (class M) at 300°C, in thicker foils. Dislocations were obstructed by subgrain walls, although the experiments were not described in substantial detail. As mentioned earlier, Mills [157] performed *in situ* deformation on an Al–Mg alloy within the three-power or viscous-drag regime and subgrain boundaries were not concluded as obstacles.

b. Prestraining experiments. Work by Kassner et al. [98,110,111], discussed earlier, utilized ambient temperature prestraining of austenitic stainless steel to, first, show that the elevated temperature strength was independent of the subgrain size and, second, that the influence of the dislocation density on strength was reasonably predicted by the Taylor equation. Ajaja and Ardell [152,153] also performed prestraining experiments on austenitic stainless steels and showed that the creep rate was influenced only by the dislocation density. The prestrains led to elevated ρ, without subgrains, and "quasi" steady-state creep rates. Presumably, this prestrain led to decreased average and critical link lengths in a Frank network. [Although, eventually, a new "genuine" steady-state may be achieved [211], this may not occur over the convenient strain/time ranges. Hence, the conclusion of a "steady-state" being independent of the prestrain may be, in some cases, ambiguous.]

Others, including Parker and Wilshire [212], have performed prestraining experiments on Cu showing that ambient temperature prestrain (cold work) reduces the elevated temperature creep rate at 410°C. This was attributed by the investigators as due to the Frank network. Well-defined subgrains did not form; rather, cell walls observed. A quantitative microstructural effect of the cold work was not clear.

2.3.4 Additional Comments on Network Strengthening

Previous work on stainless steel in Figure 25 showed that the density of dislocations within the subgrain interior or the network dislocations influence the flow stress at a given strain rate and temperature. The hardening in stainless stress is shown to be consistent with the Taylor relation if a linear superposition of "lattice" hardening (τ_o, or the stress necessary to cause dislocation motion in the absence of a dislocation substructure) and the dislocation hardening ($\alpha MGb\rho^{1/2}$) is assumed. The Taylor equation also applies to pure aluminum (with a steady-state structure), having both a much higher stacking fault energy than stainless steel and an absence of substantial solute additions.

If both the phenomenological description of the influence of the strength of dislocations in high-purity metals such as aluminum have the form of the Taylor equation and *also* have the expected values for the constants, then it would appear that the elevated temperature flow stress is actually provided by the "forest dislocations" (Frank network).

Figure 21 illustrates the well-established trend between the steady-state dislocation density and the steady-state stress. From this and from Figure 15, which plots modulus-compensated steady-state stress versus diffusion-coefficient compensated steady-state strain-rate, the steady-state flow stress can be predicted at a reference strain-rate (e.g., $5 \times 10^{-4} s^{-1}$), at a variety of temperatures, with an associated steady-state dislocation density. If equation (29) is valid for Al as for 304 stainless steel, then the values for α could be calculated for each temperature, by assuming that the annealed dislocation density and the σ_o values account for the annealed yield strength reported in Figure 36.

Figure 37 indicates, first, that typical values of α at 0.5 T_m are within the range of those expected for Taylor strengthening. In other words, the phenomenological relationship for strengthening of (steady-state) structures suggests that the strength can be reasonably predicted based on a Taylor equation. We expect the strength we observe, based only on the (network) dislocation density, completely independent of the heterogeneous dislocation substructure. This point is consistent with the observation that the elevated temperature yield strength of annealed polycrystalline aluminum is essentially independent of the grain size, and misorientation of boundaries. Furthermore, the values of α are completely consistent with the values of α in other metals (at both high and low temperatures) in which dislocation hardening is established [see Table 1]. The fact that the higher temperature α values of Al and 304 stainless steel are consistent with the low temperature α values of Table 1 is also consistent with the athermal behavior of Figure 37. The non-near-zero annealed dislocation density observed experimentally may be consistent with the Ardell *et al.* suggestion of network frustration creating a lower dislocation density.

One point to note is in Figure 37, the variation in α with temperature depends on the value selected for the annealed dislocation density. For a value of $2.5 \times 10^{11} \mathrm{m}^{-2}$

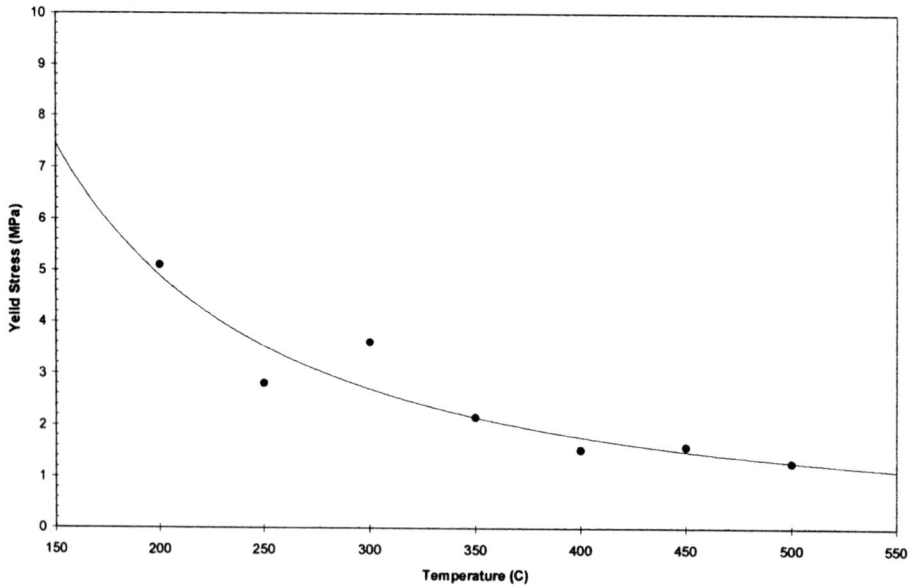

Figure 36. The yield strength of annealed 99.999% pure Al as a function of temperature. From Ref. [149]. $\dot{\varepsilon} = 5 \times 10^{-4} \mathrm{s}^{-1}$.

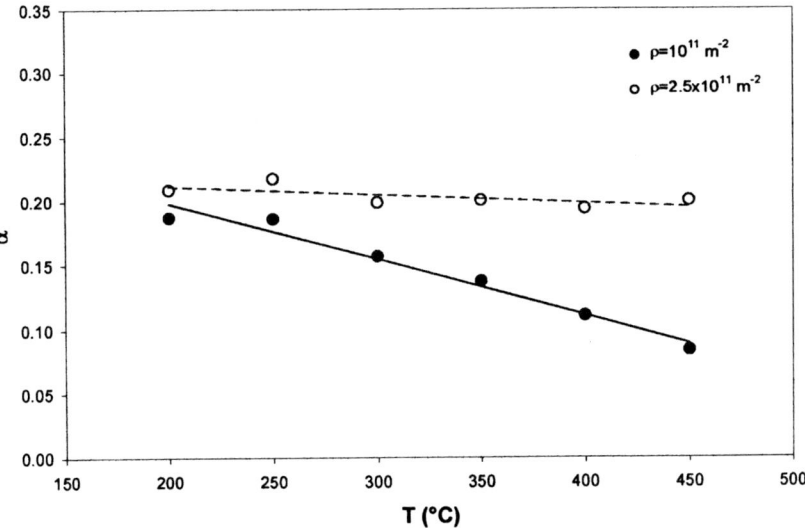

Figure 37. The values of the constant alpha in the Taylor equation (29) as a function of temperature. The alpha values depend somewhat on the assumed annealed dislocation density. Dark dots, $\rho = 10^{11}\,\mathrm{m}^{-2}$; hollow, $\rho = 10^{10}\,\mathrm{m}^{-2}$; diamond, $\rho = 2.5 \times 10^{11}\,\mathrm{m}^{-2}$.

Table 1. Taylor equation α values for various metals

Metal	T/T_m	α (Eq. 6)	Notes	Ref.
304	0.57	0.28	$\sigma_0 \neq 0$, polycrystal	[107]
Cu	0.22	0.31	$\sigma_0 = 0$, polycrystal	[144]
Ti	0.15	0.37	$\sigma_0 \cong 0.25$–0.75 flow stress, polycrystal	[151]
Ag	0.24	0.19–0.34	Stage I and II single crystal $M = 1.78 - 1$ $\sigma_0 \neq 0$	[213]
Ag	0.24	0.31	$\sigma_0 = 0$, polycrystal	[150]
Al	0.51–0.83	0.20	$\sigma_0 \neq 0$, polycrystal	[149]
Fe	—	0.23	$\sigma_0 \neq 0$, polycrystal	[144]

Note: α values of Al and 304 stainless stress are based on dislocation densities of intersections per unit area. The units of the others are not known and these α values would be adjusted lower by a factor of 1.4 if line-length per unit volume is utilized.

(or higher), the values of the α constant are nearly temperature independent, suggesting that the dislocation hardening is, in fact, theoretically palatable in that it is expected to be athermal. The annealed dislocation density for which athermal behavior is observed is that which is very close to the observed value in Figure 27(a) and by Blum [214]. The suggestion of athermal dislocation hardening is consistent with the model by Nes [215], where, as in the present case, the temperature dependence of the constant (or fixed dislocation substructure) structure flow stress is provided by the important temperature-dependent σ_0 term. It perhaps should be mentioned that if it is assumed both that $\sigma_0 = 0$ and that the dislocation hardening is athermal [i.e., equation (19) is "universally" valid] then α is about equal to 0.53, or about a factor of two larger than anticipated for dislocation hardening. Hence, apart from not including a σ_0 term which allows temperature dependence, the alpha terms appears somewhat large to allow athermal behavior.

The trends in dislocation density during primary creep have been less completely investigated for the case of constant-strain-rate tests. Earlier work by Kassner et al. [98,110,111] on 304 stainless steel found that at 0.57 T_m, the increase in flow stress by a factor of three, associated with increases in dislocation density with strain, is consistent with the Taylor equation. That is, the ρ versus strain and stress versus strain give a σ versus ρ that "falls" on the line of Figure 25. Similarly, the aluminum primary transient in Figure 30(a) can also be shown consistent with the Taylor equation. The dislocation density monotonically increases to the steady-state value under constant *strain-rate* conditions.

Challenges to the proposition of Taylor hardening for five-power-law-creep in metals and Class M alloys include the microstructural observations during primary creep under constant-*stress* conditions. For example, it has nearly always been observed during primary creep of pure metals and Class M alloys that the density of dislocations not associated with subgrain boundaries increases from the annealed value to a peak value, but then gradually decreases to a steady-state value that is between the annealed and the peak density [38,92,163–165] (e.g., Figure 26). Typically, the peak value, ρ_p, measured at a strain level that is roughly one-fourth of the strain required to attain steady-state ($\varepsilon_{ss}/4$), is a factor of 1.5–4 higher than the steady-state ρ_{ss} value. It was believed by many to be difficult to rationalize hardening by network dislocations if the overall density is decreasing while the strain-rate is decreasing. Therefore, an important question is whether the Taylor hardening, observed under constant strain-rate conditions, is consistent with this observation [169]. This behavior could be interpreted as evidence for most of these dislocations to have a dynamical role rather than a (Taylor) hardening role, since the initial strain-rates in constant stress tests may require by the equation,

$$\dot{\varepsilon} = (b/M)\rho_m v \tag{66}$$

a high mobile (nonhardening) dislocation density, ρ_m, that gives rise to high initial values of total density of dislocations not associated with subgrain boundaries, ρ (v is the dislocation velocity). As steady-state is achieved and the strain-rate decreases, so does ρ_m and in turn, ρ. [We can suggest that $\rho_h + \rho_m = \rho$, where ρ is the total density of dislocations not associated with subgrain boundaries and ρ_h are those dislocations that at any instant are part of the Frank network and are not mobile.]

More specifically, Taylor hardening during primary (especially during constant stress) creep may be valid based on the following argument.

From equation (66) $\dot{\varepsilon} = \rho_m v b / M$. We assume [216]

$$v = k_7 \sigma^1 \tag{67}$$

and, therefore, for constant strain-rate tests,

$$\dot{\varepsilon}_{ss} = [k_7 b / M] \rho_m \sigma \tag{68}$$

In a constant strain-rate test at yielding ($\dot{\varepsilon} = \dot{\varepsilon}_{ss}$), ε_p (plastic strain) is small, and there is only minor hardening, and the mobile dislocation density is a fraction f_m^0 of the total density,

$$f_m^0 \rho_{(\varepsilon_p = 0)} = \rho_{m(\varepsilon_p = 0)}$$

therefore, for aluminum (see Figure 30a)

$$\rho_{m(\varepsilon_p = 0)} = f_m^0 0.64 \rho_{ss} \text{ (based on } \rho \text{ at } \varepsilon_p = 0.03) \tag{69}$$

f_m^0 is basically the fraction of dislocations in the annealed metal that are mobile at yielding (half the steady-state flow stress) in a constant strain-rate test. Also from Figure 4 $\sigma_y/\sigma_{ss} = 0.53$. Therefore, at small strains,

$$\dot{\varepsilon}_{ss} = f_m^0 0.34 [k_7 b / M] \rho_{ss} \sigma_{ss} \tag{70}$$

(constant strain-rate at $\varepsilon_p = 0.03$)

At steady-state, $\sigma = \sigma_{ss}$ and $\rho_m = f_m^s \rho_{ss}$, where f_m^s is the fraction of the total dislocation density that is mobile and

$$\dot{\varepsilon}_{ss} = f_m^s [k_7 b / M] \rho_{ss} \sigma_{ss} \tag{71}$$

(constant strain-rate at $\varepsilon_p > 0.20$).

By combining equations (70) and (71) we find that f_m at steady-state is about 1/3 the fraction of mobile dislocations in the annealed polycrystals ($0.34 f_m^0 = f_m^s$). This suggests that during steady-state only 1/3, or less, of the total dislocations (not associated with subgrain boundaries) are mobile and the remaining 2/3, or more, participate in hardening. The finding that a large fraction are immobile is consistent with the observation that increased dislocation density is associated with increased strength for steady-state and constant strain-rate testing deformation. Of course, there is the assumption that the stress acting on the dislocations as a function of strain (microstructure) is proportional to the applied flow stress. Furthermore, we have presumed that a 55% increase in ρ over primary creep with some uncertainty in the density measurements.

For the constant stress case we again assume

$$\dot{\varepsilon}_{\varepsilon_p \cong 0} = f_m^P [k_7 b/M] \rho_p \sigma_{ss} \text{(constant stress)} \tag{72}$$

where f_m^P is the fraction of dislocations that are mobile at the peak (total) dislocation density of ρ_p, the peak dislocation density, which will be assumed equal to the maximum dislocation density observed experimentally in a ρ–ε plot of a constant stress test. Since at steady-state,

$$\dot{\varepsilon}_{ss} \cong 0.34 f_m^0 [k_7 b/M] \rho_{ss} \sigma_{ss} \tag{73}$$

by combining with (72),

$$\dot{\varepsilon}_{\varepsilon_p \cong 0}/\dot{\varepsilon}_{ss} = \left(\frac{f_m^P}{f_m^0}\right) 3\rho_p/\rho_{ss} \text{ (constant stress)} \tag{74}$$

(f_m^P/f_m^0) is not known but if we assume that at macroscopic yielding, in a constant strain-rate test, for annealed metal, $f_m^0 \cong 1$, then we might also expect at small strain levels and relatively high dislocation densities in a constant-stress test, $f_m^P \cong 1$. This would suggest that fractional decreases in $\dot{\varepsilon}$ in a constant stress test are not equal to those of ρ. This apparent contradiction to purely dynamical theories [i.e., based strictly on equation (66)] is reflected in experiments [92,162–165] where the kind of trend predicted in this last equation is, in fact, observed. Equation (74) and the observations of $\dot{\varepsilon}$ against ε in a constant stress test at the identical temperature can be used to predict roughly the expected constant-stress ρ–ε curve in aluminum at 371°C and about 7.8 MPa; the same conditions as the constant strain-rate test. If we use small plastic strain levels, $\varepsilon \cong \varepsilon_{ss}/4$ (where ρ values have been measured in constant stain-rate tests), we can determine the ratio (e.g., $\dot{\varepsilon}_{\varepsilon=(\varepsilon_{ss}/4)}/\dot{\varepsilon}_{\varepsilon=\varepsilon_{ss}}$) in constant stress tests. This value seems to be roughly 6 at stresses and temperatures comparable to the present study [92,165,212]. This ratio was applied to equation (74) [assuming $(f_m^P/f_m^c) \cong 1$]; the estimated ρ–ε tends are shown in Figure 38. This estimate, which predicts a peak dislocation density of 2.0 ρ_{ss}, is consistent with the general observations discussed earlier for pure metals and Class M alloys that ρ_p is between 1.5 and 4 ρ_{ss}(1.5–2.0 for aluminum [92]). Thus, the peak behavior observed in the dislocation density versus strain-rate trends, which at first glance appears to impugn dislocation network hardening, is actually consistent, in terms of the observed ρ values, to Taylor hardening.

Two particular imprecisions in the argument above is that it was assumed (based on some experimental work in the literature) that the stress exponent for the elevated

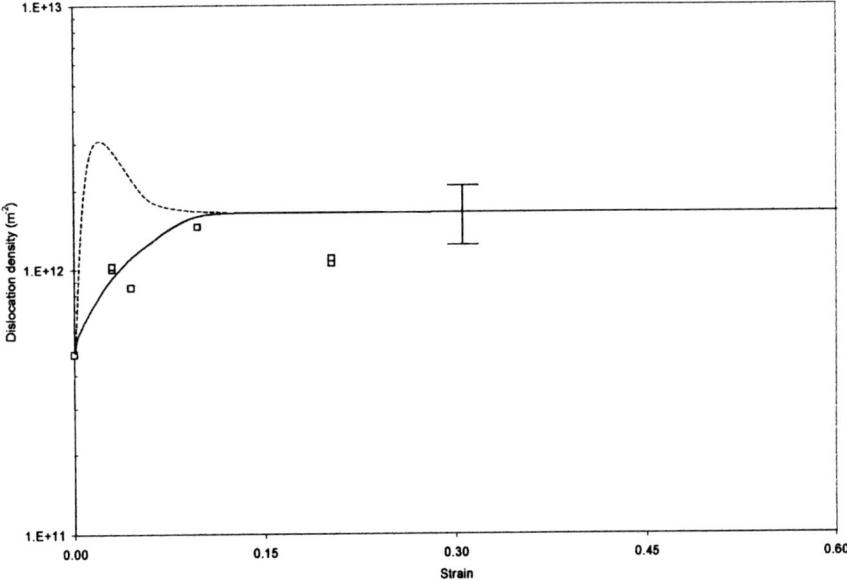

Figure 38. The predicted dislocation density (- - -) in the subgrain interior against strain for aluminum deforming under constant stress conditions is compared with that for constant strain-rate conditions (———). The predicted dislocation density is based on equation (74) which assumes Taylor hardening.

temperature (low stress) dislocation velocity, v, is one. This exponent may not be well known and may be greater than 1. The ratio ρ_p/ρ_{ss} increases from a value of 3 in equation (19) to higher values of $3\,[2^{n-1}]$, where n is defined by $v = \sigma^n$. This means that the observed strain-rate "peaks" would predict smaller dislocation peaks or even an absence of peaks for the observed initial strain-rates in constant-stress tests. In a somewhat circular argument, the consistency between the predictions of equation (74) and the experimental observations may suggest that the exponents of 1–2 may be reasonable. Also, the values of the peak dislocation densities and strain-rates are not unambiguous, and this creates additional uncertainty in the argument.

2.4 OTHER EFFECTS ON FIVE-POWER-LAW-CREEP

2.4.1 Large Strain Creep Deformation and Texture Effects

Traditionally, creep has been associated with tensile tests, and accordingly, with relatively small strains. Of course, elevated temperature creep plasticity can be observed in torsion or compression, and the phenomenological expressions presented earlier are still valid, only with modification due to different texture evolution (or changes in the average Taylor factors) with the different deformation modes. These differences in texture evolution have been discussed in detail by several investigators [38,217] for lower temperature deformation. Some lower temperature deformation texture trends may be relevant to five-power-law-creep trends. A fairly thorough review of elevated temperature torsion tests and texture measurements on aluminum is presented by Kassner et al. and McQueen et al. [218,219]. Some of the results are illustrated in Figure 39(a) and (b). Basically, the figure shows that with torsion deformation, the material hardens to a genuine steady-state that is a balance between dislocation hardening and dynamic recovery. However, with the relatively large strain deformation that is permitted by torsion, the flow stress decreases, in this case, about 17% to a new stress that is invariant even with *very* large strains to 100 or so. (Perhaps there is an increase in torque of 4% in some cases with $\varepsilon > 10$ of uncertain origin.) These tests were performed on particularly precise testing equipment. The essentially invariant stress over the extraordinarily large strains suggests a "genuine" mechanical steady-state. The cause of this softening has been carefully studied, and dynamic recrystallization and grain boundary sliding (GBS) were considered. Creep measurements as a function of strain through the "softened"

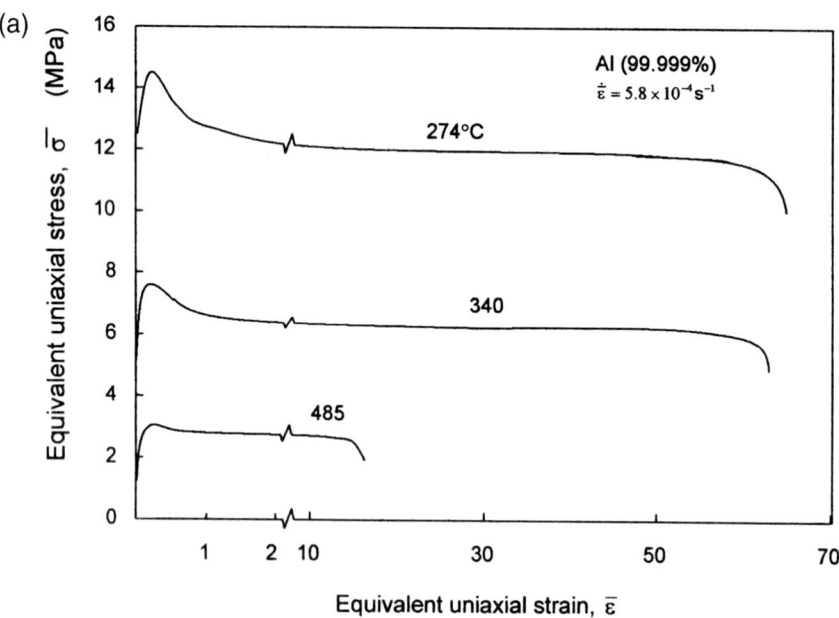

Figure 39. The stress versus strain behavior of Al deformed in torsion to very large strains at two [(a) and (b)] strain-rates. Based on Ref. [39].

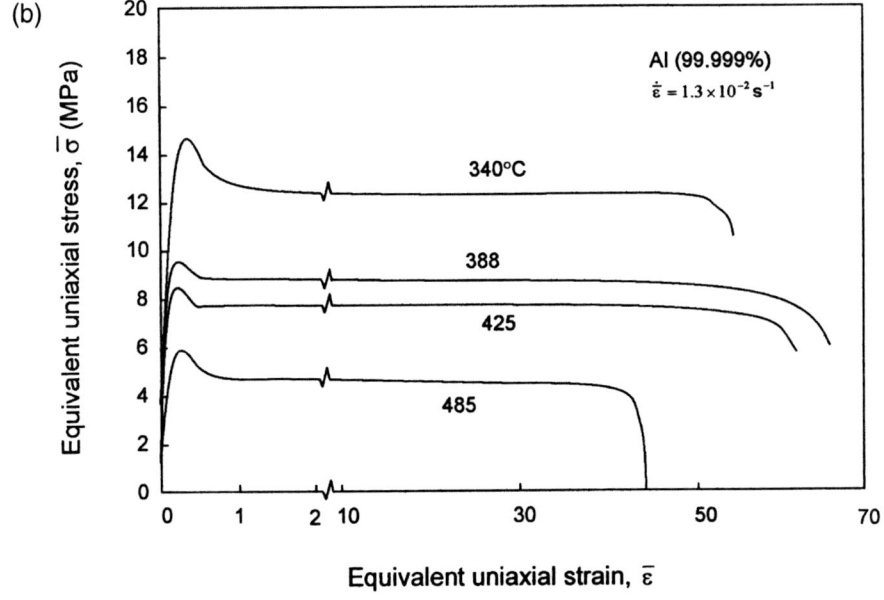

Figure 39. Continued

regime [220], and microstructural analysis using both polarized light optical (POM) and transmission electron microscopy (TEM) [218] reveal that five-power-law-creep is occurring throughout the "softened" regime and that the modest decrease in flow stress is due to a decrease in the average Taylor factor, $\overline{M}$.

This has been confirmed by X-ray texture analysis [37] and is also consistent in magnitude with theoretical texture modeling of deformation in torsion [38,217]. If compression specimens are extracted from the torsion specimen deformed into the "softened" regime, the flow stress of the compression specimen is actually higher than the torsion flow stress, again confirming the texture conclusion [218].

The microstructural evolution of specimens deformed to large strains, not achievable in tension or compression, is quite interesting and has also been extensively researched in some metals and alloys. The initial high angle grain boundaries of the aluminum polycrystalline aggregate spiral about the torsion axis with deformation within the five-power-law regime. At least initially, the total number of grains in the polycrystalline aggregate remains constant and the grains quickly fill with subgrains with low misorientation boundaries. The grains thin until they reach about twice the average subgrain diameter with increasing strain in torsion. Depending on the initial grain size and the steady-state subgrain size, this may require substantial strain, typically about 10. The high angle grain

80   Fundamentals of Creep in Metals and Alloys

boundaries of the polycrystalline aggregate are serrated (triple points) as a result of subgrain boundary formation. As the serrated spiraling grains of the polycrystalline aggregate decrease in width to about twice the subgrain size, there appears to be a pinching off of impinging serrated grains. At this point, the area of high angle boundaries, which was gradually increasing with torsion, reaches a constant value with increasing strain. Despite the dramatic increase in high angle boundaries, no change in flow properties is observed (e.g., to diffusional creep or enhanced strain-rate due to the increased contribution of grain boundary sliding). Figure 40 is a series of POM micrographs illustrating this progression. Interestingly, the

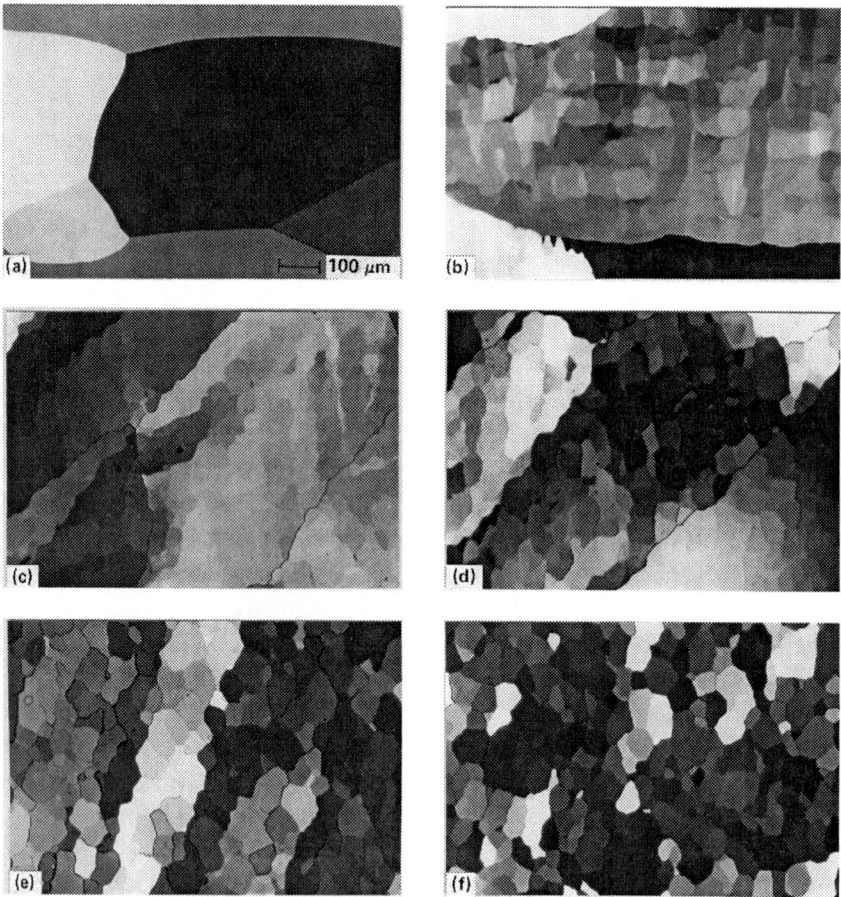

Figure 40. Polarized light optical micrographs of aluminum deformed at 371°C at $5.04 \times 10^{-4}\,\text{s}^{-1}$ [Figure 30(b)] to equivalent uniaxial strains of (a) 0, (b) 0.2, (c) 0.60, (d) 1.26, (e) 4.05, (f) 16.33. geometric dynamic recrystallization (GDX) is observed [18].

subgrain size is about constant from the "peak" stress at about 0.2 strain to the very large torsion strains. This, again, suggests that subgrain boundaries are mobile and annihilate to maintain the equiaxed structure and modest misorientation. Examination of those boundaries that form from dislocation reaction (excluding the high-angle boundaries of the starting polycrystal) reveals that the average misorientation at the onset of steady-state was, as stated earlier, only 0.5°. However, by a strain of between 1 and 1.5 it had tripled to 1.5° [also see Figure 30(b)], but appears to be fixed beyond this strain. This is, again, consistent with earlier work referenced that indicates that $\theta_{\lambda ave}$ may increase (d decreases) during at least early steady-state. Furthermore, at the onset of steady-state, nearly all of the subgrain boundaries formed are low-θ_λ dislocation boundaries. However, with very large strain deformation there is an increase in high-angle boundary area (geometric dynamic recrystallization or GDX). Nearly a third of the subgrain boundaries are high-angle boundaries, but these appear to have ancestry back to the initial, or starting, polycrystal. Notwithstanding, the flow stress is unchanged. That is, at a strain of 0.2, at about 0.7 T_m and a modest strain-rate, the average subgrain size is about 13 µm and the average misorientation angle of subgrain boundaries is about 0.5°. If we increase the plastic strain by nearly two orders of magnitude to about 16, the subgrain size and interior or network dislocation density is unchanged, but we have "replaced" nearly one-third of the subgrain facets with high-angle boundaries (through GDX) and tripled the misorientation of the remaining one-third. However, the flow stress is unchanged. This, again, suggests that the details of the subgrain boundaries are not an important consideration in the rate-controlling process for five-power-law-creep.

Other elevated temperature torsion tests on other high stacking fault energy alloys in the five-power-law regime have shown a similar softening as theoretically predicted [221]. The cause of softening was not ascribed to texture softening by those investigators but (probably incorrectly) rather to continuous reactions (continuous dynamic recrystallization) [222].

Recent work by Hughes *et al.* [141] showed that polycrystals deformed at elevated temperature may form geometrically necessary boundaries (GNBs) from dislocation reactions to accommodate differences in slip within a single grain. Whether these form in association with GDX is unclear, although it appears that the grain boundary area with large strain deformation is at least approximately consistent with grain thinning. HABs, however, have been observed to form in single crystals at elevated temperature from dislocation reaction [142] and the possibility that these form from dislocation reaction in polycrystals should also be considered.

It should be also mentioned that it has been suggested that in at least Al and same Al alloys [130], slip on {110} planes (or non-octahedral slip) can occur, leading to nontraditional textures such as the cube {(001) type}.

2.4.2 Effect of Grain Size

First, of course, it has been suggested that with fine-grain size refinement, the mechanism of plastic flow may change from five-power behavior to Coble creep [52,223], which, as discussed earlier, is a diffusion creep mechanism relying on short-circuit diffusion along grain boundaries. However, the role of high-angle grain boundaries is less clear for five-power-law-creep. Some work has been performed on the Hall–Petch relationship in copper [224] and aluminum [147] at elevated temperatures. Some results from these studies are illustrated in Figure 41. Basically, both confirm that decreasing grain size results in increased elevated temperature strength in predeformed copper and annealed aluminum (a constant dislocation density for each grain size was not confirmed in Cu). The temperature and applied strain rates correspond to five-power-law-creep in these pure metals.

Interestingly, though, the effect of diminishing grain size may decrease with increasing temperature. Figure 42 shows that the Hall–Petch constant, k_y, for high-purity aluminum significantly decreases with increasing temperature. The explanation for this in unclear. First, if the effect of decreasing grain size at elevated temperature is purely the effect of a Hall–Petch strengthening (e.g., not GBS), the, explanation for decreasing Hall–Petch constant would require knowledge of the precise strengthening mechanism. It is possible, for instance, that increased strengthening with smaller grain sizes is associated with the increased dislocation density in the grain interiors due to the activation of dislocation sources [225]. Therefore, thermal recovery may explain a decreased density and less pronounced strengthening. This is, of course, speculative and one must be careful that other effects such as grain boundary sliding are not becoming important. For example, it has been suggested that in aluminum, grain boundary sliding becomes pronounced above about 0.5 T_m [226,227]. Thus, it is possible that the decreased effectiveness of high-angle boundaries in providing elevated temperature strength may be the result of GBS, which would tend to decrease the flow stress. However, the initial Al grain size decreased from about 250 μm to only about 30 μm through GDX in Figure 40, but the flow properties at 0.7 T_m appear unchanged since the stress exponent, n, and activation energy, Q, appear to be unchanged [218,220].

The small effect of grain size changes on the elevated-temperature flow properties is consistent with some earlier work reported by Barrett *et al.* on Cu and Garafalo *et al.* on 304 stainless steel, where the steady-state creep-rate appeared at least approximately independent of the starting grain size in the former case and not substantially dependent in the latter case [228,229]. Thus, it appears that decreasing grain size has a relatively small effect on increasing the flow stress at high temperatures over the range of typical grain sizes in single phase metals and alloys.

Figure 43 plots the effect of grain size on the yield stress of annealed polycrystalline aluminum with the effect of (steady-state structure) subgrain size

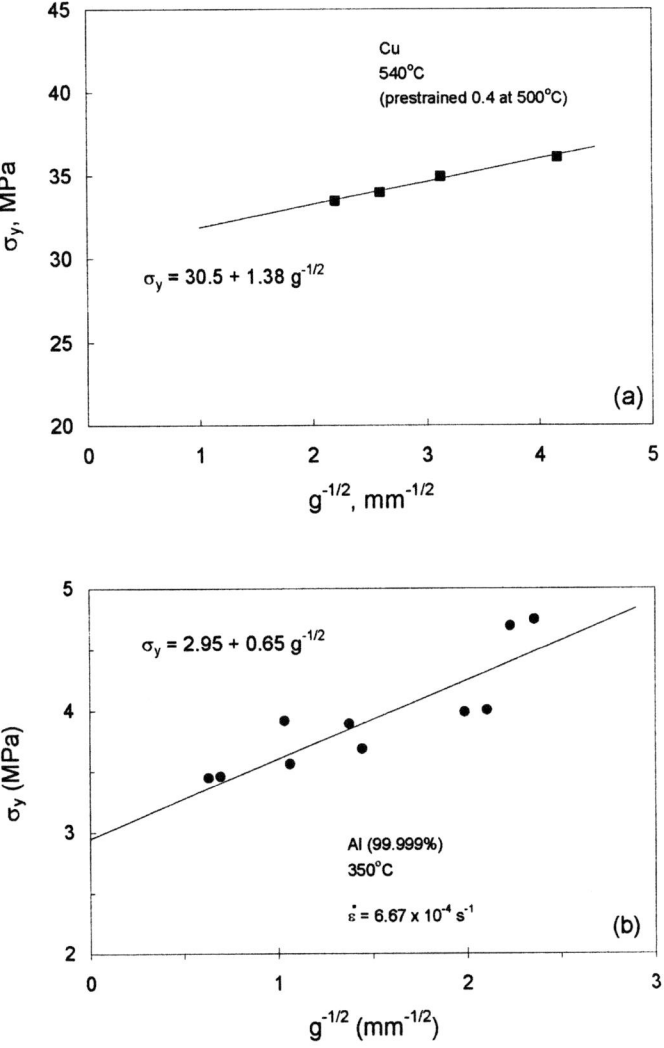

Figure 41. The effect on grain size on the elevated temperature strength of (a) pre-strained Cu and (b) annealed Al [147,224].

on the elevated flow stress all at *the same temperature and strain-rate*. Of course, while polycrystalline samples had an annealed dislocation density, the steady-state substructures with various Subgrain sizes had various elevated dislocation-densities that increased with decreasing Subgrain size. Nonetheless, the figure reveals that for identical, small-size, the subgrain substructure (typical $\theta_{\lambda_{ave}} \cong 0.5 - 1°$) had *higher* strength than polycrystalline annealed aluminum (typical $\theta = 30 - 35°$). There might

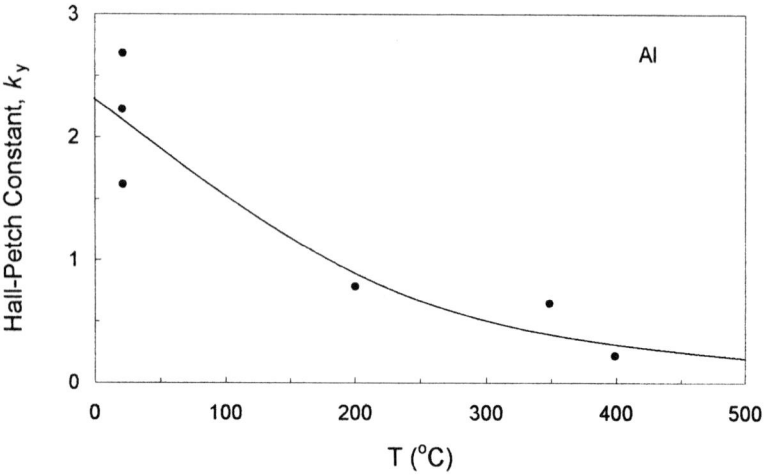

Figure 42. The variation of the Hall–Petch constant in Al with temperature. Based on Ref. [147].

be an initial inclination to suggest that subgrain boundaries, despite the very low misorientation, are more effective in hardening than high-angle grains of identical size. However, as discussed earlier, the "extra" strength may be provided by the network dislocations that are of significantly higher density in steady-state structures as compared with the annealed metal. This increase in strength appeared at least approximately predictable based solely on the equation [see equation (29)] for dislocation strengthening assuming appropriate values for the constants, such as α [107,148,231]. Wilshire [82] also recently argued that subgrains are unlikely sources for strength in metals and alloys due to the low strength provided by high-angle boundaries at elevated temperature.

2.4.3 Impurity and Small Quantities of Strengthening Solutes

It appears that the same solute additions that strengthen at ambient temperature often provide strength at five-power-law temperatures. Figure 44 shows the relationship between stress and strain-rate of high purity (99.99) and lower purity (99.5%) aluminum. The principal impurities were not specified, but probably included a significant fraction of Fe and some second phases may be present. The strength increases with decreasing purity for a fixed strain-rate. Interestingly, Figure 45 shows that the subgrain size is approximately predictable mostly on the basis of the stress, independent of composition for Al.

Staub and Blum [90] also showed that the subgram size depends only on the modulus-compensated stress in Al and several dilute Al alloys although the

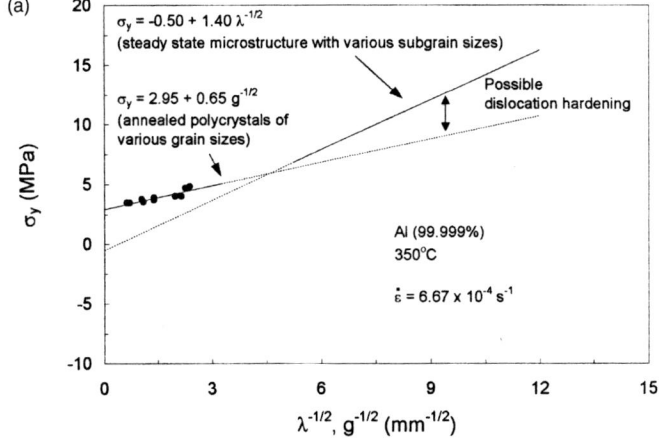

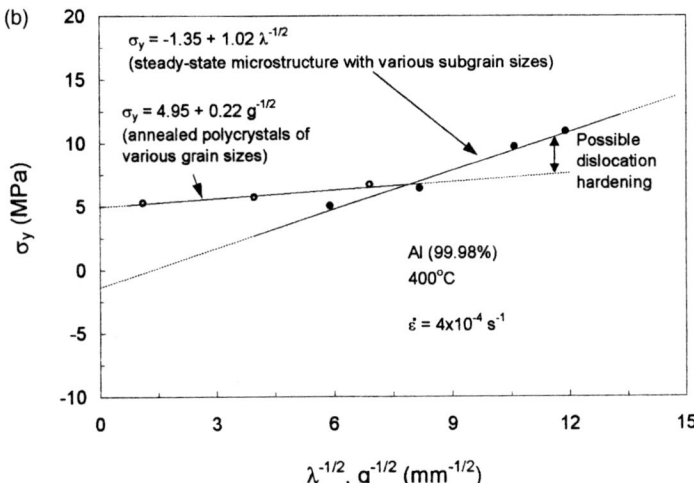

Figure 43. (a) The variation of the yield strength of annealed aluminum with various grain sizes, g, and creep deformed aluminum with various subgrain sizes, λ, at 350°C. Both λ and g data are described by the Hall–Petch equation. The annealed aluminum data is from Figure 41(b) and the "subgrain containing Al strength data at a fixed T, $\dot{\varepsilon}$ is based on interpolation of data from [4,230] and which is also summarized in [147,148,231]. (b) As in (a) but at 400°C and less pure Al, based on [147,148,232]. The subgrain containing metal here and (a), above, is stronger than expected based on Hall–Petch strengthening by the subgrains alone.

stress/strain-rate may change substantially with the purity at a specific strain-rate/ stress and temperature. If the λ_{ss} vs σ_{ss}/G in Figures 21 and 22 were placed in the same graph, it would be evident that, for identical subgrain sizes, the aluminum, curiously, would have higher strength. (The opposite is true for fixed ρ_{ss})

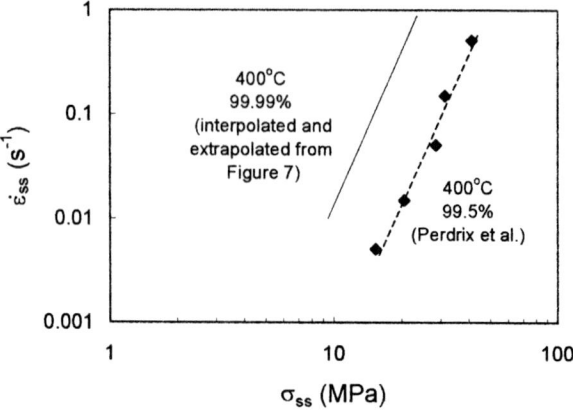

Figure 44. The steady-state strain-rate versus steady-state stress for Al of different purities. Data from Figure 21 and Perdrix et al. [233].

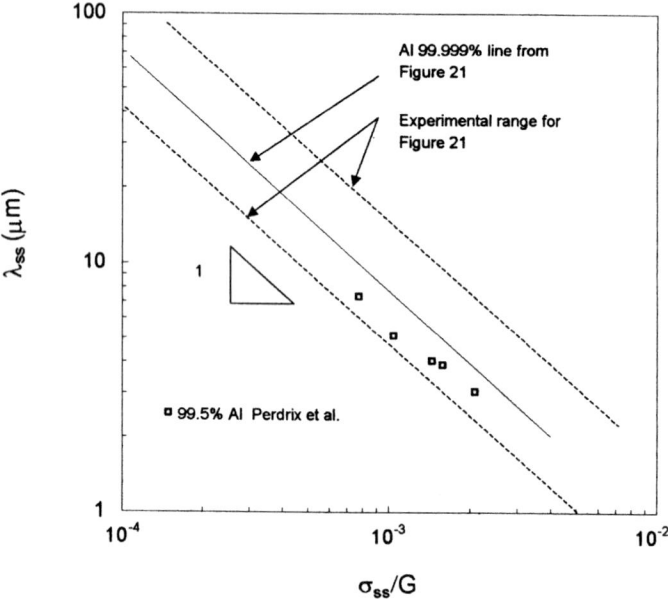

Figure 45. A plot of the variation of the steady-state subgrain size versus modulus-compensated steady-state flow stress for Al of different purities. The relationship between σ_{ss} and λ_{ss} for less pure Al is at least approximately described by the high-purity relationship.

Furthermore, it appears that λ is not predictable only on the basis of σ_{ss}/G for dispersion strengthened Al [90]. Steels do not appear to have λ_{ss} values predictable on the basis of αFe λ_{ss} versus σ_{ss}/G trends, although this failure could be a result of carbides present in some of the alloys. Thus, the aluminum trends may not be as evident in other metals. It should, however, be mentioned that for a wide range of single phase metals, there is a "rough" relationship between the subgrain size and stress [77],

$$\lambda = 23Gb/\sigma \qquad (75)$$

Of course, sometimes ambient-temperature strengthening interstitials (e.g., C in v-Fe) can *weaken* at elevated temperatures. In the case of C in v-Fe, D_{sd} increases with C concentration and $\dot{\varepsilon}_{ss}$ correspondingly increases [16].

2.4.4 Sigmoidal Creep

Sigmoidal creep behavior occurs when in a, e.g., single phase alloy, the creep-rate decreases with strain (time), but with further strain curiously increases. This increase is followed by, again, a decrease in creep-rate. An example of this behavior is

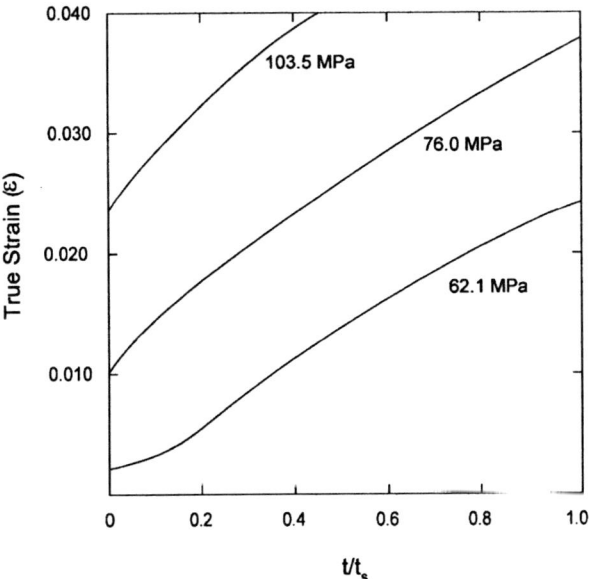

Figure 46. Transient creep curves obtained at 324°C for 70–30 α-brass, where t_s is the time to the start of steady-state creep. From Ref. [235].

illustrated in Figure 46 taken from Evans and Wilshire of 70–30 α-brass [234,235]. This behavior was also observed in the Cu–Al alloy described elsewhere [175] and also in Zr of limited purity [236]. The sigmoidal behavior in all of these alloys appears to reside within certain (temperature)/(stress/strain-rate) regimes. The explanations for sigmoidal creep are varied. Evans and Wilshire suggest that the inflection is due to a destruction in short-range order in α-brass leading to higher creep rates. Hasegawa et al. [175], on the other hand, suggest that changes in the dislocation substructure in Cu–Al may be responsible. More specifically, the increase in strain-rate prior to the inflection is associated with an increase in the total dislocation density, with the formation of cells or subgrains. The subsequent decrease is associated with cellular tangles (not subgrains) of dislocations. Evans and Wilshire suggest identical dislocation substructures with and without sigmoidal behavior, again, without subgrain formation. Warda et al. [236] attributed the behavior in Zr to dynamic strain-aging. In this case oxygen impurities give rise to solute atmospheres. Eventually, the slip bands become depleted and normal five-power behavior resumes. Dramatic increases in the activation energy are suggested to be associated with the sigmoidal behavior. Thus, the explanation for sigmoidal behavior is unclear. One common theme may be very planar slip at the high temperatures.

Chapter 3
Diffusional-Creep

Chapter 3
Diffusional-Creep

Creep at high temperatures ($T \approx T_m$) and very low stresses in fine-grained materials was attributed 50 years ago by Nabarro [237] and Herring [51] to the mass transport of vacancies through the grains from one grain boundary to another. Excess vacancies are created at grain boundaries perpendicular to the tensile axis with a uniaxial tensile stress. The concentration may be calculated using [23]

$$c = c_v \left[\exp\left(\frac{\sigma b^3}{kT}\right) - 1 \right] \qquad (76)$$

where c_v is the equilibrium concentration of vacancies. Usually $(\sigma b^3/kT) \propto 1$, and therefore equation (76) can be approximated by

$$c = \left[c_v \left(\frac{\sigma b^3}{kT}\right) \right] \qquad (77)$$

These excess vacancies diffuse from the grain boundaries lying normal to the tensile direction toward those parallel to it, as illustrated in Figure 47. Grain boundaries act as perfect sources and sinks for vacancies. Thus, grains would elongate without dislocation slip or climb. The excess concentration of vacancies per unit volume is, then, $(c_v\sigma/kT)$. If the linear dimension of a grain is "g", the concentration gradient is $(c_v\sigma/kTg)$. The steady-state flux of excess vacancies can be expressed as $(D_v c_v \sigma/kTg)$. where g is the grain size. The resulting strain-rate is given by,

$$\dot{\varepsilon}_{ss} = \frac{D_{sd}\sigma b^3}{kTg^2} \qquad (78)$$

In 1963, Coble [52] proposed a mechanism by which creep was instead controlled by grain-boundary diffusion. He suggested that, at lower temperatures ($T < 0.7\,T_m$), the contribution of grain-boundary diffusion is larger than that of self-diffusion through the grains. Thus, diffusion of vacancies along grain boundaries controls creep. The strain-rate suggested by Coble is

$$\dot{\varepsilon}_{ss} = \frac{\alpha_3 D_{gb}\sigma b^4}{kTg^3} \qquad (79)$$

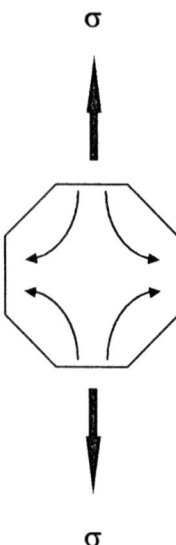

Figure 47. Nabarro–Herring model of diffusional flow. Arrows indicate the flow of vacancies through the grains from boundaries lying normal to the tensile direction to parallel boundaries. Thicker arrows indicate the tensile axis.

where D_{gb} is the diffusion coefficient along grain boundaries and α_3 is a constant of the order of unity. The strain-rate is proportional to g^{-2} in the Nabarro–Herring model whereas it is proportional to g^{-3} in the Coble model. In recent years more profound theoretical analyses of diffusional creep have been reported [238]. Greenwood [238] formulated expressions which allow an approximation of the strain-rate in materials with non-equiaxed grains under multiaxial stresses for both lattice and grain-boundary diffusional creep.

Several studies reported the existence of a threshold stress for diffusional creep below which no measurable creep is observed [239–242]. This threshold stress has a strong temperature dependence that Mishra et al. [243] suggest is inversely proportional to the stacking fault energy. They proposed a model based on grain-boundary dislocation climb by jog nucleation and movement to account for the existence of the threshold stress.

The occurrence of Nabarro–Herring creep has been reported in polycrystalline metals [244–247] and in ceramics [248–252]. Coble creep has also been claimed to occur in Mg [251], Zr and Zircaloy-2 [253], Cu [254], Cd [255], Ni [255], copper–nickel [256], copper–tin [256], iron [257], magnesium oxide [258,259], βCo [242], αFe [240], and other ceramics [260]. The existence of diffusional creep must be inferred from indirect experimental evidence, which includes agreement with the rate equations developed by Nabarro–Herring and Coble, examination of marker lines

Figure 48. Denuded zones formed perpendicular to the tensile direction in a hydrated Mg-0.5%Zr alloy at 400°C and 2.1 MPa [262].

visible at the specimen surface that lie approximately parallel to the tensile axis [261], or by the observation of some microstructural effects such as precipitate-denuded zones (Figure 48). These zones are predicted to develop adjacent to the grain boundaries normal to the tensile axis in dispersion-hardened alloys. Denuded zones were first reported by Squires et al. [263] in a Mg-0.5wt.%Zr alloy. They suggested that magnesium atoms would diffuse into the grain-boundaries perpendicular to the tensile axis. The inert zirconium hydride precipitates act as grain-boundary markers. The authors proposed a possible relation between the appearance of these zones and diffusional creep. Since then, denuded zones have been observed on numerous occasions in the same alloy and suggested as proof of diffusional creep.

The existence of diffusional creep has been questioned [264] over the last decade by some investigators [59,61,265–270] and defended by others [56–58,60,261,271,272]. One major point of disagreement is the relationship between denuded zones and diffusional creep. Wolfenstine *et al.* [59] suggest that previous studies on the Mg-0.5wt.%Zr alloy [273] are sometimes inconsistent and incomplete since they do not give information regarding the stress exponent or the grain-size exponent. By analyzing data from those studies, Wolfenstine *et al.* [59,265] suggested that the stress exponents corresponded to a higher-exponent-power-law creep regime. Wolfenstine *et al.* also suggested that the discrepancy in creep rates calculated from the width of denuded zones and the average creep rates (the former being sometimes as much as six times lower than the latter) as evidence of the absence of correlation between denuded zones and diffusional creep. Finally, the same investigators [59,265,266] claim that denuded zones can also be formed by other mechanisms including the redissolution of precipitates due to grain-boundary sliding accompanied by grain-boundary migration and the drag of solute atoms by grain-boundary migration.

Several responses to the critical paper of Wolfenstine *et al.* [59] were published defending the correlation between denuded zones and diffusional creep [57,58,271]. Greenwood [57] suggests that the discrepancies between theory and experiments can readily be interpreted on the basis of the inability of grain boundaries to act as perfect sinks and sources for vacancies. Bilde-Sørensen *et al.* [58] agree that denuded zones may be formed by other mechanisms than diffusional creep but they claim that, if the structure of the grain boundary is taken into consideration, the asymmetrical occurrence of denuded zones is fully compatible with the theory of diffusional creep. Similar arguments were presented by Kloc [271].

Recently McNee *et al.* [274] claim to have found additional evidence of the relationship between diffusional creep and denuded zones. They studied the formation of precipitate free zones in a fully hydrided magnesium ZR55 plate around a hole drilled in the grip section. The stress state around the hole is not uniaxial, as shown in Figure 49. They observed a clear dependence of the orientation of denuded zones on the direction of the stress in the region around the hole.

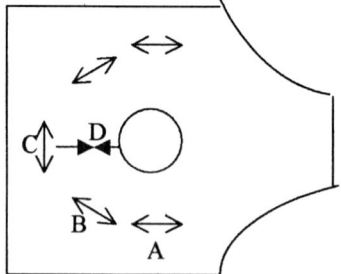

Figure 49. Orientation of stresses around a hole.

Precipitate free zones were mainly observed in boundaries perpendicular to the loading direction at each location. They claim that this relationship between the orientation of the denuded zones and the loading direction is consistent with the mechanism of formation of these zones being diffusional creep.

Ruano et al. [266–268], Barrett et al. [269], and Wang [270] suggest that the dependence of the creep-rate on stress and grain size is not always in agreement with that of diffusional creep theory. A reinterpretation of several data reported in previous studies led Ruano et al. to propose that the creep mechanism is that of Harper–Dorn creep in some cases and grain-boundary sliding in others, reporting a better agreement between experiments and theory using these models.

This suggestion has been contradicted by Burton et al. [64], Owen et al. [56], and Fiala et al. [272].

McNee et al. [275] have recently reported direct microstructural evidence of diffusional creep in an oxygen free high conductivity (OHFC) copper tensile tested at temperatures between 673 and 773 K, and stresses between 1.6 and 8 MPa. The temperature and stress dependencies were found to be consistent with diffusional creep. SEM surface examination revealed, first, displacement of scratches at grain boundaries and, second, widened grain boundary grooves on grain boundaries transverse to the applied stress in areas associated with scratch displacements. In principle, both diffusional creep, as well as some alternative mechanism involving grain-boundary sliding, could be responsible for the observed scratch displacements. The use of atomic force microscopy (AFM) to profile lines traversing boundaries both parallel and perpendicular to the tensile axis led to the conclusion that the scratch displacements originated from the deposition of material at grain boundaries transverse to the tensile axis and the depletion of material at grain boundaries parallel to the tensile axis. The investigators claimed that these features can only be attributed to the operation of a diffusional flow mechanism. However, a strain-rate with an order of magnitude higher than that predicted by Coble creep was found. Thus, the investigators questioned the direct applicability of the diffusional creep theory.

Nabarro recently suggested that Nabarro–Herring creep may be accompanied by other mechanisms (including GBS and Harper–Dorn) [276,277]. Lifshitz [278] already in 1963 pointed out the necessity of grain-boundary sliding for maintaining grain coherency during diffusional creep in a polycrystalline material More recent theoretical studies have also emphasized the essential role of grain boundary sliding for continuing steady-state diffusional creep [279–282]. The observations reported by McNee et al. [275] may, in fact, reflect the cooperative operation of both mechanisms. Many studies have been devoted to assess the separate contributions from diffusional creep and grain-boundary sliding to the total strain [283–292]. Some claim that both diffusional creep and grain-boundary sliding contribute to the overall strain and that they can be distinctly separated [284–288]; others claim that one of them is an accommodation process [289–292]. Many of these studies are based on several simplifying assumptions, such as the equal size of all grains and that the total strain is achieved in a single step. Sahay *et al.* [282] claimed that when the dynamic nature of diffusional creep is taken into account (changes in grain size, etc., that take place during deformation), separation of the strain contributions from diffusion and sliding becomes impossible.

Chapter 4
Harper–Dorn Creep

4.1.	The Size Effect	103
4.2.	The Effect of Impurities	106

Chapter 4
Harper–Dorn Creep

Another mechanism for creep at high temperatures and low stresses was proposed in 1957 by Harper and Dorn [50] based on previous studies by Mott [295] and Weertman [189]. This mechanism has since been termed Harper–Dorn creep. By performing creep tests on aluminum of high purity and large grain sizes, these investigators found that steady-state creep increased linearly with the applied stress and the activation energy was that of self-diffusion. However, the observed creep process could not be ascribed to Nabarro–Herring or Coble creep. They reported creep rates as high as a factor of 1400 greater than the theoretical rates calculated by the Nabarro–Herring or Coble models. The same observations were reported some years later by Barrett *et al.* [269] and Mohamed *et al.* [294], as summarized in Figure 50. Additionally, they observed a primary stage of creep, which would not be expected according to the Nabarro–Herring model since the concentration of vacancies immediately upon stressing cannot exceed the steady-state value. Furthermore, grain-boundary shearing was reported to occur during creep and similar steady-state creep-rates were observed in aluminum single crystals and in polycrystalline specimens with a 3.3 mm grain size. Strains as high as 0.12 were reported [199]. This evidence led these investigators to conclude that low-stress creep at high temperatures in materials of large grain sizes occurred by a dislocation climb mechanism.

The relation between the applied stress and the steady-state creep rate for Harper–Dorn creep is phenomenologically described by [295]

$$\dot{\varepsilon}_{ss} = A_{HD} \left(\frac{D_{sd} G b}{kT} \right) \left(\frac{\sigma}{G} \right)^1 \qquad (80)$$

where A_{HD} is a constant.

Since these early observations [50], Harper–Dorn creep has been reported to occur in a large number of metals and alloys (Al and Al alloys [269,295,296], Pb [294], α-Ti [297], α-Fe [240], α-Zr [298], and β-Co [299]) as well as a variety of ceramics and ice [300–315]. (The Harper–Dorn creep behavior of $MgCl_2 \cdot 6H_2O$ $(CO_{0.5}Mg_{0.5})O$ and $CaTiO_3$ was recently questioned by Berbon and Langdon [316].)

Several studies have been published over the last 20 years with the objective of establishing the detailed mechanism of Harper–Dorn creep [295,296,317]. Mohamed [317] concluded that both Harper–Dorn and Nabarro–Herring creep are independent processes and that the predominance of one or the other would depend on the grain size. The former would be rate controlling for grain sizes higher than a critical

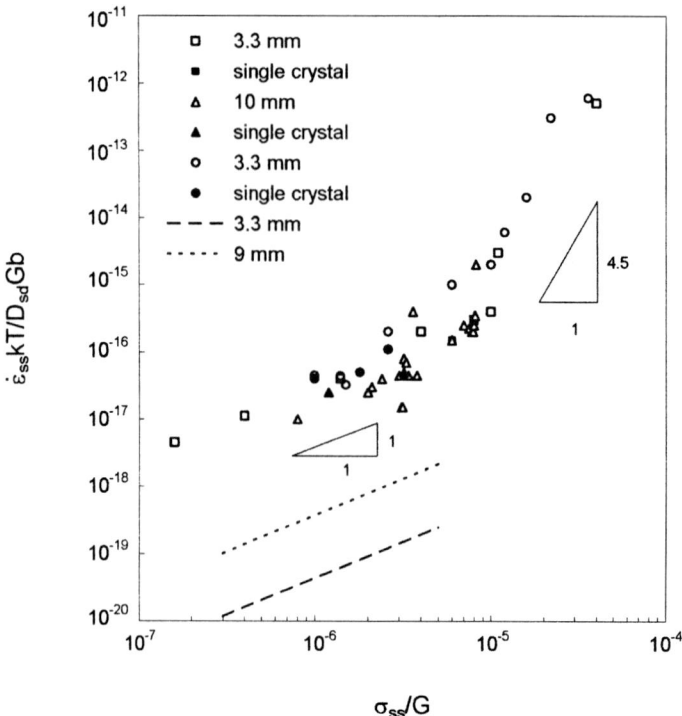

Figure 50. Comparison between the diffusion-coefficient compensated strain-rate versus modulus-compensated stress for pure aluminum based on [50,269,294], with theoretical predictions for Nabarro–Herring creep [295].

grain size (g_l) and the latter for lower grain sizes. Yavari et al. [295] provided evidence that the Harper–Dorn Creep-rate is independent of the specimen grain size. Similar rates were observed both in polycrystalline materials and in single crystals. They determined, by etch-pits, that the dislocation density was relatively low, at about $5 \times 10^7 \, \text{m}^{-2}$, and *independent* of the applied stress. Recently, Owen and Langdon [56] corrected the values of the dislocation density to near $10^9 \, \text{m}^{-2}$. (Dislocations were found to be predominantly close to edge orientation.) Nes [215], however, found that the dislocation density cannot be accurately measured by TEM due to the low values of ρ, and X-ray topography apparently showed that ρ was dependent of stress though only scaled by $\sigma^{1.3}$. Mohamed et al. [296] suggested that the transition between Nabarro–Herring and Harper–Dorn Creep may, thus, take place only when the grain size exceeds a critical value *and* the dislocation density is relatively low.

The fact that the activation energy for Harper–Dorn Creep is about equal to that of self-diffusion suggests that vacancy diffusion is rate-controlling. The failure of the Nabarro–Herring model to predict the experimental creep-rates, together with the independence of the creep-rate on grain size, suggests that dislocations may be sources and sinks of vacancies. Accordingly, Langdon et al. and Wang et al. [318,319] proposed that Harper–Dorn creep occurs by climb of edge dislocations. Weertman and Blacic [320] suggest that creep is not observed at constant temperatures, but only with low amplitude temperature fluctuations, where the vacancy concentration would not be in thermal equilibrium, thus leading to climb stresses on edge dislocations of the order 3–6 MPa. Although clever, this explanation does not appear widely accepted, partly due to activation energy disparities ($Q \neq Q_{sd}$) and that H–D creep is consistently observed by a wide assortment of investigators, presumably with different temperature control abilities [321]. The fact that both within the Harper–Dorn and the five-power-law regime, the underlying mechanism of plastic flow appeared to be diffusion controlled, led Wu and Sherby [53] to propose a unified relation to describe the creep behavior over both ranges. This model incorporates an internal stress that arises from the presence of random stationary dislocations present within subgrains. At any time during steady-state flow, half of the dislocations moving under an applied stress are aided by the internal stress field (the Internal stress adds to the applied stress), whereas the motion of the other half is inhibited by the internal stress. It is also assumed that each group of dislocations are contributing to plastic flow independent of each other. The Internal stress is calculated from the dislocation density by the dislocation hardening equation ($\tau = \alpha G b \sqrt{\rho}$ where $\alpha \cong 0.5$). The unified equation is [322]

$$\dot{\varepsilon}_{ss} = \frac{1}{2} A_{10} \frac{D_{eff}}{b^2} \left\{ \left(\frac{\sigma + \sigma_i}{E}\right)^n + \frac{|\sigma - \sigma_i|}{(\sigma - \sigma_i)} \left|\frac{\sigma - \sigma_i}{E}\right|^n \right\} \tag{81}$$

where A_{10} is a constant and σ_i is the internal stress. At high stresses, where $\sigma \gg \sigma_i$, σ_i is negligible compared to σ and equation (81) reduces to the (e.g., five-power-law) relation:

$$\dot{\varepsilon}_{ss} = A_{10} \frac{D_{eff}}{b^2} \left(\frac{\sigma}{E}\right)^n \tag{82}$$

At low stresses, where $\sigma \vee \sigma_i$ (Harper–Dorn regime), equation (81) reduces to equation (80). A reasonable agreement has been suggested between the predictions from this model and experimental data [53,322] for pure aluminum, γ-Fe and β-Co. The internal stress model was criticized by Nabarro [321], who claimed that a unified approach to both five-power-law and Harper–Dorn creep is not possible since none of these processes are, in themselves, well understood and unexplained dimensionless constants were introduced in order to match theoretical predictions with experimental data. Also, the dislocation density in Harper–Dorn creep is constant, whereas it increases with the square of the stress in the power-law regime. Thus, the physical processes occurring in both regimes must be different (although Ardell [54] attempts to rationalize this using network creep models). Nabarro [321] proposed a new mechanism for plastic flow during Harper–Dorn creep. According to this model, an equilibrium concentration of dislocations is established during steady-state creep which exerts a stress on its neighbors equal to the Peierls stress. The mechanism of plastic flow would be the motion of these dislocations controlled by climb.

The Internal stress model was also criticized by Wang [323], who proposed that the transition between power-law creep and Harper–Dorn creep takes place instead at a stress (σ) equal to the "Peierls stress (σ_p)" [324,325] (which simply appears to equal the yield stress of the annealed material). Wang [326] suggests that the steady-state dislocation density is related to the Peierls stress in the following way: in equilibrium, the stress due to the mutual interaction of moving dislocations is in balance not only with the applied stress but also with lattice friction which fluctuates with an amplitude of the Peierls stress. As a result, the steady-state dislocation density ρ in dislocation creep can be written as:

$$b\rho^{1/2} = 1.3\left[\left(\frac{\tau}{G}\right)^2 + \left(\frac{\tau_p}{G}\right)^2\right]^{1/2} \tag{83}$$

where τ is referred to as the applied shear stress. Accordingly, when $\tau \gg \tau_p$, the dislocation density is proportional to the square of the applied stress, and five- (or three-) power-law creep is observed. Conversely, when $\tau \ll \tau_p$, the dislocation density is independent of the applied stress and Harper–Dorn occurs. Wang [327] suggested that the transition between Harper–Dorn creep and Nabarro–Herring creep was influenced by σ_p since it was observed to be inversely proportional to the critical grain size, g_t. Therefore, he predicted that Harper–Dorn creep would be observed in polycrystalline materials with high σ_p values, such as ceramic oxides, carbides, and silicates, when the grain sizes are small. Conversely, Nabarro–Herring creep would predominate in materials with low σ_p values such as f.c.c. metals, even when g is very large. Wang [328] suggests that Harper–Dorn creep may be a combination of dislocation glide and climb, described by equation (61), with the dimensionless factor A_{HD} given by

$$A_{HD} = 1.4\left(\frac{\sigma_p}{G}\right)^2 \tag{84}$$

Wang compared available experimental data to equation (84) and obtained the empirical values of A_{HD}. Wang utilizes the Weertman description for climb-control with $\rho_m \propto \sigma_p = $ constant, thus a "one-power" stress dependence is derived. A different approach to Harper–Dorn is based on the dislocation network theory by Ardell and coworkers [54,86,101,329]. The dislocation link length distribution contains no segments that are long enough to glide or climb freely. Harper–Dorn is, therefore, a phenomenon in which all the plastic strain in the crystal is a consequence of stress-assisted dislocation network coarsening, during which the glide and climb of dislocations is constrained by the requirement that the forces acting on the links are balanced by the line tension on the links. Accidental collisions between these links can refine the network and stimulate further coarsening. The transition stress between power-law creep and Harper–Dorn creep (σ_T) is determined by the magnitude of a critical dislocation link length, P_c, inversely related to the stress required to activate a Frank–Read or Bardeen–Herring source [54],

$$\frac{\sigma_T}{G} = \frac{b}{g(\ell_c/3)} \frac{\{(1+\nu)(1-2\nu)\}^{1/2}}{(1-\nu)} \ell n \frac{(\ell_c/3)}{2b} \tag{85}$$

When $\sigma < \sigma_T$, P_c is larger than the maximum link length, P_m. The critical dislocation link length is related to the dislocation density by the equation

$$\frac{\ell_c}{3} = \frac{\beta_3}{\rho^{1/2}} \tag{86}$$

where β is a constant of the order of unity. The independence of ρ with σ is a consequence of the frustration of the dislocation network coarsening, which arises because of the exhaustion of Burgers vectors that can satisfy Frank's rule at the nodes. Some have suggested that it is not clear as to how ρ is self-consistently determined [277]. The correlation between the predicted and measured transition stress appears reasonable [54].

4.1 THE SIZE EFFECT

Blum et al. [55] recently questioned the existence of Harper–Dorn creep, not having been able to observe the decrease of the stress exponent to a value of 1 when performing compression tests with changes in stress in pure aluminum (99.99% purity). Nabarro [330] responded to these reservations claiming that the lowest stress used by Blum et al. (0.093 MPa) was still too high to observe Harper–Dorn. Therefore, Blum et al. [331] performed compression tests using even lower stresses (as low as 0.06 MPa), failing again to observe $n=1$ stress exponents. Instead, exponents close to 5 were measured. It is interesting to note that Blum et al. used relatively large compression specimens measuring 35 mm in length and with a cross section of about 29 × 29 mm. The strain was accurately measured using a contactless optical device consisting of a laser beam that scans the distance between two markers in the compression specimen. Blum et al. suggest that the $n=1$ exponents reported previously by other investigators might be due to a *size effect*. At the high temperatures at which Harper–Dorn occurs, the dislocation structural spacings are no longer small compared with the dimensions of a conventional creep specimen. For example, at a stress of 0.03 MPa, a stable subgrain size of about 3.7 mm is expected [331]. Also, the average spacing between dislocations inside subgrains is about 0.16 mm. Thus, the grains, usually in the cm size range, as well as many subgrains, extend up to the surface of the specimens. Under these circumstances, the average distance a mobile dislocation migrates, 1, is very large and in some cases comparable to the specimen size. Following is the analysis made by Blum *et al.* [201] to show how this effect could lead to the observation of $n = 1$ exponents when testing at very high temperatures and low stresses.

Harper–Dorn creep experiments are usually performed in high-purity specimens with a very low dislocation density. Upon loading a strain burst takes place and, thus, a significant amount of initially mobile dislocations are introduced, as demonstrated by Ardell and coworkers [54,86,101]. Subsequently, under dynamic conditions of metal plasticity, the network evolution (dislocation evolution), $\dot{\rho}$, is a consequence of the combined effect of the athermal storage of dislocations, $\dot{\rho}^+$, and dynamic recovery, $\dot{\rho}^-_{\text{dynamic}}$. Dynamic recovery consists on the local annihilation processes between mobile dislocations and network dislocations. Additionally, since H–D creep takes place at extremely high temperatures and low stresses, static recovery (or network growth due to stored line energy) also has a significant effect on the network evolution and, therefore, should be taken into account. Thus, the equation for the dislocation network evolution during H–D creep is [201]

$$\dot{\rho} = \dot{\rho}^+ + \dot{\rho}^-_{\text{dynamic}} + \dot{\rho}^-_{\text{static}} \tag{87}$$

Following the Nes and Marthinsen's model [215,332,333], dynamic recovery in a Frank network consists of the collapse of dislocation dipoles of separation l_g where the dipoles are a result of interactions between mobile dislocations and dislocations stored in the network. Thus, equation (87) can be written as:

$$\dot{\rho} = \frac{2}{b\ell}\gamma - v_g\rho - B_{\text{FN}}\rho^2\left(\frac{Gb^3}{kT}\right)D_{\text{sd}} \tag{88}$$

where ℓ is the average distance a mobile dislocation migrates from the source to the site where it is stored in the network, v_g is the dislocation collapse frequency, and B_{FN} is a parameter of order unity. In pure metals, the collapse reaction is expected to be driven by the sharp curvatures resulting from dipole pinch-off reactions. The pinched-off segments will subsequently climb due to the large curvature forces. The dislocation collapse frequency can be written as $v_g = 2v_c\sqrt{\rho}$, where v_c is the climb speed given by:

$$v_c = v_D b^2 B_\rho c_j \exp\left(\frac{-Q_{sd}}{kT}\right) 2\sinh\frac{fb^2}{kT} \approx$$
$$\approx 2v_D b^2 B_\rho c_j \left(\frac{fb^2}{kT}\right) \exp\left(\frac{-Q_{sd}}{kT}\right) \tag{89}$$

where v_D is the Debye frequency, B_ρ is a constant of order unity, f is the driving force generated by the curved network segments, and c_j represents the concentration of trailing jogs controlling the climb rate of the curved segments.

In order to consider the size effect on the substructure evolution, Blum et al. redefined equation (88) introducing a new parameter $S(S = R^2/((R+1)^2 - R^2)\lambda' + R^2$ where R = specimen radius and λ' is a parameter expected to be of the order of 1, reflecting surface conditions) in the following way:

$$\dot\rho = S\frac{2}{b\ell}\dot\gamma - S^{3/2}v_g\rho - B_{FN}\rho^2\left(\frac{Gb^3}{kT}\right)D_{sd} \tag{90}$$

The parameter S is introduced to exclude phantom sources of dislocations that are taken into account in equation (88) when the slip-length becomes comparable to, or larger than, the specimen radius. The parameter S should fulfill the following boundary conditions to account for a size effect: when $1 \ll R$, $S \to 1$ [(equation (88) would still hold in this case], and when $1 \gg R$, $S \to 0$, i.e., there is no storage of dislocations at all and the dynamic recovery rate approaches 0. In summary, almost all the dislocations would exit the specimen when the slip-length is much larger than the specimen radius. It can be noted that the static recovery rate is not affected by any size effect. Finally, Blum assumes that the thermal component (e.g., σ_0) of the flow stress is negligible and the athermal component can be written as $\tau = \alpha Gb\sqrt{\sigma}$ and the creep-rate may be expressed as [201]:

$$\dot\gamma = \frac{C_1}{2\alpha^2 b^2}\left(\frac{Gb^3}{kT}\right)\left[\frac{B_{FN}}{S}\left(\frac{\tau}{G}\right)^3 D_{SD} + \sqrt{S}\frac{13.6b^2}{\alpha}\xi^2 B_\rho\left(\frac{\tau}{G}\right)^4 v_D \exp\left(\frac{-Q_{sd}}{kT}\right)\right] \tag{91}$$

where C_1 is a constant larger than 1, ξ is a scaling parameter larger than 1, and the rest have the usual meaning. When $1 \gg R$ (and $\lambda \approx 1$), it can be shown that equation (91) reduces to a linear dependency of the strain-rate with respect to the applied stress, which is a constitutive equation for Harper–Dorn creep. Therefore, Blum et al. suggest that the observation of $n = 1$ exponents is due to the use of specimens with a cross section similar or smaller than the slip length. Thus, using larger specimens higher stress exponents would be observed. The model proposed by Blum et al. suggests static recovery as the predominant restoration mechanism during Harper–Dorn creep.

4.2 THE EFFECT OF IMPURITIES

Recently, Mohamed et al. [334–336] suggested that impurities may play an essential role in so-called "Harper–Dorn creep." They performed large strain (up to 10%) creep tests at stresses lower than 0.06 MPa in Al polycrystals of 99.99 and 99.9995 purity. They only observed H–D creep in the latter material. Accelerations in the creep curve corresponding to the high-purity grade Al are apparent, as illustrated in Figure 51. These accelerations are absent in the 99.99 Al creep curve.

Mohamed et al. report also that the microstructure of the 99.9995 Al is formed by wavy grain boundaries, an inhomogeneous dislocation density distribution, small new grains forming at the specimen surface and large dislocation density gradients across grain boundaries. Well-defined subgrains are not observed. However, the microstructure of the deformed 99.99 Al is formed by a well-defined array of subgrains.

These observations have led Mohamed et al. to conclude that the restoration mechanism taking place during Harper–Dorn creep is dynamic recrystallization. Nucleation of recrystallized grains would take place at the specimens surfaces and, due to the low amount of impurities, highly mobile boundaries would migrate toward the specimen interior. This restoration mechanism would give rise to the periodic accelerations observed in the creep curve, by which most of the strain is produced. Therefore, Mohamed et al. believed that two essential prerequisites for the occurrence of "Harper-Dorn creep" are high purity and low dislocation density (of the order of 10^7m^{-2} to $3 \times 10^7 \text{m}^{-2}$), that favor dynamic recrystallization.

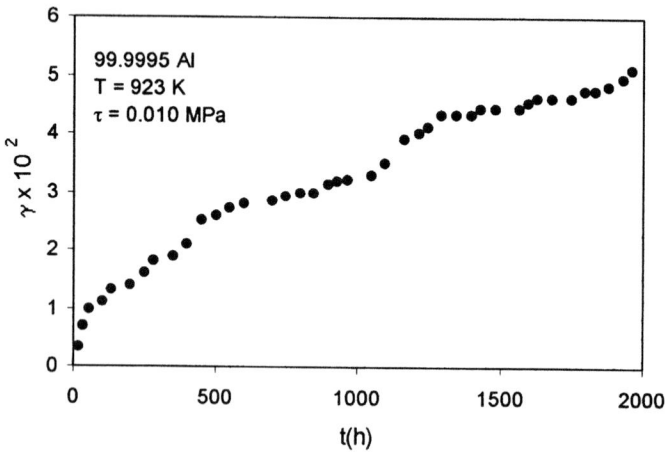

Figure 51. Creep curve corresponding to 99.9995 Al deformed at a stress of 0.01 MPa at 923 K.

It is difficult to accurately determine the stress exponent due to the appearance of periodic accelerations in the creep curves. Mohamed *et al.* [335] claim that $n=1$ exponents are only obtained if creep curves up to small strains (1–2%) are analyzed, as was done in the past. Mohamed *et al.* estimated a stress exponent of about 2.5 at larger strains.

The work by Mohamed *et al.* has received some criticism. Langdon [337] argues that the jumps in the creep curves are not very clearly defined. Also, he claims that dynamic recrystallization occurs during creep of very high purity metals at regular strain increments, whereas the incremental strains corresponding to the accelerations reported by Mohamed *et al.* tend to be relatively nonuniform. He suggests that grain growth might be a more appropriate restoration mechanism.

The restoration mechanism predominant during Harper–Dorn creep is, thus, not well established at this point. Blum *et al.* [201] support static recovery, Mohamed *et al.* [335] favor dynamic recrystallization, and Langdon [337] grain growth. Further work needs to be done in this area in order to clarify this issue.

Chapter 5
Three-Power-Law Viscous Glide Creep

Chapter 5
Three-Power-Law Viscous Glide Creep

Creep of solid solution alloys (designated Class I [16] or class A alloys [338]) at intermediate stresses and under certain combinations of materials parameters, which will be discussed later, can often be described by three regions [36,339,340]. This is illustrated in Figure 52. With increasing stress, the stress exponent, n, changes in value from 5 to 3 and again to 5 in regions I, II, and III, respectively. This section will focus on region II, the so-called Three-Power-Law regime.

The mechanism of deformation in region II is viscous glide of dislocations [36]. This is due to the fact that the dislocations interact in several possible ways with the solute atoms, and their movement is impeded [343]. There are two competing mechanisms over this stress range, dislocation climb and glide, and glide is slower

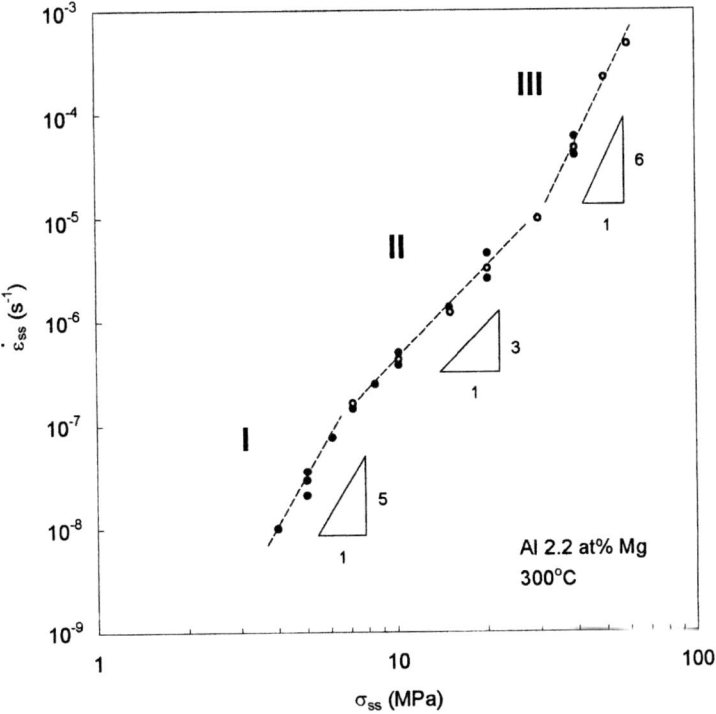

Figure 52. Steady-state creep rate vs. applied stress for an Al-2.2 at%Mg alloy at 300°C. Three different creep regimes, I, II, and III, are evident. Based on Refs. [341,342].

and thus rate controlling. A three-power-law may follow naturally then from equation (16) [24,344,345],

$$\dot{\varepsilon} = 1/2\,\bar{v}\,b\,\rho_m$$

It has been theoretically suggested that $\bar{v}$ is proportional to σ [346,347] for solute drag viscous glide. It has been determined empirically that ρ_m is proportional to σ^2 for Al–Mg alloys [76,93,118,318,341,348]. Nabarro [23], Weertman [344,345], Horiuchi et al. [349] have suggested a possible theoretical explanation for this relationship. Thus, $\dot{\varepsilon} \propto \sigma^3$. More precisely, following the original model of Weertman [344,345], viscous glide creep is described by the equation

$$\dot{\varepsilon}_{ss} \cong \frac{0.35}{A} G \left(\frac{\sigma}{G}\right)^3 \qquad (92)$$

where A is an interaction parameter which characterizes the particular viscous drag process controlling dislocation glide.

There are several possible viscous drag (by solute) processes in region II, or Three-Power-Law regime [344,350–353]. Cottrell and Jaswon [350] proposed that the dragging process is the segregation of solute atmospheres to moving dislocations. The dislocation speed is limited by the rate of migration of the solute atoms. Fisher [351] suggested that, in solid solution alloys with short-range order, dislocation motion destroys the order creating an interface. Suzuki [352] proposed a dragging mechanism due to the segregation of solute atoms to stacking faults. Snoek and Schoeck [353,354] suggested that the obstacle to dislocation movement is the stress-induced local ordering of solute atoms. The ordering of the region surrounding a dislocation reduces the total energy of the crystal, pinning the dislocation. Finally, Weertman [344] suggested that the movement of a dislocation is limited in long-range-ordered alloys since the implied enlargement of an anti-phase boundary results in an increase in energy. Thus, the constant A in equation (92) is the sum of the different possible solute–dislocation interactions described above, such as

$$A = A_{C-J} + A_F + A_S + A_{Sn} + A_{APB} \qquad (93)$$

Several investigators proposed different three-power models for viscous glide where the principal force retarding the glide of dislocations was due to Cottrell–Jaswon interaction ($A_{C-J} + A_F + A_S + A_{Sn} + A_{APB}$) [87,118,345,355]. In one of the first theories, Weertman [345] suggested that dislocation loops are emitted by sources and sweep until they are stopped by the interaction with the stress field of loops on different planes, and dislocation pile-ups form. The leading dislocations can, however, climb and annihilate dislocations on other slip planes. Mills et al. [118] modeled the dislocation substructure as an array of elliptical loops, assuming that no drag force exists on the pure screw segments of the loops. Their model intended to explain transient three-power creep behavior. Takeuchi and Argon [355] proposed a dislocation glide model based on the assumption that once dislocations are emitted from the source, they can readily disperse by climb and cross-slip, leading to a homogeneous dislocation distribution. They suggested that both glide and climb are controlled by solute drag. The final relationship is similar to that by Weertman. Mohamed and Langdon [357] derived the following relationship that is frequently referenced for three-power-law viscous creep when only a Cottrell–Jaswon dragging mechanism is considered

$$\dot{\varepsilon}_{ss} \cong \frac{\pi(1-v)kT\tilde{D}}{6e^2Cb^5G}\left(\frac{\sigma}{G}\right)^3 \qquad (94)$$

where e is the solute–solvent size difference, C is the concentration of solute atoms and $\tilde{D}$ is the diffusion coefficient for the solute atoms, calculated using Darken's [358] analysis. Later Mohamed [359] and Soliman et al. [360] suggested that Suzuki and Fischer interactions are necessary to predict the Three-Power-Law creep behavior of several Al–Zn, Al–Ag, and Ni–Fe alloys with accuracy. They also suggested that the diffusion coefficients defined by Fuentes-Samaniego and coworkers [361] should be used, rather than Darken's equations.

Region II has been reported to occur preferentially in materials with a relatively large atom size mismatch [362,363]. Higher solute concentrations also favor the occurrence of three-power-law creep [338,340,344,357]. As illustrated in Figure 53, for high enough concentrations, region III can even be suppressed. The difference between the creep behaviors corresponding to Class I (A) and Class II (M) is evident in Figure 54 [349] where strain-rate increases with time with the former and decreases with the latter. Others have observed even more pronounced primary creep features in Class I (A) Al–Mg [156,364]. Alloys with 0.6% and 1.1 at.%Mg are Class II (M) alloys and those with 3.0%, 5.1%, and 6.9% are Class I (A) alloys. Additionally, inverse Creep transient behavior is observed in Class I (A) alloys [118,349,365,366] and illustrated in Figure 55 [349]. A drop in stress is followed by a decrease in the strain rate in pure aluminum, which then increases with a recovering dislocation substructure until steady-state at the new, lower, stress. However, with a stress decrease in Class I (A) alloys [Figures 55(b) and (c)], the strain rate continually decreases until the new steady-state. Analogous disparities are observed with stress increases [i.e., decreasing strain-rate to steady-state in Class II (M) while increasing rates with Class I (A).] Horiuchi et al. [349] argued that this is explained by the strain-rate being proportional to the dislocation density and the dislocation velocity. The latter is proportional to the applied stress while the square root of the former is proportional to the stress. With a stress drop, the dislocation velocity decreases to

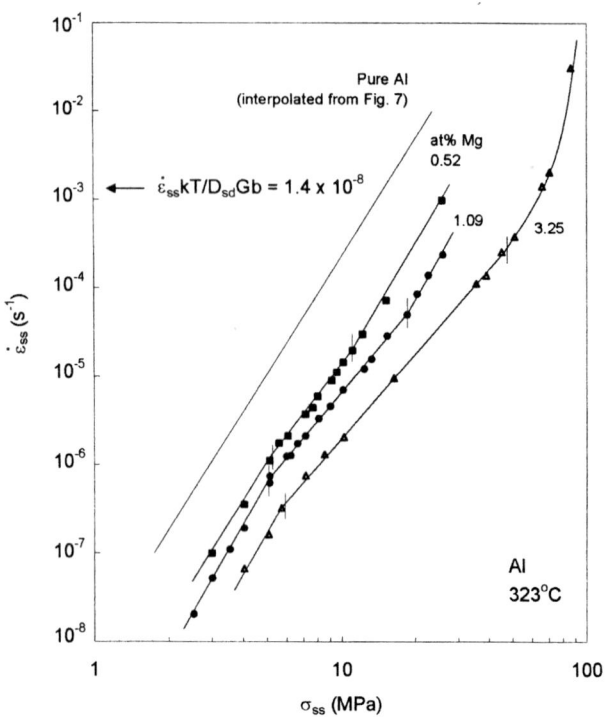

Figure 53. Steady-state creep-rate vs. applied stress for three Al–Mg alloys (Al-0.52 at.%Mg, ν; Al-1.09 at.%Mg, λ Al-3.25 at.%Mg, σ) at 323°C [356].

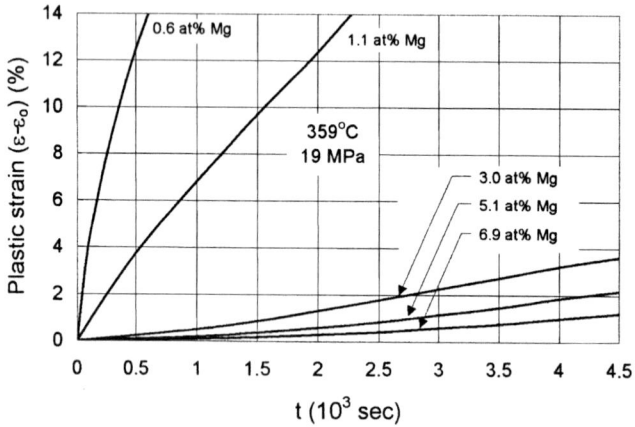

Figure 54. Creep behavior of several aluminum alloys with different magnesium concentrations: 0.6 at.% and 1.1 at.% (class II (M)) and 3.0 at.%, 5.1 at.%, and 6.9 at.% (class I (A)). The tests were performed at 359°C and at a constant stress of 19 MPa [349].

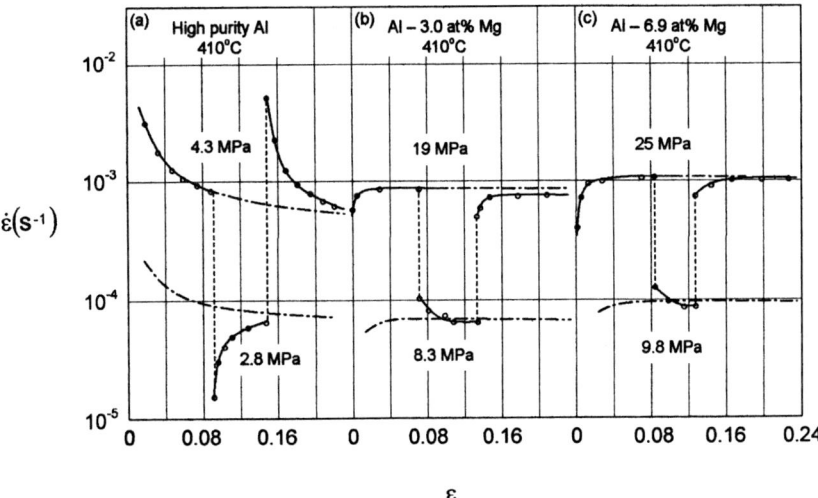

Figure 55. Effect of changes in the applied stress to the creep rate in (a) high-purity aluminum, (b) Al-3.0 at.%Mg (class I (A) alloy), and (c) Al-6.9 at.%Mg (class I (A)alloy), at 410°C [349].

the value corresponding to the lower stress. The dislocation density continuously decreases, also leading to a decrease in strain-rate. It is presumed that nearly all the dislocations are mobile in Class I (A) alloys while this may not be the case for Class II (M) alloys and pure metals. Sherby et al. [367] emphasize that the transition from strain softening to strain hardening at lower stresses in class I alloys is explained by taking into account that, within the viscous glide regime, the mobile dislocation density controls the creep-rate. Upon a stress drop, the density of mobile dislocations is higher than that corresponding to steady-state, and thus it will be lowered by creep straining, leading to a gradual decrease in strain-rate (strain hardening). If the stress is increased, the initial mobile dislocation density will be low, and, thus, the creep strength will be higher than that corresponding to steady-state. More mobile dislocations will be generated as strain increases, leading to an increase in the creep-rate, until a steady-state structure is achieved (strain softening). The existence of internal stresses during three-power-law creep is also not clearly established. Some investigators have reported internal stresses as high as 50% of the applied stress [368]. Others, however, suggest that Internal stresses are negligible compared with the applied stress [121,349].

The transitions between regions I and II and between regions II and III are now well established [362]. The condition for the transition from region I ($n = 5$, climb-controlled creep behavior) to region II ($n = 3$, viscous glide) with increasing applied stress is, in general, represented by [359]

$$\frac{kT}{D_g b A} = C\left(\frac{\chi}{Gb}\right)^3 \left(\frac{D_c}{D_g}\right)\left(\frac{\tau}{G}\right)_t^2 \quad (95)$$

where D_c and D_g are the diffusion coefficients for climb and glide, respectively, C is a constant and $(\tau/G)_t$ is the normalized transition stress. If only the Cottrell-Jaswon interaction is considered [357,359], equation (95) reduces to

$$\left(\frac{kT}{ec^{1/2}Gb^3}\right)^2 = C\left(\frac{c}{Gb}\right)^3 \left(\frac{D_c}{D_g}\right)\left(\frac{\tau}{G}\right)_t^2 \quad (96)$$

The transition between regions II ($n=3$, viscous glide) and III ($n=5$, climb-controlled creep) has been the subject of several investigations [36,344,345,369–372]. It is generally agreed that it is due to the breakaway of dislocations from solute atmospheres and are thus able to glide at a much faster velocity. The large difference in dislocation speed between dislocations with and without clouds has been measured experimentally [373]. Figure 56 (from Ref. [373]) shows the σ vs. ε and $\dot{\varepsilon}$ vs. ε curves corresponding to Ti of commercial purity, creep tested at 450°C at different stresses after initial loading at $10^{-3}\,\text{s}^{-1}$. Instead of a smooth transition to steady-state creep, a significant drop in strain-rate takes place upon the beginning of the creep test, which is associated with the significant decrease in dislocation speed due to the formation of solute clouds around dislocations. Thus, after a critical breakaway stress is exerted on the material, glide becomes faster than climb and the latter is, then, rate-controlling in region III. Friedel [87] predicted that the breakaway stress for unsaturated dislocations may be expressed as

$$\tau_b = A_{11}\left(\frac{W_m^2}{kTb^3}\right)C \quad (97)$$

where A_{11} is a constant, W_m is the maximum interaction energy between a solute atom and an edge dislocation and C is the solute concentration. Endo et al. [342] showed, using modeling of mechanical experiments, that the critical velocity, v_{cr}, at breakaway agrees well with the value predicted by the Cottrell relationship

$$v_{cr} = \frac{DkT}{eGbR_S^3} \quad (98)$$

where R_S is the radius of the solvent atom and e is the misfit parameter.

It is interesting that the extrapolation of Stage I (five-power) predicts significantly lower Stage III (also five-power) stress than observed (see Figure 52). The explanation for this is unclear. TEM and etch-pit observations within the

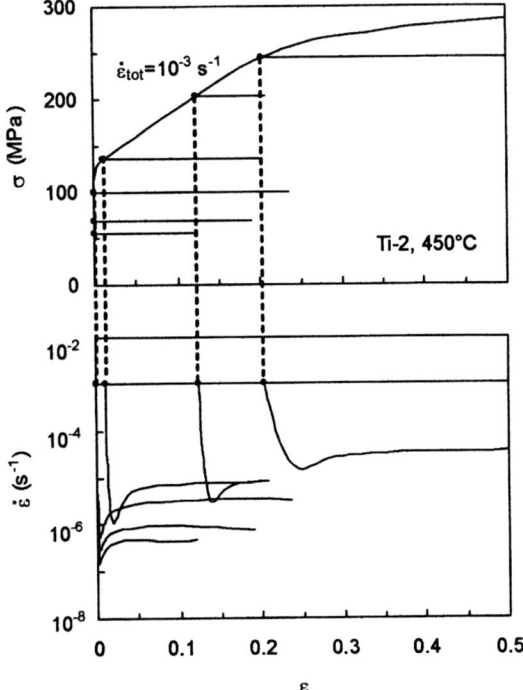

Figure 56. σ vs. ε and $\dot{\varepsilon}$ vs. ε curves corresponding to Ti of commercial purity, creep tested at 450°C at different stresses after initial loading at 10^{-3} s^{-1}. The circles mark the end of loading/beginning of creep testing. From Refs. [373].

three-power-law creep regime [118,338,374–377] show a random distribution of bowed long dislocation lines, with only a sluggish tendency to form subgrains. However, subgrains eventually form in Al–Mg [121,156,378], as illustrated in Figure 57 for Al-5.8%Mg.

Class I (A) alloys have an intrinsically high strain-rate sensitivity ($m = 0.33$) within regime II and therefore are expected to exhibit high elongations due to resistance to necking [379–381]. Recent studies by Taleff et al. [382–385] confirmed an earlier correlation between the extended ductility achieved in several binary and ternary Al–Mg alloys and their high strain-rate sensitivity. The elongations to failure for single phase Al–Mg can range from 100% to 400%, which is sufficient for many manufacturing operations, such as the warm stamping of automotive body panels [385]. These elongations can be achieved at lower temperatures than those necessary for conventional superplasticity in the same alloys and still at reasonable strain rates. For example, enhanced ductility has been reported to occur at 10^{-2} s^{-1} in a

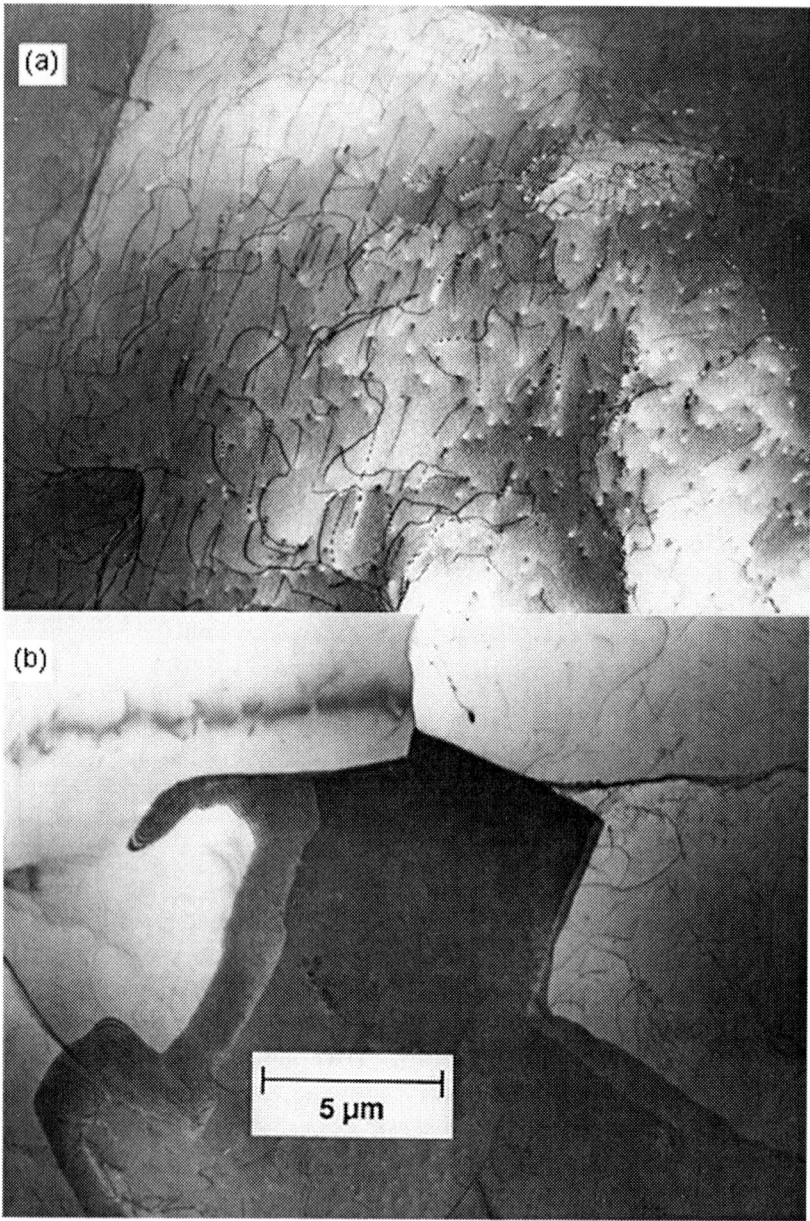

Figure 57. Al-5.8 at.%Mg deformed in torsion at 425°C to (a) 0.18 and (b) 1.1 strain in the three-power regime.

coarse-grained Al–Mg alloy [385] at a temperature of 390°C, whereas a temperature of 500°C is necessary for superplasticity in the same alloy with a grain size ranging from 5 to 10 µm. The expensive grain refinement processing routes necessary to fabricate superplastic microstructures are unnecessary. The solute concentration in a binary Al–Mg alloy does not affect significantly mechanical properties such as tensile ductility, strain-rate sensitivity, or flow stress [386]. For example, under conditions of viscous-glide creep, variations in Mg concentration ranging from 2.8 to 5.5 wt. pct. only change the strain-rate sensitivity from 0.29 to 0.32, which does not have a substantial effect on the elongation to failure [382]. McNelley *et al.* [378] attributed this observation to the saturation effect of Mg atoms in the core of the moving dislocation. However, ternary additions of Mn, Fe, and Zr seem to significantly affect the mechanical behavior of Al–Mg alloys [382]. The stress exponent increases and ductility decreases significantly, especially for Mn concentrations higher than 0.46 wt. pct. Ternary additions above the solubility limit favor the formation of second phase particles around which cavities tend to nucleate preferentially. Thus, a change in the failure mode from necking-controlled to cavity-controlled may occur, accompanied by a decrease in ductility [382]. Also, Mn atoms may interfere with the solute drag. The hydrostatic stress by which an atom interacts with a dislocation is determined by the volumetric size factor (Ω). The Ω values corresponding to Mg and Mn in Al are [363]: $\Omega^{\text{Al-Mg}} = +40.82$ and $\Omega^{\text{Al-Mn}} = -46.81$. Both factors are nearly equal in magnitude and of opposite sign. Therefore, each added Mn atom acts as a sink for one atom of Mg, thus reducing the effective Mg concentration [382]. If Mn is added in sufficient quantities that the effective Mg concentration is lower than that required for viscous-drag creep, the stress exponent would increase and the ductility would consequently decrease significantly. This effect seems to be less important than the change in failure mode described above [382].

Class I behavior has been reported to occur in a large number of metallic alloys. These include Al–Mg [16,357,379,387], Al–Zn [362], Al–Cu [388],Cu–Al [389], Au–Ni [372], Mg- [324, 386, 390-392], Pb- [372], In- [372], and Nb- [393,394] based alloys. Viscous-glide creep has also been observed in dispersion strengthened alloys. Sherby *et al.* [367] attribute the differences in the creep behavior between pure Al–Mg and DS Al–Mg alloys to the low mobile dislocation density of the latter. Due to the presence of precipitates, dislocations are pinned and their movement is impeded. Therefore, at a given strain-rate ($\dot{\gamma} = b\rho v$) the velocity of the dislocations is very high. Thus, much higher temperatures are required for solute atoms to form clouds around dislocations. The viscous glide regime is, therefore, observed at higher temperatures than in the pure Al–Mg alloys. A value of $n = 3$ has been observed in intermetallics with relatively coarse grain sizes ($g > 50\,\mu m$), such as Ni_3Al [395], Ni_3Si [396], TiAl [397], Ti_3Al [398], Fe_3Al [399,400], FeAl [401], and CoAl [402]. The mechanism of creep, here, is still not clear. Yang [403] has argued that, since the glide of dislocations introduces disorder, the steady-state velocity is limited by the rate at which chemical diffusion can reinstate order behind the gliding dislocations. Other explanations are based on the non-stoichiometry of intermetallics, and the frequent presence of interstitial impurities, such as oxygen, nitrogen, and carbon. Finally, dislocation drag has also been attributed to lattice friction effects [402]. Viscous glide has also been reported to occur in metal matrix composites [404], although this is still controversial [405].

Chapter 6
Superplasticity

6.1.	Introduction	123
6.2.	Characteristics of Fine Structure Superplasticity	123
6.3.	Microstructure of Fine Structure Superplastic Materials	127
	6.3.1. Grain Size and Shape	127
	6.3.2. Presence of a Second Phase	127
	6.3.3. Nature and Properties of Grain Boundaries	127
6.4.	Texture Studies in Superplasticity	128
6.5.	High Strain-Rate Superplasticity	128
	6.5.1. High Strain-Rate Superplasticity in Metal–Matrix Composites	129
	6.5.2. High Strain-Rate Superplasticity in Mechanically Alloyed Materials	134
6.6.	Superplasticity in Nano and Submicrocrystalline Materials	136

Chapter 6
Superplasticity

6.1 INTRODUCTION

Superplasticity is the ability of a polycrystalline material to exhibit, in a generally isotropic manner, very high tensile elongations prior to failure ($T > 0.5\,T_\text{m}$) [406]. The first observations of this phenomenon were made as early as 1912 [407]. Since then, superplasticity has been extensively studied in metals. It is believed that both the arsenic bronzes, used in Turkey in the Bronze Age (2500 B.C.), and the Damascus steels, utilized from 300 B.C. to the end of the nineteenth century, were already superplastic materials [408]. One of the most spectacular observations of superplasticity is perhaps that reported by Pearson in 1934 of a Bi–Sn alloy that underwent nearly 2000% elongation [409]. He also claimed then, for the first time, that grain-boundary sliding was the main deformation mechanism responsible for superplastic deformation. The interest in superplasticity has increased due to the recent observations of this phenomenon in a wide range of materials, including some materials (such as nanocrystalline materials [410], ceramics [411,412], metal matrix composites [413], and intermetallics [414]) that are difficult to form by conventional forming. Recent extensive reviews on superplasticity are available [415–418].

There are two types of superplastic behavior. The best known and studied, fine-structure superplasticity (FSS), will be briefly discussed in the following sections. The second type, internal stress superplasticity (ISS), refers to the development of internal stresses in certain materials, which then deform to large tensile strains under relatively low externally applied stresses [418].

6.2 CHARACTERISTICS OF FINE STRUCTURE SUPERPLASTICITY

Fine structure superplastic materials generally exhibit a high strain-rate sensitivity exponent (m) during tensile deformation. Typically, m is larger than 0.33. Thus, n in equation (3), is usually smaller than 3. In particular, the highest elongations have been reported to occur when $m \sim 0.5$ ($n \sim 2$) [418]. Superplasticity in conventional materials usually occurs at low strain rates ranging from $10^{-3}\,\text{s}^{-1}$ to $10^{-5}\,\text{s}^{-1}$. However, it has been reported in recent works that large elongations to failure may also occur in selected materials at strain rates substantially higher than $10^{-2}\,\text{s}^{-1}$ [419]. This phenomenon, termed high-strain-rate superplasticity (HSRSP), has been observed in some conventional metallic alloys, in metal matrix composites and in mechanically alloyed materials [420], among others. This will be discussed in Section 6.5. Very recently, HSRSP has been observed in cast alloys prepared by ECA (equal channel angular) extrusion [421–424]. In this case, very high temperatures are not required and the grain size is very small (< 1 μm). The activation energies for fine structure superplasticity tend to be low, close to the value for grain boundary diffusion, at intermediate temperatures. At high temperatures, however, the activation energy for superplastic flow is about equal to that for lattice diffusion.

The microscopic mechanism responsible for superplastic deformation is still not thoroughly understood. However, since Pearson's first observations [409], the most widely accepted mechanisms involve grain-boundary sliding (GBS) [425–432]. GBS is generally modeled assuming sliding takes place by the movement of extrinsic dislocations along the grain boundary. This would account for the observation that the amount of sliding is variable from point to point along the grain boundary [433]. Dislocation pile-ups at grain boundary ledges or triple points may lead to stress concentrations. In order to avoid extensive cavity growth, GBS must be aided by an accommodation mechanism [434]. The latter must ensure rearrangement of grains during deformation in order to achieve strain compatibility and relieve any stress concentrations resulting from GBS. The accommodation mechanism may include grain boundary migration, recrystallization, diffusional flow or slip. The accommodation process is generally believed to be the rate-controlling mechanism.

Over the years a large number of models have emerged in which the accommodation process is either diffusional flow or dislocation movement [435]. The best known model for GBS accommodated by diffusional flow, depicted schematically in Figure 58, was proposed by Ashby and Verral [436]. This model

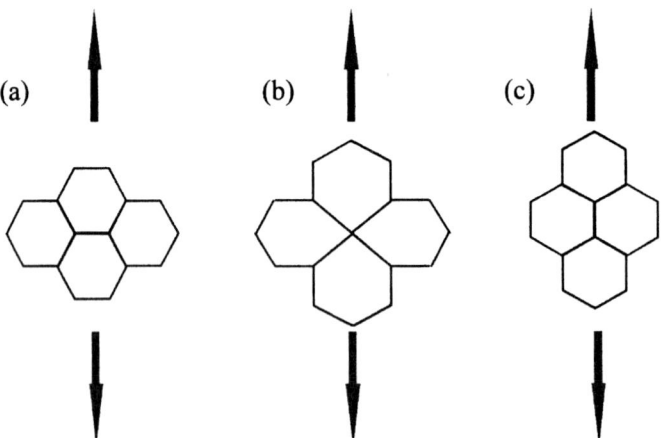

Figure 58. Ashby–Verral model of GBS accommodated by diffusional flow [436].

explains the experimentally observed switching of equiaxed grains throughout deformation. However, it fails to predict the stress dependence of the strain-rate. According to this model,

$$\dot{\varepsilon}_{ss} = K_1(b/g)^2 D_{eff}(\sigma - \sigma_{TH_s}/E) \tag{99}$$

where $D_{eff} = D_{sd}$ 9 $[1 + (3.3 w/g)(D_{gb}/D_{sd})]$, K_1 is a constant, σ_{TH_s} is the threshold stress and w is the grain boundary width. The threshold stress arises since there is an increase in boundary area during grain switching when clusters of grains move from the initial position (Figure 58(a)) to the intermediate one (Figure 58(b)).

Several criticisms of this model have been reported [437–442]. According to Spingarn and Nix [437] the grain rearrangement proposed by Ashby–Verral cannot occur purely by diffusional flow. The diffusion paths are physically incorrect. The first models of GBS accommodated by diffusional creep were proposed by Ball and Hutchison [443], Langdon [444], and Mukherjee [445]. Among the most cited are those proposed by Mukherjee and Arieli [446] and Langdon [447]. According to these authors, GBS involves the movement of dislocations along the grain boundaries, and the stress concentration at triple points is relieved by the generation and movement of dislocations within the grains (Figure 59). Figure 60 illustrates the model proposed by Gifkins [448], in which the accommodation process, which also consists of dislocation movement, only occurs in the "mantle" region of the grains, i.e., in the region close to the grain boundary. According to all these GBS accommodated by slip models, $n = 2$ in a relationship such as

$$\dot{\varepsilon}_{ss} = K_2(b/g)^{p'} D(\sigma/E)^2 \tag{100}$$

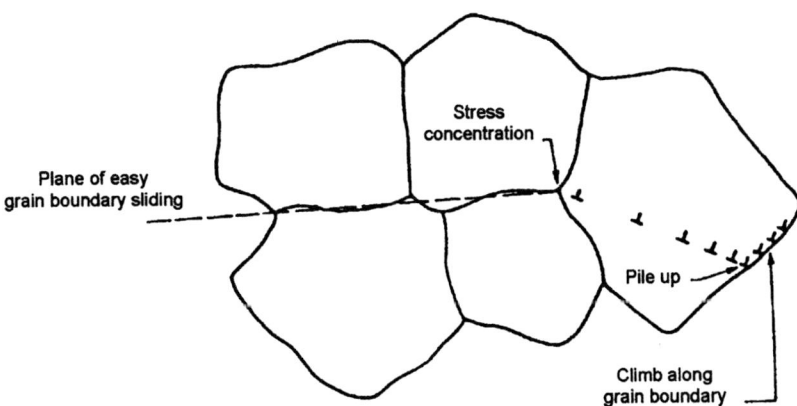

Figure 59. Ball–Hutchinson model of GBS accommodated by dislocation movement [443].

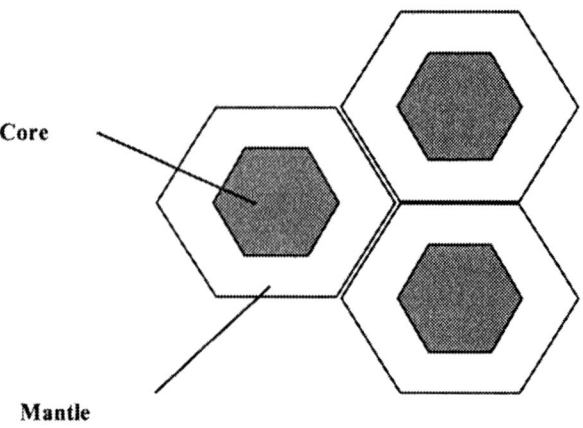

Figure 60. Gifkins "core and mantle" model [448].

where $p' = 2$ or 3 depending on whether the dislocations move within the lattice or along the grain boundaries, respectively. K_2 is a constant, which varies with each of the models and the diffusion coefficient, D, can be D_{sd} or D_{gb}, depending on whether the dislocations move within the lattice or along the grain boundaries to accommodate stress concentrations from GBS. In order to rationalize the increase in activation energy at high temperatures, Fukuyo et al. [449] proposed a model based on the GBS mechanism in which the dislocation accommodation process takes place by sequential steps of climb and glide. At intermediate temperatures, climb along the grain boundaries is the rate-controlling mechanism due to the pile-up stresses. Pile-up stresses are absent and the glide of dislocations within the grain is the rate-controlling mechanism at high temperatures. It is believed that slip in superplasticity is accommodating and does not contribute to the total strain [450]. Thus, GBS is traditionally believed to account for all of the strain in Superplasticity [451]. However, recent studies, based on texture analysis, indicate that slip may contribute to the total elongation [452–467].

The proposed mechanisms predict some behavior but have not succeeded in fully predicting the dependence of the strain rate on σ, T, and g during superplastic deformation. Ruano and Sherby [468,469] formulated the following phenomenological equations, which appear to describe the experimental data from metallic materials,

$$\dot{\varepsilon}_{ss} = K_3(b/g)^2 D_{sd}(\sigma/E)^2 \qquad (101)$$

$$\dot{\varepsilon}_{ss} = K_4(b/g)^3 D_{gb}(\sigma/E)^2 \qquad (102)$$

where K_3 and K_4 are constants. These equations, with $n=2$, correspond to a mechanism of GBS accommodated by dislocation movement. Equation (101) corresponds to an accommodation mechanism in which the dislocations would move within the grains (g^2) and equation (102) corresponds to an accommodation mechanism in which the dislocations would move along the grain boundaries (g^3). Only the sliding of individual grains has been considered. However, currently the concept of cooperative grain-boundary sliding (CGBS), i.e., the sliding of blocks of grains, is gaining increasing acceptance. Several deformation models that account for CGBS are described in Ref. [421].

6.3 MICROSTRUCTURE OF FINE STRUCTURE SUPERPLASTIC MATERIALS

The microstructures associated with fine structure superplasticity are well established for conventional metallic materials. They are, however, less clearly defined for intermetallics, ceramics, metal matrix composites, and nanocrystalline materials.

6.3.1 Grain Size and Shape
GBS in metals is favored by the presence of equiaxed small grains that should generally be smaller than 10 µm. Consistent with equations (99–102), the strain-rate is usually inversely proportional to grain size, according to

$$\dot{\varepsilon}_{ss} = K_5 g^{-p'} \quad (103)$$

where $p' = 2$ or 3 depending, perhaps, on the accommodation mechanism and K_5 is a constant. Also, for a given strain-rate, the stress decreases as grain size decreases. grain size refinement is achieved during the thermomechanical processing by successive stages of warm and cold rolling [468–473]. However, the present understanding of microstructural control in engineering alloys during industrial processing by deformation and recrystallization is still largely empirical.

6.3.2 Presence of a Second Phase
The presence of small second-phase particles uniformly distributed in the matrix prevents rapid grain growth that can occur in single-phase materials within the temperature range over which superplasticity is observed.

6.3.3 Nature and Properties of Grain Boundaries
GBS is favored along high-angle disordered (not CSL) boundaries. Additionally, sliding is influenced by the grain boundary composition. For example, a heterophase

boundary (i.e., a boundary which separates grains with different chemical composition) slides more readily than a homophase boundary. Stress concentrations develop at triple points and at other obstacles along the grain boundaries as a consequence of GBS. mobile grain boundaries may assist in relieving these stresses. Grain boundaries in the matrix phase should not be prone to tensile separation.

6.4 TEXTURE STUDIES IN SUPERPLASTICITY

Texture analysis has been utilized to further study the mechanisms of superplasticity [442–467,474], using both X-ray texture analysis and computer-aided EBSP techniques [475]. Commonly, GBS, involving grain rotation, is associated with a decrease in texture [417], whereas crystallographic slip leads to the stabilization of certain preferred orientations, depending on the number of slip systems that are operating [476,477].

It is interesting to note that a large number of investigations based on texture analysis have led to the conclusion that crystallographic slip (CS) is important in superplastic deformation. According to these studies, CS is not merely an accommodation mechanism for GBS, but also operates in direct response to the applied stress. Some investigators [453–458,474] affirm that both GBS and CS coexist at all stages of deformation; other investigators [459–461] conclude that CS only operates during the early stages of deformation, leading to a microstructure favoring GBS. Others [462–467] even suggest that CS is the principal deformation mechanism responsible for superplastic deformation.

6.5 HIGH STRAIN-RATE SUPERPLASTICITY

High strain-rate superplasticity (HSRS) has been defined by the Japanese Standards Association as superplasticity at strain-rates equal to or greater than $10^{-2} s^{-1}$ [418,478,479]. This field has awakened considerable interest in the last 15 years since these high strain-rates are close to the ones used for commercial applications ($10^{-2} s^{-1}$ to $10^{-1} s^{-1}$). Higher strain-rates can be achieved by reducing the grain size [see equation (100)] or by engineering the nature of the interfaces in order to make them more suitable for sliding [478,479]. High strain-rate Superplasticity was first observed in a 20%SiC whisker reinforced 2124 Al composite [419]. Since then, it has been achieved in several metal–matrix composites, mechanically alloyed materials, conventional alloys that undergo continuous reactions or (continuous dynamic recrystallization), alloys processed by power consolidation, by physical vapor deposition, by intense plastic straining [480] (for example, equal channel angular pressing (ECAP), equal channel angular extrusion (ECAE), torsion straining under high pressure), or, more recently, by friction stir processing [481]. The details of the microscopic mechanism responsible for high strain-rate superplasticity are not yet well understood, but some recent theories are reviewed below.

6.5.1 High Strain-Rate Superplasticity in Metal–Matrix Composites

High strain-rate superplasticity has been achieved in a large number of metal–matrix composites. Some of them are listed in Table 2 and more complete lists can be found elsewhere [478,482]. The microscopic mechanism responsible for high strain-rate superplasticity in metal–matrix composites is still a matter of controversy. Any theory must account for several common features of the mechanical behavior of metal–matrix composites that undergo high strain-rate superplasticity, such as [491]:

(a) Maximum elongations are achieved at very high temperatures, sometimes even slightly higher than the incipient melting point.
(b) The strain-rate sensitivity exponent changes at such high temperatures from ~0.1 (n~10) (low strain-rates) to ~0.3 (n~3) (high strain-rates).
(c) High apparent activation energy values are observed. Values of 920 kJ/mol and 218 kJ/mol have been calculated for $SiC_w/2124Al$ at low and high strain-rates, respectively. These values are significantly higher than the activation energy for self-diffusion in Al (140 kJ/mol).

Both grain-boundary sliding and interfacial sliding have been proposed as the mechanisms responsible for HSRS. The significant contribution of interfacial sliding is evidenced by extensive fiber pullout apparent on fracture surfaces [491]. However, an accommodation mechanism has to operate simultaneously in order to avoid

Table 2. Superplastic characteristics of some metal–matrix composites exhibiting high strain-rate superplasticity.

Material	Temperature (°C)	Strain rate (s^{-1})	Elongation (%)	Reference
$SiC_w/2124\ Al$	525	0.3	~300	[419]
$SiC_w/2024\ Al$	450	1	150	[483]
$SiC_w/6061\ Al$	550	0.2	300	[484]
$SiC_p/7075\ Al$	520	5	300	[485]
$SiC_p/6061\ Al$	580	0.1	350	[486]
$Si_3N_{4w}/6061\ Al$	545	0.5	450	[487]
$Si_3N_{4w}/2124\ Al$	525	0.2	250	[488]
$Si_3N_{4w}/5052\ Al$	545	1	700	[489]
$AlN/6061\ Al$	600	0.5	350	[490]

(w = whisker; p = particle).

cavitation at such high strain-rates. The nature of this accommodation mechanism, that enables the boundary and interface mobility, is still uncertain.

A fine matrix grain size is necessary but not sufficient to explain high strain-rate superplasticity. In fact, HSRS may or may not appear in two composites having the same fine-grained matrix and different reinforcements. For example, it has been found that a 6061 Al matrix with β-Si$_3$N$_4$ whiskers does experience HSRS, whereas the same matrix with β-SiC does not [491]. The nature, size, and distribution of the reinforcement are critical to the onset of high strain-rate superplasticity.

a. Accommodation by a Liquid Phase. Rheological Model. Nieh and Wadsworth [491] have proposed that the presence of a liquid phase at the matrix-reinforcement interface and at grain boundaries within the matrix is responsible for accommodation of interface sliding during HSRS and thus for strain-rate enhancement. The presence of this liquid phase would be responsible for the observed high activation energies. A small grain size would favor HSRS since the liquid phase would then be distributed along a larger surface area and thus can have a higher capillarity effect, preventing decohesion. The occurrence of partial melting even during tests at temperatures slightly below solvus has been explained in two different ways. First, as a consequence of solute segregation, a low melting point region could be created at the matrix-reinforcement interfaces. Alternatively, local adiabatic heating at the high strain-rates used could contribute to a temperature rise that may lead to local melting.

It has been suggested [492] that high strain-rate superplasticity with the aid of a liquid phase can be modeled in rheological terms in a similar way to semi-solid metal forming. A fluid containing a suspension of particles behaves like a non-Newtonian fluid, for which the strain-rate sensitivity and the shear strain-rate are related by

$$\tau = K_7 \cdot \dot{\gamma}^m \tag{104}$$

where τ is the shear stress, and K_7 and m are both materials constants, m being the strain-rate sensitivity of the material. The shear stress and strain rate of a semi-solid that behaves like a non-Newtonian fluid are related to the shear viscosity by the following equations:

$$\eta = K_7 \cdot \dot{\gamma}^{-u} \tag{105}$$

$$\eta = \tau/\dot{\gamma} \tag{106}$$

where η is the shear viscosity and u is a constant of the material, related to the strain-rate sensitivity by the expression m = 1 − u. The viscosity of several Al-6.5%Si metal matrix composites was measured experimentally [418] at 700°C as a function of shear rate. High strain sensitivity values similar to those reported for MMCs (~0.3–0.5) in the HSRS regime were obtained at very high shear strain rates (200–1000 s^{-1}). These data support the rheological model. The temperature used, however, is higher than the temperatures at which high strain-rate superplasticity is observed.

The role of a liquid phase as an accommodation mechanism for interfacial and grain-boundary sliding has been supported by other authors [493–498]. It is suggested that the liquid phase acts as an accommodation mechanism, relieving stresses originated by sliding and thus preventing cavity formation. However, in order to avoid decohesion, it is emphasized that the liquid phase must either be distributed discontinuously or be present in the form of a thin layer. The optimum amount of liquid phase may depend on the nature of the grain boundary or interface. Direct evidence of local melting at the reinforcement–matrix interface was obtained using *In situ* transmission electron microscopy by Koike *et al.* [495] in a Si_3N_{4p}/6061 Al. The rheological model was criticized by Mabuchi *et al.* [493], arguing that testing the material at a temperature within the solid–liquid region is not sufficient to achieve high strain-rate superplasticity. For example, an unreinforced 2124 alloy fails to exhibit high tensile ductility when tested at a temperature above solvus. Additionally, it has been observed experimentally that ductility decreases when testing above a certain temperature.

b. Accommodation by Interfacial Diffusion. Mishra *et al.* [499–502] rationalized the mechanical behavior of HSRS metal–matrix composites by taking into account the presence of a threshold stress. This analysis led them to conclude that the mechanism responsible for HSRS in metal–matrix composites is grain-boundary sliding accommodated by interfacial diffusion along matrix–reinforcement interfaces. It is important to note that the particle size is often comparable to grain size, and therefore interfacial sliding is geometrically necessary, as illustrated in Figure 61. Partial melting, especially if it is confined to triple points, may be beneficial for superplastic deformation, but it is not necessary to account for the superplastic elongations observed.

Threshold stresses are often used to explain the variation of the strain-rate sensitivity exponent with strain-rate in creep deformation studies. The presence of a threshold stress would explain the transition to a lower strain-rate sensitivity value (and, thus, to a higher n) at low strain rates that takes place during HSRS in metal–matrix composites. Calculating threshold stresses and a (true$^-$) stress exponent, n_{hsrs}, that describes the predominant deformation mechanism is a non-trivial process, as explained in Ref. [500]. Mishra *et al.* [500,501] concluded that a true stress exponent of 2 would give the best fit for their data, suggesting the predominance of grain-boundary sliding as a deformation mechanism responsible for HSRS in metal–matrix composites. Additionally, activation energies (Q_{hsrs}) of the order of

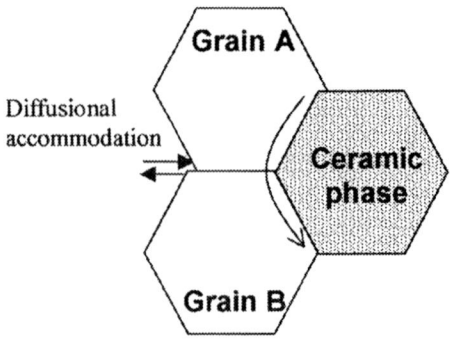

Figure 61. Interfacial diffusion-controlled grain-boundary sliding. The ceramic phase would not allow slip accommodation.

300 kJ/mol were obtained from this analysis. Both parametric dependencies ($n_{hsrs} = 2$ and $Q_{hsrs} \cong 300$ kJ/mol) are best predicted by Arzt's model for "interfacial diffusion-controlled diffusional creep" [503]. Mishra et al. [502] suggested that, since both diffusional creep and grain-boundary sliding are induced by the movement of grain-boundary dislocations, and the atomic processes involved are similar for both processes, it is reasonable to think that the parametric dependencies would be similar in both interfacial diffusion-controlled diffusional creep and interfacial diffusion-controlled superplasticity. Thus, the latter is invoked to be responsible for HSRS. Figure 61 illustrates this deformation mechanism.

The phenomenological constitutive equation proposed by Mishra et al. for HSRS in metal–matrix composites is the following:

$$\dot{\varepsilon}_{ss} = A_{12} \frac{D_i G b}{kT} \left(\frac{b^2}{g_m g_p}\right) \left(\frac{\sigma - \sigma_{TH_{hsrs}}}{E}\right)^2 \qquad (107)$$

where D_i is the coefficient for interfacial diffusion, g_m is the matrix grain size and g_p is the particle/reinforcement size, $\sigma_{TH_{hsrs}}$ is the threshold stress for high strain-rate superplasticity, and A_{12} is a material constant. An inverse grain size and reinforcement size dependence is suggested.

According to this model [501], as temperature rises, the accommodation mechanism would change from slip accommodation (at temperatures lower than the optimum) to interfacial diffusion accommodation. The need for very high temperatures to attain HSRS is due to the fact that grain-boundary diffusivity increases with temperature. Therefore, the higher the temperature, the faster interface diffusion, which leads to less cavitation and thus, higher ductility.

Mabuchi et al. [504,505] claim the importance of a liquid phase in HSRS arguing that, when introducing threshold stresses, the activation energy for HSRS at temperatures at which no liquid phase is present is similar to that corresponding to lattice self-diffusion in Al. However, at higher temperatures, at which partial melting has taken place, the activation energy increases dramatically. It is at these temperatures that the highest elongations are observed. The origin of the threshold stress for superplasticity is not well known. Its magnitude depends on the shape and size of the reinforcement and it generally decreases with increasing temperature.

c. Accommodation by Grain-Boundary Diffusion in the Matrix. The Role of Load Transfer. The two theories described above were critically examined by Li and Langdon [506–508]. First, the rheological model was questioned, since HSRS had been recently found in Mg–Zn metal–matrix composites at temperatures below the incipient melting point, where no liquid phase is present [509]. Second, Li *et al.* [506] claim that it is hard to estimate interfacial diffusion coefficients at ceramic–matrix interfaces, and therefore validation of the interfacial diffusion-controlled grain-boundary sliding mechanism is difficult. These investigators used an alternative method for computing threshold stresses, described in detail in Ref. [506], which does not require an initial assumption of the value of n. This methodology also rendered a true stress exponent of 2 and true activation energy values that were higher than that for matrix lattice self-diffusion and grain boundary self-diffusion. These results were explained by the occurrence of a transfer of load from the matrix to the reinforcement. Following this approach, that was used before to rationalize creep behavior in metal–matrix composites [510], a temperature-dependent load-transfer coefficient α' was incorporated in the constitutive equation as follows:

$$\dot{\varepsilon}_{ss} = \frac{A'''' DGb}{kT} \left(\frac{b}{g}\right)^{p'} \left[\frac{(1-\alpha')(\sigma - \sigma_{TH_{hsrs}})}{G}\right]^{n} \quad (108)$$

where A'''' is a dimensionless constant. In their calculations, Li *et al.* assumed that D is equal to D_{gb} and the remaining constants and variables have the usual meaning. Load-transfer coefficients are expected to vary between 0 (no load-transfer) and 1 (all the load is transferred to the reinforcement). It was found that the load-transfer coefficients obtained decreased with increasing temperature, becoming 0 at temperatures very close to the incipient melting point. This indicates that load transfer would be inefficient in the presence of a liquid phase. The effective activation energies Q^* calculated by introducing the load-transfer coefficient into the rate equation for flow are similar to those corresponding to grain-boundary diffusion within the matrix alloys (until up to a few degrees from the incipient melting point). Therefore, Li and Langdon proposed that the mechanism responsible for HSRS is grain-boundary sliding controlled by grain-boundary diffusion in the matrix. This mechanism, that is characteristic of conventional superplasticity at high temperatures, would be valid up to temperatures close to the incipient melting point.

The origin of the threshold stress is still uncertain. It has been shown that it decreases with increasing temperature, and that it depends on the shape and size of the reinforcement [502]. The temperature dependence of the threshold stress may be expressed by an Arrhenius-type equation of the form:

$$\frac{\sigma_{TH_{hsrs}}}{G} = B \exp\left(\frac{Q_{TH_{hsrs}}}{RT}\right) \qquad (109)$$

where $\sigma_{TH_{hsrs}}$ is the threshold stress for high strain-rate superplasticity, B is a constant, and $Q_{TH_{hsrs}}$ is an energy term which seems to be associated with the process by which the mobile dislocations surpass the obstacles in the glide planes. (The threshold stress concept will be discussed again in Chapter 8.)

Li and Langdon [508] claim that the threshold stress values obtained in metal–matrix composites tested under HSRS and under creep conditions may have the same origin. They showed that similar values of $Q_{TH_{hsrs}}$ are obtained under these two conditions when, in addition to load transfer, substructure strengthening is introduced into the rate equation for flow. Substructure strengthening may arise, for example, from an increase in the dislocation density due to the thermal mismatch between the matrix and the reinforcement or to the resistance of the reinforcement to plastic flow. The "effective stress" acting on the composite in the presence of load-transfer and substructure strengthening is given by,

$$\sigma_e = (1 - \Phi)\sigma - \sigma_{TH_{hsrs}} \qquad (110)$$

where Φ is a temperature-dependent coefficient. At low temperatures at which creep tests are performed, the value of Φ may be negligible, but since HSRS takes place at very high temperatures, often close to the melting point, the temperature dependence of Φ must be taken into account to obtain accurate values of $Q_{TH_{hsrs}}$. In fact, when the temperature dependence of Φ is considered, $Q_{TH_{hsrs}}$ values close to 20–30 kJ/mol, typical of creep deformation of MMCs, are obtained under HSRS conditions.

6.5.2 High Strain-Rate Superplasticity in Mechanically Alloyed Materials

HSRS has also been observed in some mechanically alloyed (MA) materials that are listed in Table 3.

As can be observed in Table 3, mechanically alloyed materials attain superplastic elongations at higher strain-rates than metal–matrix composites. Such high strain rates are often attributed to the presence of a very fine microstructure (with average grain size of about 0.5 µm) and oxide and carbide dispersions approximately 30 nm in diameter that have an interparticle spacing of about 60 nm [418]. These particle dispersions impart stability to the microstructure. The strain-rate sensitivity exponent (m) increases with temperature reaching values usually higher than 0.3 at the temperatures where the highest elongations are observed. Optimum superplastic elongations are often obtained at temperatures above solvus.

After introducing a threshold stress, $n = 2$ and the activation energy is equal to that corresponding to grain-boundary diffusion. These values are similar to those obtained for conventional superplasticity and would indicate that the main deformation mechanism is grain-boundary sliding accommodated by dislocation slip. The rate-controlling mechanism would be grain-boundary diffusion [502,506,517]. Mishra et al. [502] claim that the small size of the precipitates allows for diffusion relaxation of the stresses at the particles by grain-boundary sliding, as illustrated in Figure 62. Higashi et al. [517] emphasize the importance of the presence of a small amount of liquid phase at the interfaces that contributes to stress relaxation and thus enhanced superplastic properties at temperatures above solvus. Li et al. [506] state that, given the small size of the particles, no load transfer

Table 3. Superplastic properties of some mechanically alloyed materials.

Material	Temperature (°C)	Strain-rate (s^{-1})	Elongation (%)	Reference
IN9021	450	0.7	300	[511]
IN90211	475	2.5	505	[512,513]
IN9052	590	10	330	[514]
IN905XL	575	20	190	[515]
SiC/IN9021	550	50	1250	[515]
MA754	1100	0.1	200	[516]
MA6000	1000	0.5	308	[516]

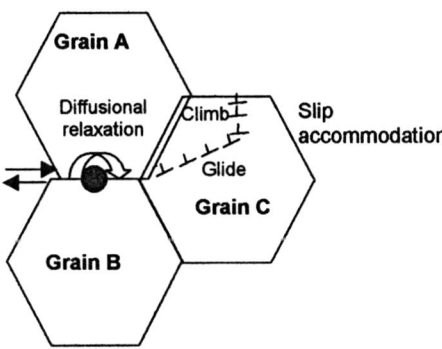

Figure 62. Grain-boundary sliding accommodated by boundary diffusion-controlled dislocation slip.

takes place and thus the values obtained for the activation energy after introducing a threshold stress are the true activation energies. According to Li et al., the same mechanism (GBS rate controlled by grain-boundary diffusion) predominates during HSRS in both metal–matrix composites and mechanically alloyed materials.

6.6 SUPERPLASTICITY IN NANO AND SUBMICROCRYSTALLINE MATERIALS

The development of grain size reduction techniques in order to produce microstructures capable of achieving superplasticity at high strain rates and low temperatures has been the focus of significant research in recent years [480,518–522]. Some investigations on the mechanical behavior of sub-microcrystalline (1 μm > g > 100 nm) and nanocrystalline (g < 100 nm) materials have shown that superplastic properties are enhanced in these materials, with respect to microcrystalline materials of the same composition [518–530]. Improved superplastic properties have been reported in metals [518–522,524–529], ceramics [523], and intermetallics [527,529,530]. The difficulties in studying superplasticity in nanomaterials arise from (a) increasing uncertainty in grain size measurements, (b) difficulty in preparing bulk samples, (c) high flow stresses may arise, that may approach the capacity of the testing apparatus, and (d) the mechanical behavior of nanomaterials is very sensitive to the processing, due to their metastable nature.

The microscopic mechanisms responsible for superplasticity in nanocrystalline and sub-microcrystalline materials are still not well understood. Together with superior superplastic properties, significant work hardening and flow stresses larger than those corresponding to coarser microstructures have often been observed [526–528]. Figure 63 shows the stress–strain curves corresponding to Ni_3Al deformed at 650 °C and 725°C at a strain rate of $1 \times 10^{-3} s^{-1}$ [Figure 63(a)] and to Al-1420 deformed at 300°C at $1 \times 10^{-2} s^{-1}$, $1 \times 10^{-1} s^{-1}$, and $5 \times 10^{-1} s^{-1}$ [Figure 63(b)]. It is observed in Figure 63(a) that nanocrystalline Ni_3Al deforms superplastically at temperatures which are more than 400°C lower than those corresponding to the microcrystalline material [532]. The peak flow stress, that reaches 1.5 GPa at 650°C, is the highest flow stress ever reported for Ni_3Al. Significant strain hardening can be observed. In the same way, Figure 63(b) shows that the alloy Al-1420 undergoes superplastic deformation at temperatures about 150°C lower than the microcrystalline material [533], and at strain-rates several orders of magnitude higher ($1 \times 10^{-1} s^{-1}$ vs. $4 \times 10^{-4} s^{-1}$). High flow stresses and considerable strain hardening are also apparent.

The origin of these anomalies is still unknown. Mishra et al. [528,531] attributed the presence of high flow stresses to the difficulty in slip accommodation in nanocrystalline grains. Islamgaliev et al. [529] support this argument. The difficulty

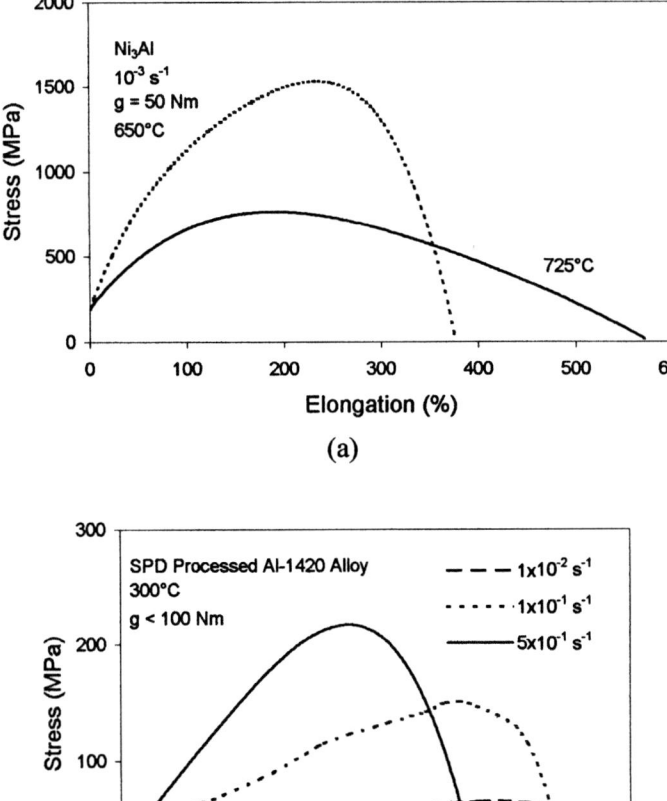

Figure 63. Stress–strain curves corresponding to (a) Ni$_3$Al deformed at 650°C (dotted line) and 725°C (full line) at a strain rate of 1×10^{-3} s^{-1} and (b) to Al-1420 deformed at 300°C at 1×10^{-2} s^{-1}, 1×10^{-1} s^{-1}, and 5×10^{-1} s^{-1}. From Refs. [529] and [531].

of dislocation motion in nanomaterials has also been previously reported in Ref. [534]. The stress necessary to generate the dislocations responsible for dislocation accommodation is given by Ref. [528]:

$$\tau = \frac{Gb}{4\pi\lambda(1-\nu)}\left(\ln\left(\frac{\lambda_p}{b}\right) - 1.67\right) \quad (111)$$

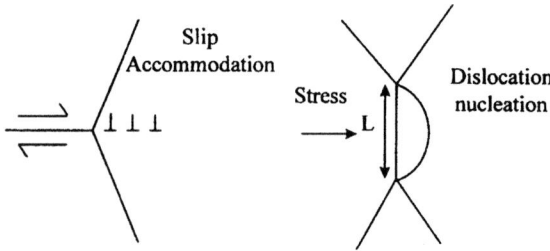

Figure 64. Generation of dislocations for slip accommodation of GBS. From Ref. [531].

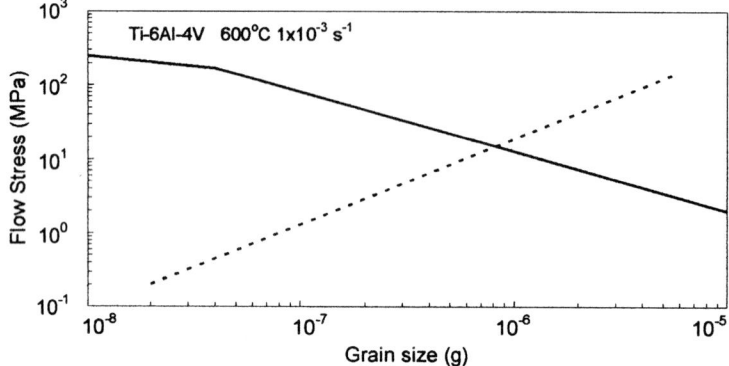

Figure 65. Theoretical stress for slip accommodation and flow stress for overall superplasticity vs. grain size in a Ti-6Al-4V alloy deformed at $1 \times 10^{-3}\,\mathrm{s}^{-1}$. From Ref. [531]. (Full line: theoretical stress for slip accommodation; dashed line: predicted stress from empirical correlation $\dot{\varepsilon}_{ss} = 5 \times 10^9 (\sigma/E)2(D_{sd}/g^2)$).

where λ_p is the distance between the pinning points and τ is the shear stress required to generate the dislocations (see Figure 64). Figure 65 is a plot showing the variation with grain size of the stress calculated from equation (111) and the flow stress required for overall superplastic deformation [obtained from equation (100) assuming the main deformation mechanism is GBS accommodated by lattice diffusion-controlled slip]. It can be observed that, for coarser grain sizes, the flow stress is high enough to generate dislocations for the accommodation of grain-boundary sliding. For submicrocrystalline and nanocrystalline grain sizes, however, the stress required for slip accommodation is higher than the overall flow stress. This is still a rough approximation to the problem, since equation (111) does not include strain-rate dependence, temperature dependence other than the modulus, as well as the details for dislocation generation from grain boundaries. However, Mishra et al. use this argument to emphasize that the microcrystalline behavior can apparently not be extrapolated to nanomaterials. Instead, there may be a transition between both kinds of behavior. The large strain hardening found during superplasticity of nanocrystalline materials has still not been thoroughly explained.

A classification of nanomaterials according to the processing route has been made by the same authors [528,531]. Nanomaterials processed by mechanical deformation (such as ECAP) are denoted by "D" (for deformation) and nanomaterials processed by sintering of powders are denoted by "S". In the first, a large amount of dislocations are already generated during processing, which can contribute to deformation by an "exhaustion plasticity" mechanism. Thus, the applied stress, which at the initial stage of deformation is not enough to generate new dislocations for slip accommodation, moves the previously existing dislocations. As the easy paths of grain-boundary sliding become exhausted, the flow stress increases until it is high enough to generate new dislocations. "S" nanomaterials may not be suitable for obtaining large tensile strains, due to the absence of pre-existing dislocations.

A significant amount of grain growth takes place during deformation even when superplasticity occurs at lower temperatures. In fact, the transition from low plasticity to superplasticity in nanomaterials is often accompanied by the onset of grain growth. This seems unavoidable, since both grain growth and grain-boundary sliding are thermally activated processes. It has been found that a reduction of the superplastic temperature is usually offset by a reduction of grain-growth temperature [527]. As the grain size decreases, the surface area of grain boundaries increases, and thus the reduction of grain-boundary energy emerges as a new driving force for grain growth. This force is much less significant for coarser grain sizes (which, in turn, render higher superplastic temperatures). Thus, the possibility of observing superplasticity in nanomaterials, that remain nanoscale after deformation, seems small.

Chapter 7
Recrystallization

7.1.	Introduction	143
7.2.	Discontinuous Dynamic Recrystallization (DRX)	145
7.3.	Geometric Dynamic Recrystallization	146
7.4.	Particle-Stimulated Nucleation (PSN)	147
7.5.	Continuous Reactions	147

Chapter 7
Recrystallization

7.1 INTRODUCTION

The earlier chapters have described creep as a process where dislocation hardening is accompanied by dynamic recovery. It should be discussed at this point that dynamic recovery is not the only (dynamic) restoration mechanism that may occur with dislocation hardening. Recrystallization can also occur and this process can also "restore" the metal and reduce the flow stress. Often, recrystallization during deformation (dynamic recrystallization) is observed at relatively high strain rates which is outside the common creep realm. However, any complete discussion of elevated temperature creep, and particularly, a discussion of high-temperature plasticity must include this restoration mechanism. An understanding of the hotworking (high strain-rates and high temperatures) requires an appreciation of both dynamic recovery and recrystallization processes. Some definitions are probably useful, and we will use those definitions adopted by Doherty et al. [222].

During deformation, energy is stored in the material mainly in the form of dislocations. This energy is released in three main processes, those of recovery, recrystallization, and grain coarsening (subsequent to recyrstallization). The usual definition of recrystallization [222] is the formation and migration of high-angle boundaries, driven by the stored energy of deformation. The definition of recovery includes all processes releasing stored energy that do not require the movement of high-angle boundaries. In the context of the processes we have discussed, creep is deformation accompanied only by dynamic recovery. Typical recovery processes involve the rearrangement of dislocations to lower their energy, for example by the formation of low-angle subgrain boundaries, and annihilation of dislocation line length in the subgrain interior, such as by Frank network coarsening. Grain coarsening is the growth of the mean grain size driven by the reduction in grain-boundary area.

It is now recognized that recrystallization is not a Gibbs I transformation that occurs by classic nucleation and growth process as described by Turnbull [535] and Christian [536]. ΔG^* and r^*, the critical Gibb's free energy and critical-sized embryos are unrealistically large if the proper thermodynamic variables are used. As a result of this disagreement, it is now universally accepted [537], as first proposed by Cahn [538], that the new grains do not nucleate as totally new grains by the atom by atom construction assumed in the classic kinetic models. Rather, new grains grow from small regions, such as subgrains, that are already present in the deformed microstructure. Special grains do not have to form. These embryos are present in the starting structure. Only subgrains with a high misorientation angle to the adjacent deformed material appear to have the necessary mobility to evolve into new recrystallized grains. Typical nucleation sites include pre-existing high-angle boundaries, shear bands, and highly misoriented deformation zones around hard particles. Misoriented "transition" bands (or geometric necessary boundaries) inside grains are a result of different parts of the grain having undergone different lattice rotations due to different slip systems being activated. Figure 66 (from Ref. [222]) illustrates an example of recrystallization in 40% compressed pure aluminum. New grains 3 and 17 are only growing into the deformed regions A and B, respectively, with which they are strongly misoriented and not into regions with which they share a common misorientation; 17 has a low-angle misorientation with A and 3 with B. It should be mentioned that recrystallization often leads to a characteristic texture(s),

Figure 66. Static recrystallization in aluminum cold worked 40%. A large grain has fragmented into two regions, A and B. From Doherty et al. [222].

usually different that any texture developed as a consequence of the prior deformation that is the driving force for any recrystallization.

7.2 DISCONTINUOUS DYNAMIC RECRYSTALLIZATION (DRX)

Recrystallization can occur under two broad conditions; static and dynamic. Basically, static occurs in the absence of external plasticity during the recrystallization. The most common case for static is heating cold-worked metal leading to a recrystallized microstructure. Dynamic recrystallization occurs with concomitant plasticity. This distinction is complicated, somewhat, by the more recent suggestion of metadynamic recrystallization (MRDX) [539] that can follow dynamic recrystallization, generally at elevated temperature. Although it occurs without external plasticity, it can occur, quickly. It is distinguished from static recrystallization (SRX) in that MDRX is relatively sensitive to prior strain-rate but insensitive to prestrain and temperature. Static recrystallization depends on prestrain and temperature, but only slightly on strain-rate. The recrystallization remarks in the previous section are equally valid for these two cases (although Figure 66 was a Static recrystallization example) although differences are apparent. Dynamic recrystallization is more important to discuss in the context of creep plasticity.

A single broad stress peak, where the material hardens to a peak stress, followed by significant softening is often evidenced in DRX. The softening is largely attributable to the nucleation of growing, "new," grains that annihilate dislocations during growth. This is illustrated in Figure 67. The restoration is contrasted by dynamic recovery, where the movement of, and annihilation of, dislocations at, high-angle boundaries is not important. DRX may commence well before the peak stress. This becomes evident without microstructural examination by examining the hardening rate, θ, as a function of flow stress. For customary Stage III hardening, θ decreases at a constant or decreasing "rate" with stress. DRX, on the other hand, causes an "acceleration" of the decrease in hardening rate.

Sometimes the single peak in the stress versus strain behavior in DRX is not observed; rather multiple peaks may be evident leading to the appearance of undulations in the stress versus strain behavior that "dampen" into an effective "steady-state." This is also illustrated in Figure 67. It has been suggested that the cyclic behavior indicates that grain coarsening is occurring while a single peak is associated with grain refinement [540].

Although DRX is frequently associated with commercial metal-forming strain rates (e.g., $1 s^{-1}$ and higher), Figure 67 illustrates that DRX can occur at more modest rates that approach those in ordinary creep conditions. This explains why some ambiguity has been experienced in interpreting creep deformation where both

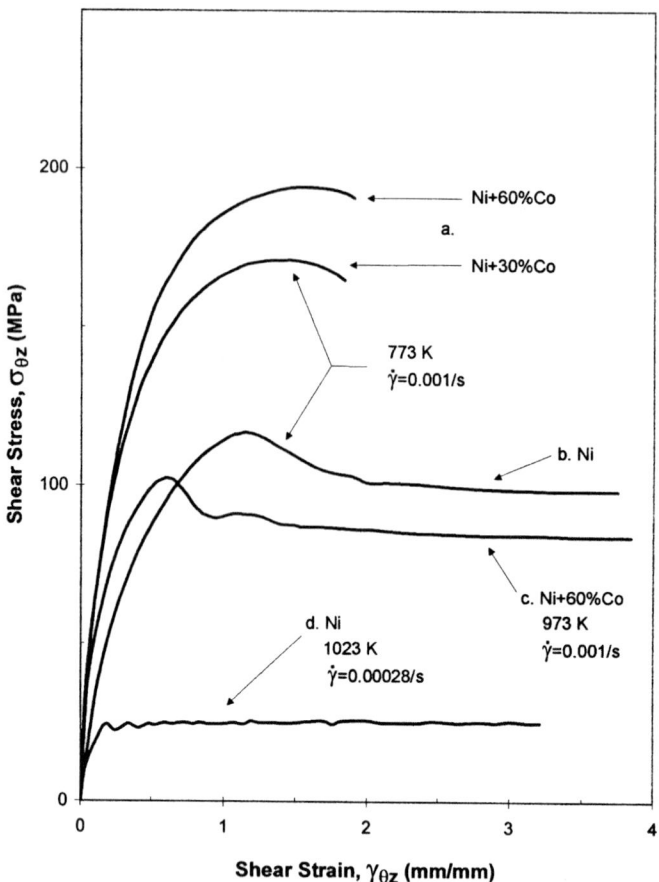

Figure 67. DRX in Ni and Ni–Co alloys in torsion [13].

dynamic recovery and recrystallization are both occurring. As some recent analysis has indicated, some of the creep data of Figure 17 of Zr may include some data for which some DRX may be occurring. Metals such as pure Ni and Cu frequently exhibit DRX [7,78].

7.3 GEOMETRIC DYNAMIC RECRYSTALLIZATION

The starting grains of the polycrystalline aggregate distort with relatively large strain deformation. These boundaries may thin to the dimensions of the subgrain diameter with strain approaching 2–10 (depending on the starting grain size), achievable in torsion or compression. In the case of Al, the starting HABs (typically 35° misorientation) are serrated as a result of subgrain-boundary formation, in association with DRV, where the typical misorientations are about a degree, or so. As the grains thin to about twice the subgrain diameter, nearly 1/3–1/2 of the subgrain facets have been replaced by high-angle boundaries, which have ancestry to the starting polycrystal. The remaining two-thirds are still of low misorientation polygonized boundaries typically of a degree or so. As deformation continues, "pinching off" may occur which annihilates HABs and the high-angle boundary area remains constant. Thus, with GDX, the HAB area can dramatically increase but not in the same discontinuous way as DRX. GDX has been confused with DRX, as well and CR (discussed in a later section), but has been confirmed in Al and Al–Mg alloys [146,156], and may occur in other alloys as well, including Fe-based and Zr [221,541]. Figure 40 showed the progression in Al at elevated temperature in torsion [18].

7.4 PARTICLE-STIMULATED NUCLEATION (PSN)

As pointed out by Humphreys [222], an understanding of the effects of second-phase particles on recrystallization is important since most industrial alloy contain second-phase particles and such particles have a strong influence on the recrystallization kinetics, microstructure, and texture. Particles are often known for their ability to impede the motions of high-angle boundaries during high-temperature annealing or deformation (Zener pinning). During the deformation of a particle-containing alloy, the enforced strain gradient in the vicinity of a non-deforming particle creates a region of high dislocation density and large orientation gradient, which is an ideal site for the development of a recrystallization nucleus. The mechanisms of recrystallization in two-phase alloys do not differ from those in single-phase alloys. There are not a great deal of systematic measurements of these zones, but it appears that the deformation zone may extend a diameter of the particle, or so, into the matrix and lead to misorientations of tens of degrees from the adjacent matrix.

7.5 CONTINUOUS REACTIONS

According to McNelley [222], it is now recognized that refined grain structures may evolve homogeneously and gradually during the annealing of deformed metals, either with or without concurrent straining. This can occur even when the heterogeneous nucleation and growth stages of primary recrystallization do not occur. "Continuous reactions" is a term that is sometimes used in place of others that imply at least similar process such as "continuous recrystallization", "*In situ* recrystallization", and "extended recovery." It is often commonly observed that deformation textures sharpen and components related to the stable orientations within the prior deformation textures are retained [542]. These observations are consistent with recovery as the sole restoration mechanism, suggesting that the term "continuous reactions" may be more meaningful a description than "continuous recrystallization."

Mechanisms proposed to explain the role of recovery in high-angle boundary formation include subgrain growth via dislocation motion [540], the development of higher-angle boundaries by the merging of lower-angle boundaries during subgrain coalescence [542] and the increase of boundary misorientation through the accumulation of dislocations into the subgrain boundaries [540]. These processes have been envisioned to result in a progressive buildup of boundary misorientation during (static or dynamic) annealing, resulting in a gradual transition in boundary character and formation of high-angle grain boundaries.

Chapter 8
Creep Behavior of Particle-Strengthened Alloys

8.1.	Introduction	151
8.2.	Small Volume-Fraction Particles that are Coherent and Incoherent with the Matrix with Small Aspect Ratios	151
	8.2.1. Introduction and Theory	151
	8.2.2. Local and General Climb of Dislocations over Obstacles	155
	8.2.3. Detachment Model	158
	8.2.4. Constitutive Relationships	162
	8.2.5. Microstructurals Effects	166
	8.2.6. Coherent Particles	168

Chapter 8
Creep Behavior of Particle-Strengthened Alloys

8.1 INTRODUCTION

This chapter will discuss the behavior of two types of materials that have creep properties enhanced by second phases. These materials contain particles with square or spherical aspect ratios that are both coherent and, but generally, incoherent with the matrix, and of relatively low volume fractions. This chapter is a review of work in this area, but it must be recognized that other reviews have been published and this chapter reflects the particularly high-quality reviews by Reppich and coworkers [543–545] and Arzt [546,547] and others [548–554]. It should also be mentioned that this chapter will emphasize those cases where there are relatively wide separations between the particles, or, in otherwords, the volume fraction of the precipitate discussed here is nearly always less than 30% and usually less than 10%. This contrasts the case of some γ/γ' alloys where the precipitate, γ' occupies a substantial volume fraction of the alloys.

8.2 SMALL VOLUME-FRACTION PARTICLES THAT ARE COHERENT AND INCOHERENT WITH THE MATRIX WITH SMALL ASPECT RATIOS

8.2.1 Introduction and Theory
It is well known that second-phase particles provide enhanced strength at lower temperatures and there have been numerous discussions on the source of this strength. A relatively recent review of, in particular, low-temperature strengthening by second-phase particles was published by Reppich [544]. Although a discussion of the mechanisms of lower temperature second-phase strengthening is outside the scope of this chapter, it should be mentioned that the strength by particles has been believed to be provided in two somewhat broad categories of strengthening, Friedel cutting or Orowan bypassing. Basically, the former involves coherent particles and the flow stress of the alloy is governed by the stress required for the passage of the dislocation through particle. The Orowan stress is determined by the bypass stress based on an Orowan loop mechanism. In the case of oxide dispersion strengthened (ODS) alloys, in which the particles are incoherent, the low-temperature yield stress is reasonably predicted by the Orowan

loop mechanism [544,555]. The Orowan bowing stress is approximated by the classic equation,

$$\tau_{or} = Gb/L = \left(\frac{2T_d}{bL}\right) \tag{112}$$

where τ_{or} is the bowing stress, L is the average separation between particles, and T_d is the dislocation line tension. This equation, of course, assumes that the elastic strain energy of a dislocation can be estimated by $Gb^2/2$, which, though reasonable, is not firmly established.

This chapter discusses how the addition of second phases leads to enhanced strength (creep resistance) at elevated temperature. This discussion is important for at least two reasons: First, as Figure 68 illustrates, the situation at elevated

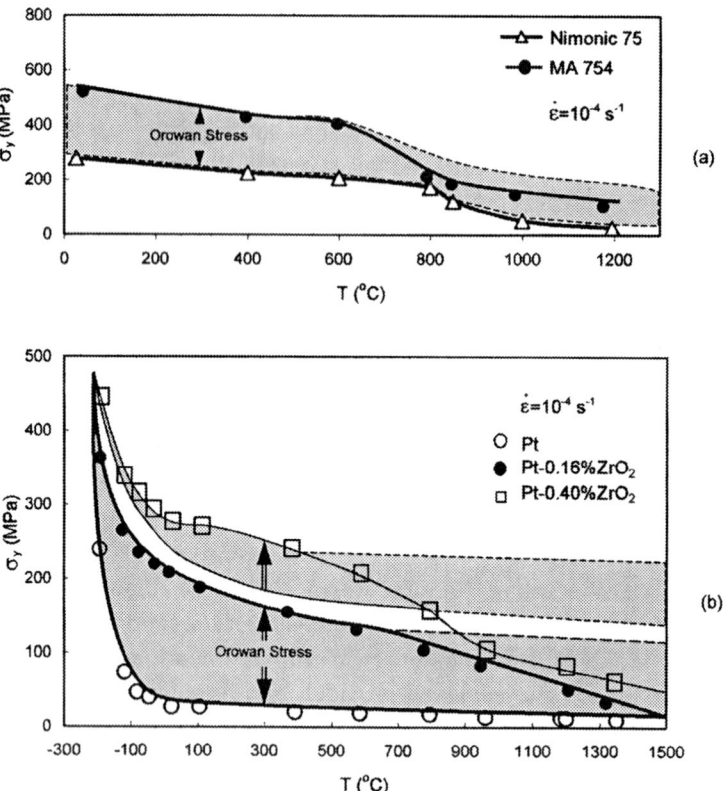

Figure 68. Compressive 0.2% yield stress versus temperature. Shaded: Orowan stress given as low-temperature yield-stress increment due to oxide dispersoids. (a) ODS Superalloy MA 754, (b) Pt-based ODS alloys. From Ref. [544].

temperature is different than at lower temperatures. This figure illustrates that the yield stress of the single-phase matrix is temperature dependent, of course, but there is a superimposed strengthening (suggested in the figure to be approximately athermal) by Orowan bowing [555]. It is generally assumed that the bowing process cannot be thermally activated, but the non-shearable particles can be negotiated by climb. At higher temperatures, it is suggested that the flow stress becomes only a fraction of this superimposed stress and an understanding of the origin is a significant focus of this chapter. Second, in the previous chapters it was illustrated how solute additions, basically obstacles, lead to increased creep strength. There is, essentially, a roughly uniform shifting of the power law, power-law-breakdown and low-stress exponent regimes to higher stresses. This is evident in Figure 69 where the additions of Mg to Al are described. In the case of alloys with second-phase particles, however, there is often a lack of this uniform shift and sometimes the appearance what many investigators have termed a "threshold stress," σ_{TH}. The intent of this term is illustrated in Figure 70, based on the data of Lund and Nix and additional interpretations by Pharr and Nix [557,558]. Figure 70 reflects "classic" particle

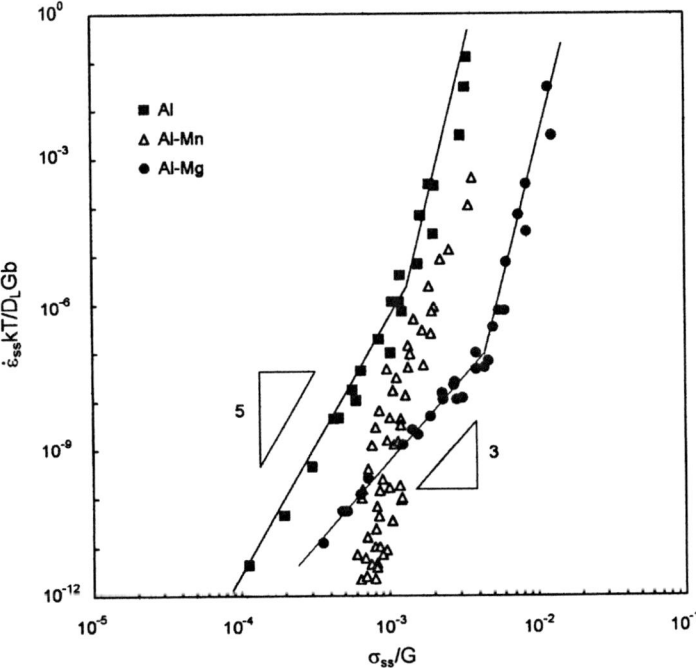

Figure 69. Steady-state relation between strain-rate $\dot{\varepsilon}$ and flow stress for the alloys of this work compared to literature data from slow tests (Al, Al–Mg, Al–Mn). Adapted from Ref. [556].

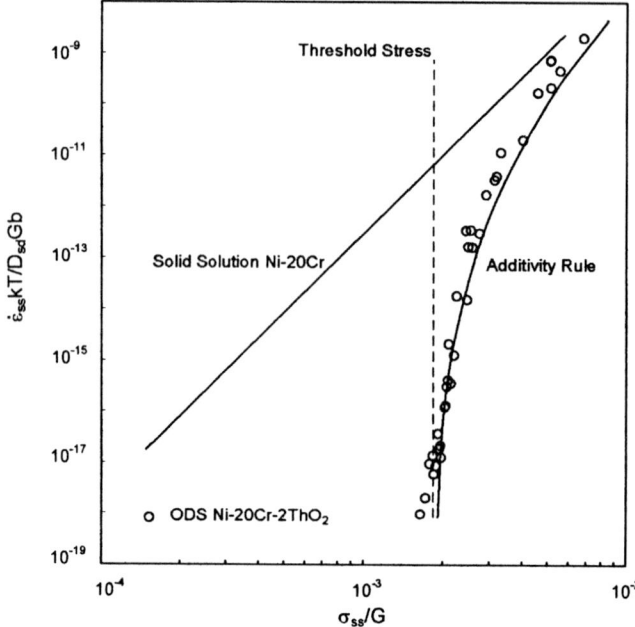

Figure 70. The normalized steady-state creep-rate versus modulus-compensated steady-state stress. Adapted from Ref. [557,558].

strengthening by oxide dispersoids (ThO$_2$) in a Ni–Cr solid solution matrix. These particles, of course, are incoherent with the matrix. The "pure" solid solution alloy behavior is also indicated. Particle strengthening is evident at all steady-state stress levels, but at the lower stresses there appears a modulus-compensated stress below which creep does not appear to occur or is at least very slow. Again, this has been termed the "threshold stress," σ_{TH}. In otherwords, there is not a uniform shift of the strengthening on logarithmic axes; there appears a larger fraction of the strength provided by the particles at lower stress (high temperatures) than at higher stresses (lower temperatures). In fact, the concept of a threshold stress was probably originally considered one of athermal strengthening. This will be discussed more later, but this "coarse" description identifies an aspect of particle strengthening that appears generally different from dislocation substructure strengthening and solution strengthening (although the latter, in certain temperature regimes, may have a nearly athermal strengthening character). The "threshold stress" of Figure 70 is only about half the Orowan bowing stress, suggesting that Orowan bowing may not be the basis of the threshold. Activation energies appear relatively high (greater than lattice self-diffusion of the matrix) as well as the stress exponents being relatively high ($n \gg 5$) in the region where a threshold is apparent.

Particle strengthening is also illustrated in Figure 69, based on a figure from Ref. [556], where there is, again, a non-uniform shift in the behavior of the particle-strengthened Al. The figure indicates the classic creep behavior of high-purity Al. Additionally, the behavior of Mg (4.8 wt.%)-solute strengthened Al is plotted (there is additionally about 0.05 wt.% Fe and Si solute in this alloy). The Mg atoms significantly strengthen the Al. The strengthening may be associated with viscous glide in some temperature ranges in this case. However, an important point is that at higher modulus-compensated stresses, the Al–Mn alloy (strengthened by incoherent Al_6Mn particles) has slightly greater strength than pure Al (there is also an additional 0.05 wt.% Fe and Si solute in this alloy). It does not appear to have as high a strength as the solute strengthened Al-4.8 Mg alloy. However, at lower applied stresses (lower strain rates) or higher temperatures, the second-phase strengthened alloy has *higher* strength than both the pure matrix and the solution-strengthened alloy. As with the ODS alloy of Figure 70, a "threshold" behavior is evident. The potential technological advantage of these alloys appears to be provided by this threshold-like behavior. As the temperature is increased, the flow stress does not experience the magnitudes of decreases as by the other (e.g., solute and dislocation) strengthening mechanisms.

Basically, the current theories for the threshold stress fall into one of two main categories; a threshold arising due to increased dislocation line length with climb over particles and the detachment stress to remove the dislocation from the particle matrix interface after climb over the particle.

8.2.2 Local and General Climb of Dislocations over Obstacles

It was presumed long ago, by Ansell and Weertman [559], that dislocation climb allowed for passage at these elevated temperatures and relatively low stresses. The problem with this early climb approach is that the creep-rate is expected to have a low stress-dependence with an activation energy equivalent to that of lattice self-diffusion. As indicated in the figures just presented in this chapter, the stress dependence in the vicinity of the "threshold" is relatively high and the activation energy in this "threshold" regime can be much higher than that of lattice self-diffusion. More recent analysis has attempted to rationalize the apparent threshold. One of the earlier approaches suggested that for stresses below the cutting, σ_{ct} (relevant for some cases of coherent precipitates) or Orowan bowing stress, σ_{or}, the dislocation must, as Ansell and Weertman originally suggested, climb over the obstacle. This climbing process could imply an increase in dislocation line length and, hence, total elastic strain energy, which would act as an impediment to plastic flow [560–565]. The schemes by which this has been suggested are illustrated in Figure 71, adapted from Ref. [543]. Figure 72, also from Ref. [543], shows an edge

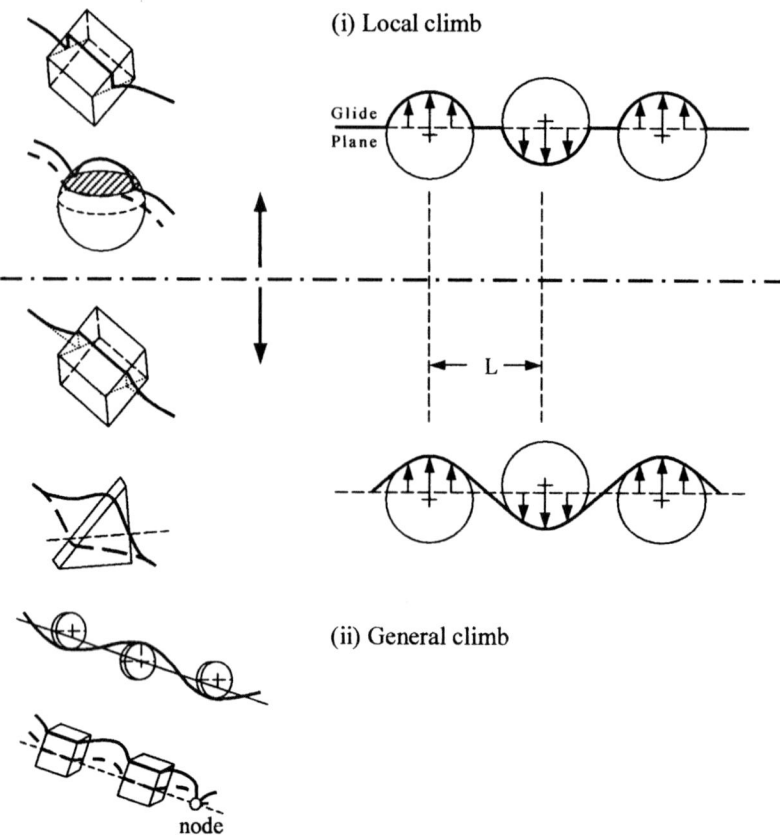

Figure 71. Compilation by Blum and Reppich [543] of models for dislocation climb over second-phase particles.

dislocation climbing, with concomitant slip, over a spherical particle. As the dislocation climbs, work is performed by the applied shear. The total energy change can be described by

$$dE \cong (Gb^2/2)dL - \tau bLdx - \sigma_n bLdy - dE_{el} \tag{113}$$

The first term is the increase in elastic strain energy associated with the increase in dislocation line length. This is generally the principal term giving rise to the (so-called or apparent) threshold stress. The second term is the work done by the applied stress as the dislocation glides. The third term is the work done by the normal component of the stress as the dislocation climbs. The fourth term accounts for any elastic interaction between the dislocation and the particle [566],

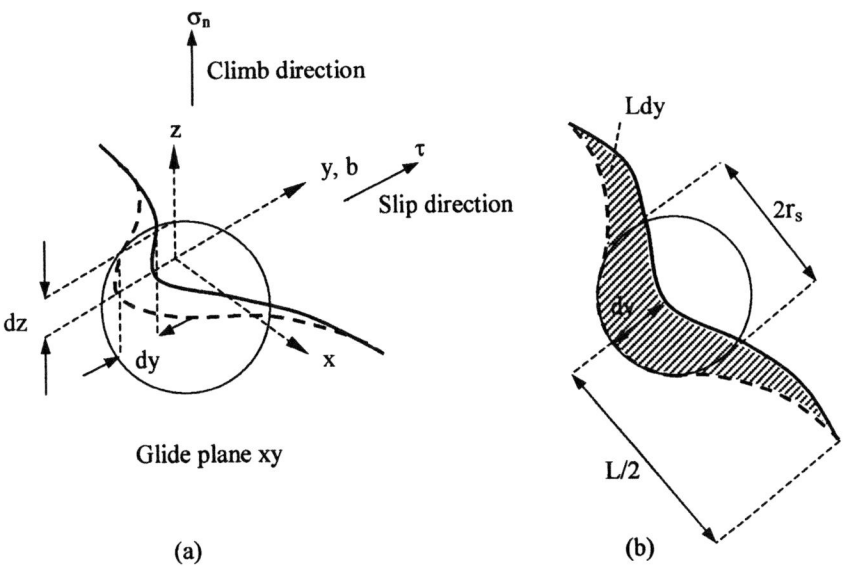

Figure 72. (a) Climb of an edge dislocation over a spherical particle; (b) top view. From Ref. [543].

which does not, in its original formulation, appear to include coherency stresses, although this would be appropriate for coherent particles. This equation is often simplified to

$$dE = (Gb^2/2)dL - \tau bLdx \tag{114}$$

The critical stress for climb of the dislocation over the particle is defined under the condition where

$$(dE/dx) = 0$$

or

$$\tau_c = Gb/L(0.5\alpha') \tag{115}$$

Arzt and Ashby defined the α' parameter $= (dL/dx)_{max}$ as the climb resistance and τ_c can be regarded as the apparent threshold stress. Estimates have been made of α' by relating the volume fraction of the particles, and the particle diameter to the value of L in equation (115). Furthermore, tyre is a statistical distribution of particle spacings and Arzt and Ashby suggest,

$$\tau_c/(Gb/L) = \alpha'/(1.68 + \alpha') \tag{116}$$

while Blum and Reppich use a similar relationship which includes the so-called "Friedel correction"

$$\tau_c/(Gb/L) = \alpha'^{1.5}/(2\sqrt{2} + \sqrt{\alpha'^3}) \qquad (117)$$

These equations (115–117), together with the values of α', allow a determination of the threshold stress. Note that for general climb there is the suggestion that τ_c is particle size independent.

The determination of α' will depend on whether climb is local or general; both cases are illustrated in Figure 71. The portion of the dislocation that climbs can be either confined to the particle–matrix interfacial region (local), or the climbing region can extend beyond the interfacial region, well into the matrix (general). This significantly affects the α' calculation.

First, equation (115) suggests that a maximum value for α' which corresponds to the Orowan bowing stress. It has been suggested that the Orowan stress can be altered based on randomness and elastic interaction considerations [561,564], giving values of $0.5 < \alpha' < 1.0$. For local climb, the value of α' depends on the shape of the particle, with $0.77 < \alpha' < 1.41$, from spherical to square shapes [560,561,564]. For extended or general climb, which is a more realistic configuration in the absence of any particular attraction to the particle [567], the α' is one order of magnitude, or so, smaller. Additionally, the value of α' will be dependent on the volume fraction, f, as $f^{1/2}$ and values of α' range from 0.047 to 0.14 from $0.01 < f < 0.10$ [543]. Blum and Reppich suggested that for these circumstances,

for local climb $\qquad \tau_c = 0.19\ [Gb/L]$
and for general climb $\qquad \tau_c = 0.004$ to $0.02\ [Gb/L]$

This implies that there is a threshold associated with the simple climbing of a dislocation over particle-obstacles, without substantial interaction. This "threshold" stress is a relatively small fraction of the Orowan stress. There is the implicit suggestion in all of this analysis that the stress calculated from the above equations is athermal in nature and this will be discussed subsequently.

8.2.3 Detachment Model

In connection with the above, however, there has been evidence that dislocations may interact with incoherent particles. This was observed by Nardone and Tien [568] and later by Arzt and Schroder [569] and others [571] using TEM of creep-deformed ODS alloys. Figures 73 and 74 illustrate this. The dislocations must undergo local climb over the precipitate and then the dislocation must undergo "detachment". Srolovitz et al. [571] suggested that incoherent particles have interfaces that may slip

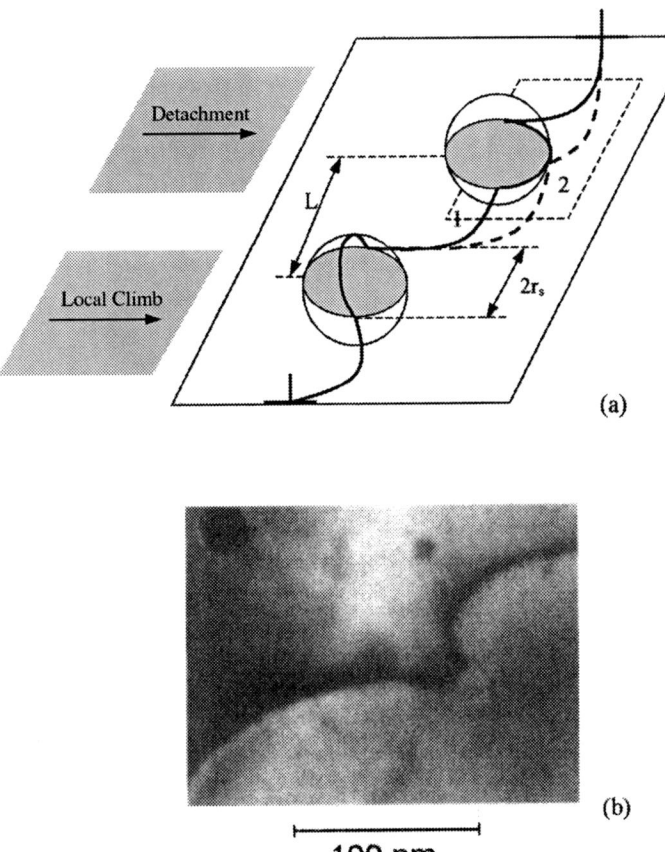

Figure 73. The mechanism of interfacial pinning. (a) Perspective view illustrating serial local climb over spherical particles of mean (planar) radius r_s and spacing l and subsequent detachment. (b) Circumstantial TEM evidence in the creep-exposed ferritic ODS superalloy PM 2000 [570].

and can attract dislocations by reducing the total elastic strain energy. Thus, there is a detachment stress that reflects the increase in strain energy of the dislocation on leaving the interface. Basically, Arzt and coworkers suggest that the incoherent dispersoids strengthen by acting as, essentially, voids. Arzt and coworkers [562,573–575] analyzed the detachment process in some detail and estimated τ_d as,

$$\tau_d = [1 - k_R^2]^{1/2}(Gb/L) \tag{118}$$

where k_R is the relaxation factor described by

$$(Gb^2)_p = k_R(Gb^2)_m \tag{119}$$

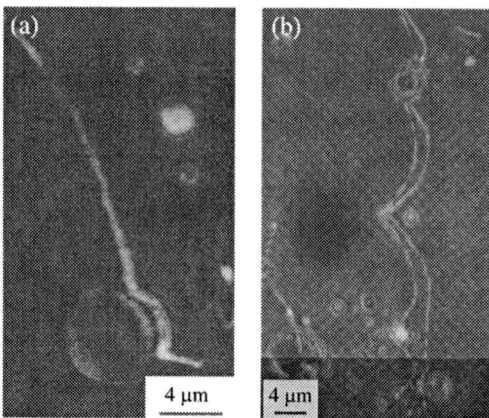

Figure 74. TEM evidence of an attractive interaction between dislocation and dispersoid particles: (a) dislocation detachment from a dispersoid particle in a Ni alloy; (b) dissociated superdislocation detaching from dispersoid particles in the intermetallic compound Ni_3Al. From Refs. [547,569,572].

where "p" refers to the particle interface and "m" the matrix. In the limit that $k_R = 1$, there is no detachment process.

Reppich [532] modified the Arzt et al. analysis slightly, using Fleischer–Friedel obstacle approximation and suggested that:

$$\tau_d = 0.9(1 - k_R^2)^{3/4} / [1 + (1 - k_R^2)^{3/4}](Gb/L) \qquad (120)$$

This decreases the values of equation (118) roughly by a factor of two. Note that as with general climb, τ_d and τ_c are independent of the particle size. It was suggested, by both of the above groups, that this detachment process could be thermally activated. This equation suggests the τ_d is roughly $Gb/3L$, substantially higher than τ_c for general climb, as later illustrated in Figure 80. Arzt and Wilkinson [562] showed that if k_R is such that there is just a 6%, or less, reduction in the elastic strain energy, then local climb becomes the basis of the threshold stress instead of detachment. For general climb, the transition point is k_R about equal to one.

Rosler and Arzt [575] extended the detachment analysis to a "full kinetic model" and suggested a constitutive equation for "detachment-controlled" creep,

$$\dot{\varepsilon} = \dot{\varepsilon}_0 \exp[(-Gbr_s^2/kT)(1 - k_R)^{3/2}(1 - \sigma/\sigma_d)^{3/2}] \qquad (121)$$

where

$$\dot{\varepsilon}_0 = C_2 D_v L \rho_m / 2b,$$

and r_s is the particle radius. This was shown to be valid for random arrays of particles. Figure 75 plots this equation for several values of k_R as a function of

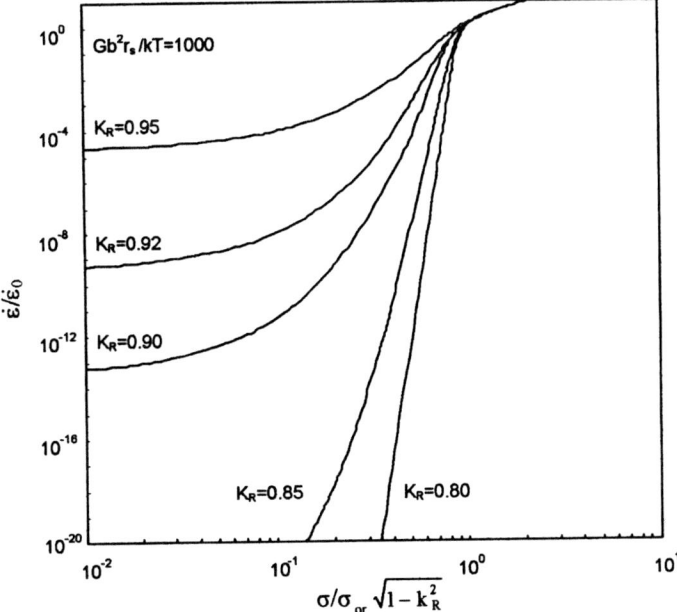

Figure 75. Theoretical prediction of the creep-rate (normalized) as a function of stress (normalized) on the basis of thermally activated dislocation detachment from attractive dispersoids, as a function of interaction parameter k. The change of curvature at high strain rates (broken line) indicates the transition to the creep behavior of dispersoid-free material and does not follow from the equation [574].

strain-rate. Threshold behavior is apparent for modest values of k_R. This model appears to reasonably predict the creep behavior of various dispersion-strengthened Al alloys [546,576] with reasonable k_R values (0.75–0.95). As Arzt points out, this model, however, does not include the effects of dislocation substructure. Arzt noted from this equation that an optimum particle size is predicted. This results from the probability of thermally activated detachment being raised for small dispersoids and that large particles (for a given volume fraction) have a low Orowan stress and, hence, small detachment stress [546]. It should be noted that Figure 75 does not suggest a "pure" threshold stress (below which plasticity does not occur). Rather, thermally activated detachment is suggested, and this will be discussed more later.

More recently, Reppich [545] reviewed the reported *In situ* straining experiments in an ODS alloys at elevated temperature and concluded that the observations of the detachment are essentially in agreement with the above description (thermal activation aside). *In situ* straining experiments by Behr *et al.* at 1000°C in the TEM also appear to confirm this detachment process in dispersion-strengthened intermetallics [572], as shown in Figure 74.

8.2.4 Constitutive Relationships

The suggestion of the above is that particle-strengthened alloys can be approximately described by relationships that include the threshold stress. A common relationship that is used to describe the steady-state behavior of second-phase strengthened alloys (at a fixed temperature) is,

$$\dot{\varepsilon}_{ss} = A'(\sigma - \sigma_{th})^{n_m} \qquad (122)$$

where σ_{th} is the threshold stress and n_m is the steady-state stress exponent of the matrix. Figure 76 [577] graphically illustrates this superposition strategy. As will be discussed subsequently, this equation is widely used to assess the value of the threshold stress. Additional data that illustrates the value of equation (122) for superalloys is illustrated in Figure 77 adapted from Ajaja et al. [578]. Figure 78 (adapted from Ref. [579]) illustrates that at higher stresses, above τ_{or}, decreases in stress (and strain-rate) illustrate a threshold behavior. That is, a plot of $\dot{\varepsilon}^{1/n}$ versus

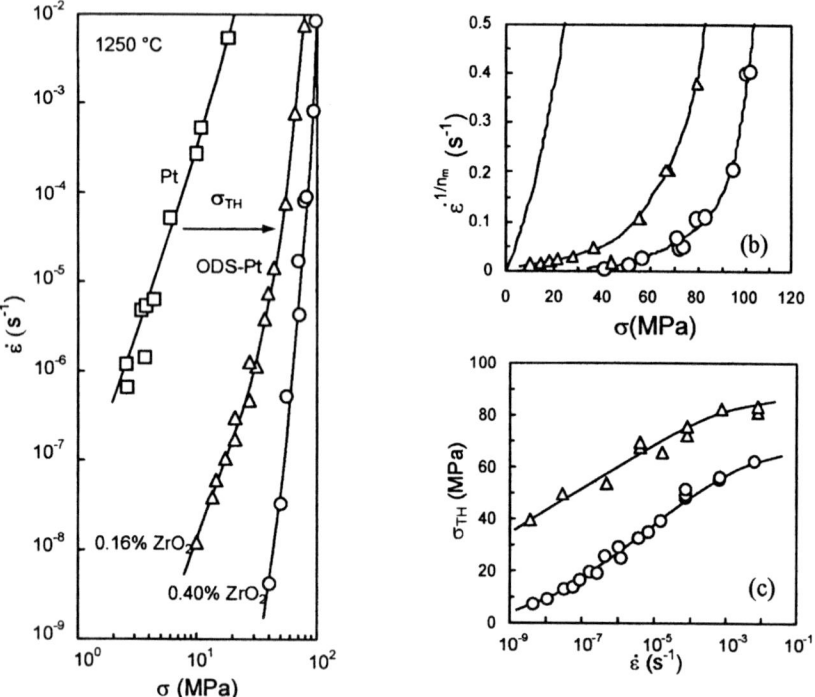

Figure 76. Comparison of the creep behavior of ZrO_2 dispersion strengthened Pt-based alloys at 1250°C: (a) Double-logarithmic Norton plot of creep rate $\dot{\varepsilon}$ versus stress σ; (b) Lagneborg–Bergman plot; (c) dependence of σ_{TH} on $\dot{\varepsilon}$. The shaded bands denote the ratio σ_p/σ_{or}. From Ref. [572].

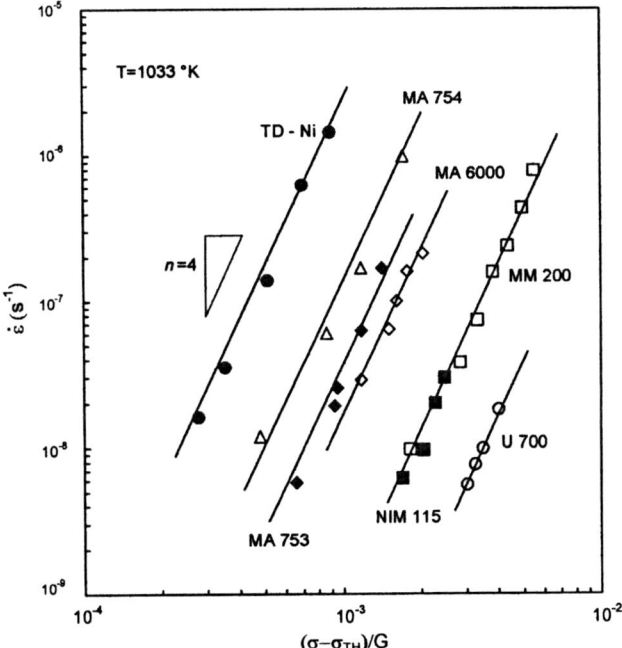

Figure 77. Double-logarithmic plot of creep-rate versus reduced stress $\sigma - \sigma_{TH}$ for various superalloys. After Ajaja et al. [578].

σ extrapolates to τ_{or}, an "apparent σ_{TH}." However, as σ decreases below τ_{or}, a new threshold appears and this, then, is the σ_{TH} relevant to creep plasticity. A σ_{TH} can be estimated with low-stress plots such as Figure 78 [580,581] (also see Figure 80). These give rise to estimates of σ_{TH} and allow plots such as Figure 77. However, Figure 76(b) and (c) illustrate that the high temperature σ_{TH} is not a true threshold and creep occurs below σ_{TH}. This is why σ_{TH} estimates based on plots such as Figure 78 decrease with increasing temperature.

An activation energy term can be included in the form of,

$$\dot{\varepsilon}_{ss} = A'' \exp(-Q/kT)\left(\frac{\sigma - \sigma_{th}}{G}\right)^{n_m} \quad (123)$$

or

$$\dot{\varepsilon}_{ss} = \frac{A'' DGb}{kT}\left(\frac{\sigma - \sigma_{th}}{G}\right)^{n_m} \quad (124)$$

where D is the diffusion coefficient. However, the use of D works in some cases while not in others for both coherent and incoherent particles. Figure 79, taken from

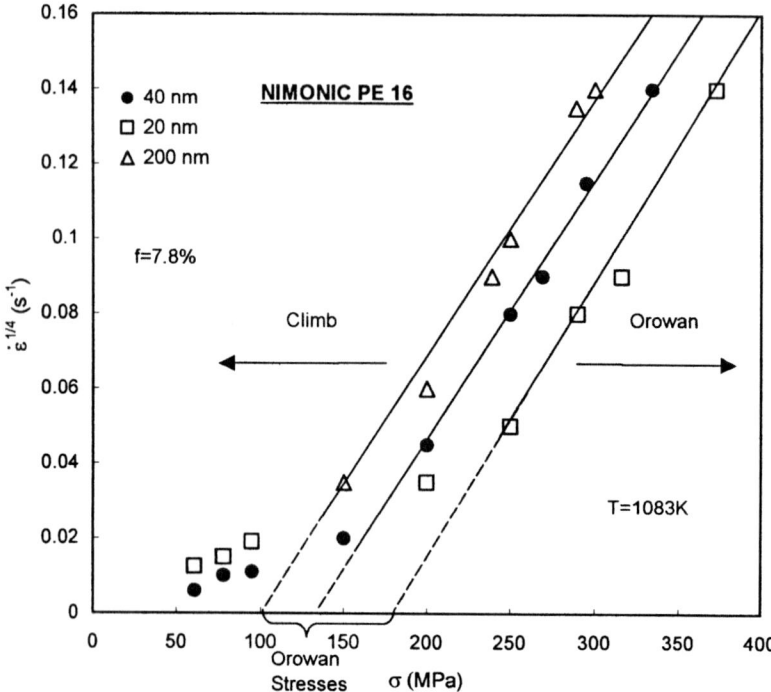

Figure 78. $\dot{\varepsilon}^{1/n}$ versus σ-plot for determination of σ_{TH} in γ'-hardened Nimonic PE 16 by back extrapolation (arrows). From Ref. [579].

Ref. [582], illustrates somewhat different behavior from Figure 70 in that the data do not appear to reduce to a single line when the steady-state stress is modulus-compensated and the steady-state creep-rates are lattice self-diffusion compensated. The activation energy for creep is reported to be relatively high at 537 kJ/mol (as compared to 142 kJ/mol for lattice self-diffusion). Cadek and coworkers [583,584] illustrate that for experiments on ODS Cu, the (modulus-compensated) threshold stress, determined by the extrapolation procedure described in Figure 79, is temperature-dependent. They propose that the activation energies should be determined using the usual equation but at constant $(\sigma - \sigma_{th})/G$ rather than σ/G as used in typical (especially five-power-law for single-phase metals and alloys) creep activation energy calculations. The activation energies calculated using this procedure reasonably correspond to lattice or dislocation-core self-diffusion. Thus, the investigators argued that the activation energy for diffusion could reasonably be used as the activation energy term such as in equation (123).

As discussed earlier, Figure 80 is an idealized plot by Blum and Reppich that illustrates many of the features and parameters for particle strengthening. This is

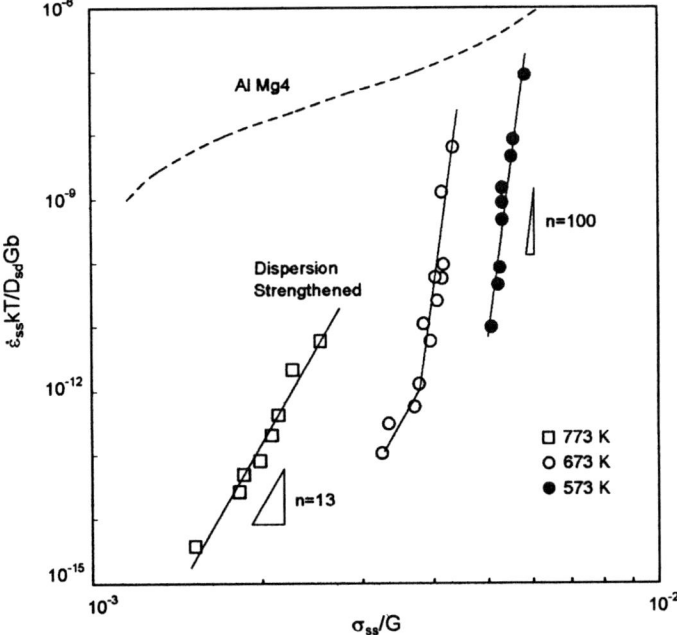

Figure 79. Lack of convergence of the different dispersion-strengthened creep curves with Q_{sd} compensation.

a classic logarithmic plot of the steady-state creep- (strain-) rate versus the steady-state stress. σ_{or} is indicated and apparent threshold behavior is observed above this stress. A second threshold-like behavior is evident below the Orowan stress; one for incoherent particles and another for coherent particles at particularly low stresses (high temperature). The incoherent particles evince interfacial pinning and the more effective (or higher) threshold-like behavior is observed in the absence of a detachment stress. There has been some discussion as to what mechanism may be applicable at stresses below the apparent threshold, and it appears that grain-boundary sliding and even diffusional creep have been suggested [585,586]. These, however, are speculative as even single crystals appear to show sub-threshold plasticity. Figure 75 was basically an attempt to explain creep below the apparent σ_{TH} through thermally activated detachment. This approach has become fairly popular [587], although recently it has been applied below, but not above, an apparent threshold. A difficulty with this approach is that Figures 73 and 74 suggest that a considerable length of dislocation is trapped in the interface which would appear to imply a very large activation energy for detachment, much larger than that for D_v, in equation (121). In at least some instances, the plasticity below the apparent threshold is due to a change in deformation mechanism [588]. Dunand and Jansen

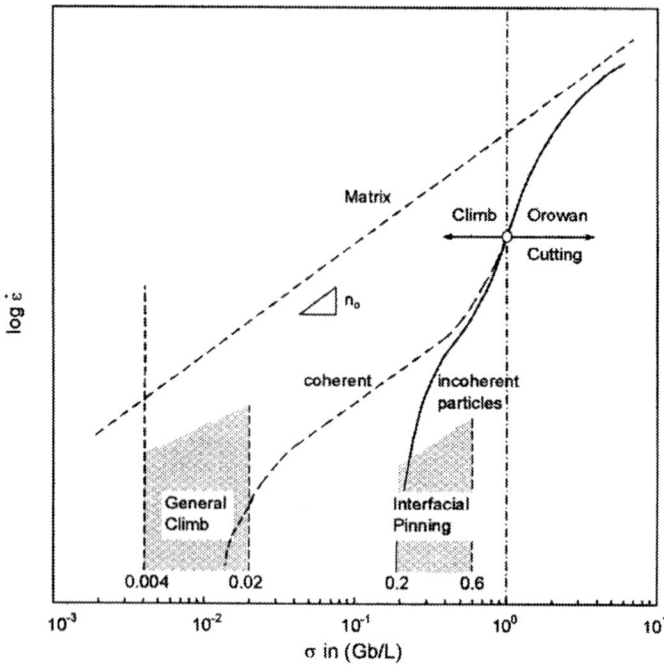

Figure 80. Creep behavior of particle-strengthened materials (schematic). The stress is given in units of the classical Orowan stress. From Ref. [543].

[589,590] suggested that for larger volume fractions of second-phase particles (e.g., 25%) dislocation pile-ups become relevant and additional stress terms must be added to the conventional equations. However, those considerations do not appear relevant to the volume fractions being typically considered here.

8.2.5 Microstructural Effects

a. Transient Creep Behavior and Dislocation Structure. The strain versus time behavior of particle-strengthened alloys during primary and transient creep is similar to that of single-phase materials in terms of the strain-rate versus strain trends as illustrated in Figure 81. Figure 81(a) illustrates Incoloy 800H [549] and (b) Nimonic PE 16 [579]. Both generally evince Class M behavior although the carbide-strengthened Incoloy shows an inverted transient (such as a Class A alloy) but this was suggested to be due to particle structure changes, which must be considered with prolonged high-temperature application. The Nimonic alloy at the lowest stress also shows such an inverted transient and this was suggested as possibly being due to particle changes in this initially coherent γ'-strengthened alloy.

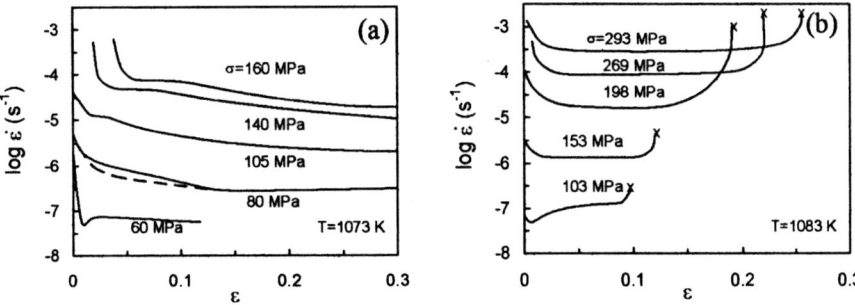

Figure 81. Half-logarithmic plot of creep rate $\dot{\varepsilon}$ versus (true) strain of a single-phase material and particle-strengthened material (a) Incoloy 800 H; (b) Nimonic PE 16. Adapted from Refs. [543,579,591].

There has been relatively little discussed in the literature regarding creep transients. Blum and Reppich suggest that the transients between steady-states with stress-drops and stress-jumps are analogous to the single phase metals both in terms of the nature of the mechanical (e.g., strain-rate versus strain) trends and the final steady-state strain-rate values, as well as the final substructural dimensions.

The dislocation structure of particle-strengthened alloys has been examined, most particularly by Blum and coworkers [543,556]. The Subgrain size in the particle-strengthened Al–Mn alloy are essentially identical to those of high purity Al at the same modulus-compensated stresses. Similar findings were reported for Incoloy 800H [591] and TD–Nichrome [563]. These results show that the total stress level affects the subgrain substructure, even if there is an interaction between the particles and the subgrain boundaries. Blum and Reppich suggest, however, that the density of dislocations within the subgrains seem to depend on $\sigma - \sigma_{TH}$ rather than σ, suggesting that particle hardening diminishes the network dislocation density compared to the single-phase alloy at the same value of σ_{ss}/G. Some of these trends are additionally evident from Figure 82 taken from Straub et al. [90]. One interpretation of this observation is that the Subgrain size reflects the stress but does not determine the strength. In contrast, Arzt suggests that only at higher stresses, where the alloys approach the behavior of dispersoid-free matrix, have dislocation substructures been reported [582]. Blum, however, suggests that this may be due to insufficient strain to develop the substructure that would ultimately form in the absence of interdiction by fracture [592].

b. Effect of Volume Fraction. As expected [585], higher volume fractions, for identical particle sizes, are associated with greater strengthening and, of course, threshold behavior. Others [593] have also suggested that the volume fraction of the second-phase particles can affect the value of the threshold stress.

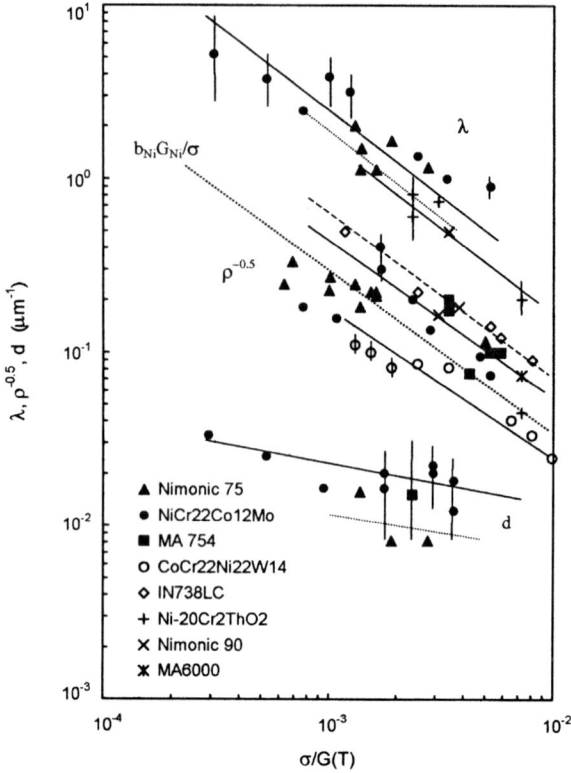

Figure 82. Steady-state dislocation spacings of Ni-based alloys. Adapted from Ref. [90].

c. Grain Size Effects. Lin and Sherby [553], Stephens and Nix [588], and Gregory et al. [594] examined the effects of grain size on the creep properties of dispersion-strengthened Ni–Cr alloys and found that smaller grain size material may not exhibit a threshold behavior and evince stress exponents more typical of single-phase polycrystalline metals with high elongations [595]. In fact, Nix [596] suggests that this decrease in stress exponent may be due to grain-boundary sliding and possibly superplasticity [596]. Again, however, Arzt [546] reports this sigmoidal behavior occurs in single crystals as well and the loss in strength (presumably below that of thermally activated detachment) involves other poorly understood processes including changes in the size or number of particles.

8.2.6 Coherent Particles

Strengthening from coherent particles can occur in a variety of ways that usually involves particle cutting. This cutting can be associated with (a) the creation of

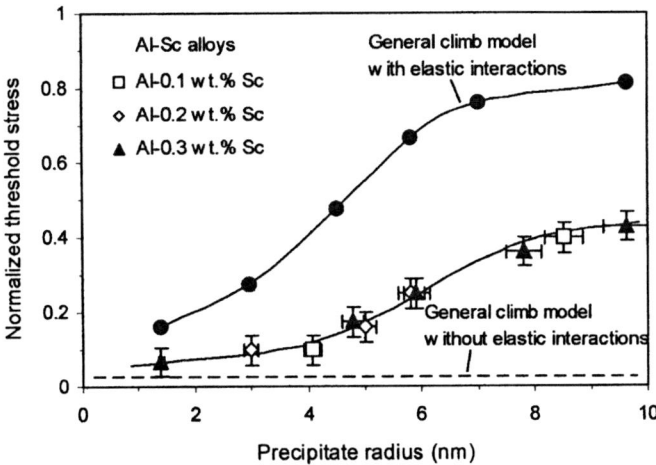

Figure 83. Normalized threshold stress versus coherent precipitate radius. From Ref. [598].

antiphase boundaries [e.g., $\gamma - \gamma'$ superalloys], (b) the creation of a step in the particle, (c) differences in the stacking fault energy between the particle and the matrix, (d) the presence of a stress field around the particle, and (e) other changes in the "lattice friction stress" to the dislocation and the particle [133].

Most of the earlier work referenced was relevant to incoherent particles. This is probably in part due to the fact that coherent particles are often precipitated from the matrix, as opposed to added by mechanical alloying, etc. Precipitates may be unstable at elevated temperatures and, as a consequence, the discussion returns to that of incoherent particles. Of course, exceptions include the earlier referenced $\gamma - \gamma'$ of superalloys and, more recently, the AlSc$_3$ precipitates in Al–Sc alloys by Seidman et al. [597]. In this latter work, coherent particles are precipitated. The elevated temperature strengths are much less than the Orowan bowing stress and also less than expected based on the shearing mechanism. Thus, it was presumed that the rate-controlling process is general climb over the particle, consistent with other literature suggestions. This, as discussed in the previous section, is associated with a relatively low threshold stress that is a small fraction of the Orowan bowing stress at about 0.03 σ_{or}, and, as discussed, is independent of the particle size. Seidman et al. found that the normalized threshold stress increases significantly with the particle size and argued that this could only be rationalized by elastic interaction effects, such as coherency strain and modulus effects. Detachment is not important. The results are illustrated in Figure 83. They also found that subgrains may or may not form. They do appear to obey the standard equations that relate the steady-state stress to subgrain size when they are observed. Seidman et al. do appear to suggest that steady-state was achieved without the formation of subgrains.

Chapter 9
Creep of Intermetallics

9.1.	Introduction	173
9.2.	Titanium Aluminides	175
	9.2.1. Introduction	175
	9.2.2. Rate Controlling Creep Mechanisms in FL TiAl Intermetallics During "Secondary" Creep	178
	9.2.3. Primary Creep in FL Microstructures	186
	9.2.4. Tertiary Creep in FL Microstructures	188
9.3.	Iron Aluminides	188
	9.3.1. Introduction	188
	9.3.2. Anomalous Yield Point Phenomenon	190
	9.3.3. Creep Mechanisms	194
	9.3.4. Strengthening Mechanisms	197
9.4.	Nickel Aluminides	198
	9.4.1. Ni_3Al	198
	9.4.2. NiAl	208

Chapter 9
Creep of Intermetallics

9.1 INTRODUCTION

The term "intermetallics" has been used to designate the intermetallic phases and compounds which result from the combination of various metals, and which form a large class of materials [598]. There are mainly three types of superlattice structures based on the f.c.c. lattice, i.e. $L1_2$ (with a variant of $L'1_2$ in which a small interstitial atom of C or N is inserted at the cube center), $L1_0$, and $L1_2$-derivative long-period structures such as DO_{22} or DO_{23}. The b.c.c.-type structures are B2 and DO_3 or $L2_1$. The DO_{19} structure is one of the most typical superlattices based on h.c.p. symmetry. Table 4 lists the crystal structure, lattice parameter and density of selected intermetallic compounds [599]. A comprehensive review on the physical metallurgy and processing of intermetallics can be found in [600].

Intermetallics often have high melting temperatures (usually higher than 1000°C), due partly to the strong bonding between unlike atoms, which is, in general, a mixture between metallic, ionic and covalent to different extents. The presence of these strong bonds also results in high creep resistance. Another factor that contributes to the superior strength of intermetallics at elevated temperature is the high degree of long-range order [602]. The effect of order is, first, to slow diffusivity. The reason for this is that the number of atoms per unit cell is large in a material with long-range order. Therefore in alloys in which dislocation climb is rate-controlling, a decrease in the diffusion rate would result in a drop in the creep rate and therefore in an increase

Table 4. Crystal structure, lattice parameters and density of selected intermetallic compounds.

Alloy	Structure (Bravais lattice)	Lattice Parameters		Density (g/cm^3)
		a (nm)	c (nm)	
Ni$_3$Al	$L1_2$ (simple cubic)	0.357	–	7.40
NiAl	B2 (simple cubic)	0.288	–	5.96
Ni$_2$AlTi	DO_3	0.585	–	6.38
Ti$_3$Al	DO_{19}	0.577	0.464	4.23
TiAl	$L1_0$	0.398	0.405	3.89
Al$_3$Ti	DO_{22}	0.395	0.860	3.36
FeAl	B2 (simple cubic)			5.4–6.7 [601]
Fe$_3$Al	DO_3			5.4–6.7 [601]
MoSi$_2$	C11b			6.3

of the creep resistance. Secondly, the presence of a high degree of long-range order may retard the viscous motion of dislocations. This is due to the fact that, when a dislocation moves in a non-perfect lattice, the long-range order is damaged and this leads to an increase in energy or a dragging force. Thus, the presence of order results in a decrease of the creep rate in the intermetallic alloys in which the mechanism of viscous glide of dislocations is rate-controlling.

One major disadvantage of these materials, that is limiting their industrial application, is brittleness [603]. This is attributed to several factors. First, the strong atomic bonds as well as the long-range order give rise to high Peierls stresses. Transgranular cleavage will occur in a brittle manner if the latter are larger than the stress for nucleation of a crack. Second, grain boundaries are intrinsically weak. The low boundary cohesion results in part from the directionality of the distribution of the electronic charge in ordered alloys [600]. The strong atomic bonding between the two main alloy constituents is related to the p-d orbital hibridization, which leads to a strong directionality in the charge distribution. In grain boundaries the directionality is reduced and thus the bonding becomes much weaker. Other factors that may contribute to the brittleness in intermetallics are the limited number of operative slip systems, segregation of impurities at grain boundaries, a high work hardening rate, planar slip, and the presence of constitutional defects. The latter may be, for example, atoms occupying sites of a sublattice other than their own sublattice (antisites) or vacancies of deficient atomic species (constitutional vacancies). The planar faults, dislocation dissociations, and dislocation core structures typical of intermetallics were summarized by Yamaguchi and Umakoshi [604]. Other so-called extrinsic factors that cause brittleness are the presence of segregants, interstitials, moisture in the environment, poor surface finish, and hydrogen [605]. It appears that those intermetallics with more potential as high-temperature structural materials, i.e., those which are less brittle, are compounds with high crystal symmetry and small unit cells. Thus, nickel aluminides, titanium aluminides, and iron aluminides have been the subject of the most activities in research and development over the last two decades. These investigations were stimulated by both the possibility of industrial application and scientific interest [598–607].

Creep resistance is a critical property in materials used for high temperature structural applications. Some intermetallics may have the potential to replace nickel superalloys in parts such as the rotating blades of gas turbines or jet engines [608] due to their higher melting temperatures, high oxidation and corrosion resistance, high creep resistance, and, in some cases, lower density. The creep behavior of intermetallics is more complicated than that of pure metals and disordered solid solution alloys due to their complex structures together with the varieties of chemical composition [609–610]. The rate-controlling mechanisms are still not fully understood in spite of significant efforts over the last couple of decades [598,604,611–619].

In the following, the current understanding of creep of intermetallics will be reviewed, placing special emphasis on investigations published over the last decade and related to the compounds with potential for structural applications such as titanium aluminides, iron aluminides and nickel aluminides.

9.2 TITANIUM ALUMINIDES

9.2.1 Introduction

Titanium aluminide alloys have the potential for replacing heavier materials in high-temperature structural applications such as automotive and aerospace engine components. This is due, first, to their low density (lower than that of most other intermetallics), high melting temperature, good elevated temperature strength and high modulus, high oxidation resistance and favorable creep properties [620,621]. Second, they can be processed through conventional manufacturing methods such as casting, forging and machining [607]. In fact, TiAl turbocharger turbine wheels have recently been used in automobiles [607]. Table 5 (from [621]) compares the properties between titanium aluminides, titanium-based conventional alloys and superalloys (see phase diagram below for phase compositions).

Many investigations have attempted to understand the creep mechanisms in titanium aluminides over the last two decades. Several excellent reviews in this area include [607,622,623]. The creep behavior of titanium aluminides depends strongly

Table 5. Properties of titanium aluminides, titanium-based conventional alloys and superalloys.

Property	Ti-based alloys	Ti$_3$Al-based α_2 alloys	TiAl-based γ alloys	superalloys
Density (g/cm^3)	4.5	4.1–4.7	3.7–3.9	8.3
RT modulus (GPa)	96–115	120–145	160–176	206
RT Yield strength (MPa)	380–1115	700–990	400–630	250–1310*
RT Tensile strength (MPa)	480–1200	800–1140	450–700	620–1620*
Highest temperature with high creep strength (°C)	600	750	1000	1090
Temperature of oxidation (°C)	600	650	900–1000	1090
Ductility (%) at RT	10–20	2–7	1–3	3–5
Ductility (%) at high T	High	10–20	10–90	10–20
Structure	hcp/bcc	DO19	L1$_0$	Fcc/L1$_2$

*Data added to the table provided in [621].

on alloy composition and microstructure. The different Ti-Al microstructures are briefly reviewed in the following sections.

Figure 84, the Ti-Al phase diagram, illustrates the following phases: γ-TiAl (ordered face-centered tetragonal, $L1_0$), α_2-Ti_3Al (ordered hexagonal, DO_{19}), α-Ti (h.c.p., high-temperature disordered), and β-Ti (b.c.c., disordered). Gamma (γ) or near γ-TiAl alloys have compositions with 49–66 at.% Al, depending on temperature. α_2 alloys contain from 22 at.% to approximately 35 at.% Al. Two-phase (γ-TiAl+α_2-Ti_3Al) alloys contain between 35 at.% and 49 at.% Al. The morphology of the two phases depends on thermomechanical processing [607]. Alloys, for example, with nearly stoichiometric or Ti-rich compositions that are cast or cooled from the β phase, going through the α single phase region and $\alpha \rightarrow \alpha + \gamma$ and $\alpha + \gamma \rightarrow \alpha_2 + \gamma$ reactions, have fully lamellar (FL) microstructures, as illustrated in Fig. 85. An FL microstructure consists of "lamellar grains" or colonies, of size g_l, that are equiaxed grains composed of thin alternating lamellae of γ and α_2. The average thickness of the lamellae, termed the lamellar interface spacing, is denoted by λ_l. The γ and α_2 lamellae are stacked such that {111} planes of the γ lamellae are parallel to (0001) planes of the α_2 lamellae and the closely packed directions on the former are parallel to those of the latter. The lamellar structure is destroyed if an FL microstructure is annealed or hot worked at temperatures higher than 1150°C within

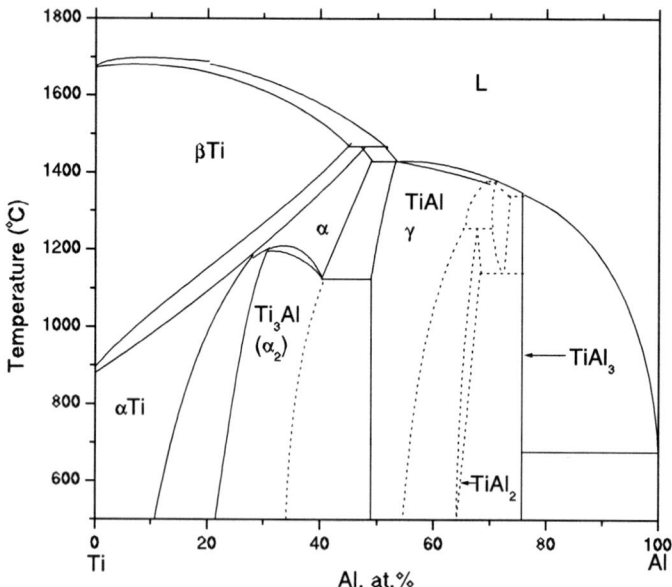

Figure 84. Ti-Al phase diagram [600].

Figure 85. Fully lamellar (FL) microstructure corresponding to a Ti-Al based alloy with a nearly stoichiometric composition [607].

the $(\alpha+\gamma)$ phase fields. A bimodal microstructure develops, consisting of lamellar grains alternating with γ grains (or grains exclusively of γ-phase). Depending on the amount of γ-grains, the microstructure is termed "nearly lamellar" (NL) when the fraction of γ-grains is small, or duplex (DP) when the fractions of lamellar and γ-grains are comparable. Detailed studies of the microstructures of TiAl alloys are described in [624–625].

Overall, two-phase γ-TiAl alloys have more potential for high-temperature applications than α-Ti$_3$Al alloys due to their higher oxidation resistance and Elastic Modulus [621]. Simultaneously, two-phase γ-TiAl alloys have comparatively lower creep strength at high temperature than α-Ti$_3$Al alloys and therefore significant efforts have been devoted in recent years to improve the creep behavior of γ-TiAl. It is now well established that the optimum microstructure for creep resistance in two-phase TiAl alloys is fully lamellar (FL) [626–628]. As will be explained in the following sections, this microstructure shows the highest creep resistance, the lowest minimum creep rate, and the best primary creep behavior (i.e., longer times to attain a specified strain). Figure 86 (from [626]) illustrates the creep curves at 760°C and 240 MPa corresponding to a Ti-48%Al alloy with several different microstructures. The FL microstructure shows superior creep resistance. Lamellar microstructures also have superior fracture toughness and fatigue resistance in comparison to duplex (DP) structures, although the latter have, in general, better ductility [607]. This section will review the fundamentals of creep deformation in FL Ti-Al alloys. Emphasis will be placed on describing prominent recent creep models, rather than on compiling the extensive experimental data [622,623,626].

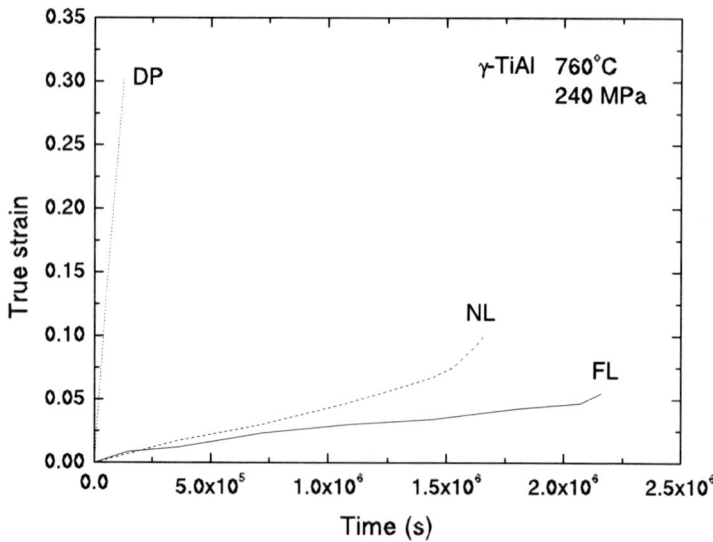

Figure 86. Creep curves at 760°C and 240 MPa corresponding to several near γ-TiAl alloys with different microstructures. (a) Ti-48Al alloy with a fully lamellar (FL) microstructure; (b) Ti-48Al alloy with a nearly lamellar (NL) microstructure; (c) Ti-48Al alloy with a duplex microstructure [626].

9.2.2 Rate Controlling Creep Mechanisms in FL TiAl Intermetallics During "Secondary" Creep

Several investigations have attempted to determine the rate-controlling mechanisms during creep of fully lamellar TiAl intermetallics [622–623,626–633]. Most creep studies were performed in the temperature range 676°C–877°C [623] and the stress range 80–500 MPa, relevant to the anticipated service conditions [629]. Clarifying the rate controlling creep mechanisms in FL TiAl alloys is difficult for several reasons. First, rationalization of creep data by conventional methods such as analysis of steady-state stress exponents is controversial, since usually an unambiguous secondary creep stage is not observed. Instead, a minimum strain rate ($\dot{\varepsilon}_{min}$) is measured, and the "secondary creep rate" or "steady-state rate" is presumed close to the minimum rate. Second, a continuous increase in the slope of the curve is observed (i.e., the stress exponent increases steadily as stress increases) when minimum strain-rates are plotted versus modulus-compensated stress over a wide stress-range. Figure 87 illustrates a minimum strain rate vs. stress plot for a FL Ti-48Al-2Cr-2Nb at 760°C. The stress exponent varies from $n=1$, at low stresses, to $n=10$ at high stresses. Stress exponents as high as 20 have been measured at elevated stresses [621]. Third, the analysis of creep data is a very difficult task because of the complex microstructures of FL TiAl alloys. Microstructural parameters such as

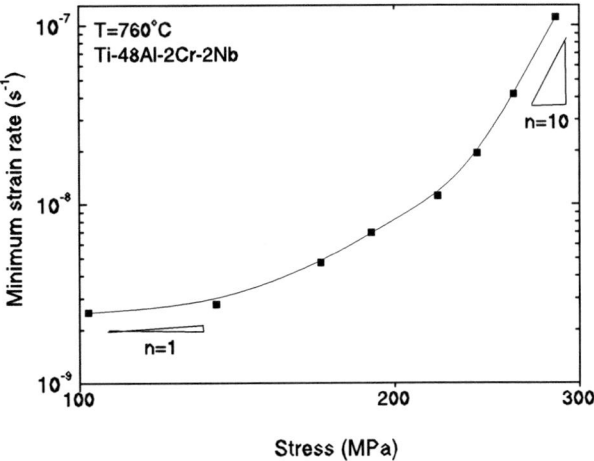

Figure 87. Minimum strain rate vs. stress curve of a Ti-48Al-2Cr-2Nb alloy deformed at 760°C (from [626]).

lamellar grain size (g_l), lamellar interface spacing (λ_l), lamellar orientation, precipitate volume fraction and grain boundary morphology all have significant influences on the creep properties that are difficult to incorporate into the traditional creep models that have been discussed earlier. Nevertheless, a variety of rate-controlling mechanisms of FL TiAl alloys has been proposed over the last few years.

a. High Stress-High Temperature Regime. Beddoes et al. [622] suggested, based on their own results and data from other investigators, that the gradual increase in the stress exponent with increasing stress might be due to changes in the creep mechanisms from diffusional creep at low stresses, to dislocation climb as the stress increases, and finally to power-law breakdown at very high stresses. (A note should be made of the fact that the creep data analyzed by Beddoes et al. originated from strain-rate change tests, rather than from independent creep tests). Thus, these investigators claim that dislocation climb would most likely be rate-controlling during creep of FL microstructures at stresses higher than about 200 MPa and temperatures higher than about 700°C. They suggested that this argument is consistent with the previous work on the creep of single phase γ-TiAl alloys by Wolfenstine and González-Doncel [634]. These investigators analyzed the creep data of Ti-50 at.%Al, Ti-53.4 at.%Al and Ti-49 at.%Al tested from 700°C to 900°C, and concluded that the creep behavior of these materials could be described by a single mechanism by incorporating a threshold stress. The stress exponent was found to be close to 5 and the activation energy equal to 313 kJ/mol, a value close to that for lattice diffusion of Ti in γ-TiAl (291 kJ/mole) [635]. Additionally, several other studies on creep of FL TiAl alloys reported activation energies of roughly 300 kJ/mole [635]. Therefore the creep of FL TiAl alloys appears to be controlled by lattice diffusion of Ti. The activation energy for lattice diffusion of Al in γ-TiAl has not been measured but it is believed to be significantly higher than that of Ti [636]. Es-Souni et al. [627] also suggested the predominance of a recovery-type dislocation-climb mechanism based on microstructural observations of the formation of dislocation arrangements (similar to subgrains) during creep. Several possible explanations have been suggested to reconcile the proposition of dislocation climb and the observation of high stress exponents ($n > 5$). First, it has been suggested [622] that back-stresses may arise within lamellar microstructures due to the trapping of dislocation segments at the lamellar interfaces, which leads to bowing of dislocations between interfaces. The shear stress required to cause bowing, which was suggested as a source of backstresses during creep, is inversely proportional to the lamellar interface spacing (λ_l). Second, it has been proposed [637] that the occurrence of microstructural instabilities such as dynamic recrystallization during deformation may contribute to a rise in the strain rate, thus rendering stress exponents with less physical meaning in terms of a single, rate-controlling restoration mechanism. Finally, it has been suggested [622] that the subgrain size corresponding to a specific creep stress if dislocation climb were rate-controlling could be larger than the lamellar spacing, λ_l, which remains constant with stress. Thus, λ_l may actually become the "effective subgrain size". These circumstances are similar to constant structure creep, which is associated with a relatively high stress exponent of 8 or higher.

Beddoes et al. [626] later more precisely delineated the stress range in which dislocation climb was rate controlling by performing stress reduction tests on FL TiAl alloys. Figure 88 illustrates the results of the reduction tests on a Ti-48Al-2Cr-2Nb alloy at 760°C, with an initial stress of 277 MPa. Data are illustrated for two different FL microstructures, both with a (lamellar) grain size of 300 μm, but with different lamellar interface spacing (120 nm and 450 nm). Deformation at a lower rate was observed upon reduction of the stress. An incubation period was observed for reduced stresses lower than a given stress (indicated with a dotted line) before deformation would continue. This was attributed [626] to the predominance of dislocation climb in the low stress regime (below the dotted line), and to the predominance of dislocation glide in the high stress regime (above the dotted line). The stress at which the change in mechanism occurs depends on the lamellar interface spacing. It was suggested that a decrease in the lamellar interface spacing results in an increase of the stress below which dislocation climb becomes rate-controlling.

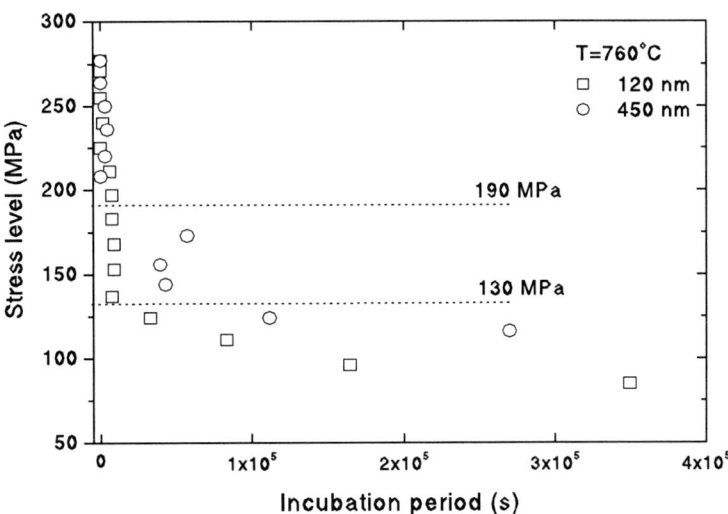

Figure 88. Incubation period following stress reduction tests to different final stresses (from [626]). Tests performed at 760°C in two Ti-48Al-2Cr-2Nb alloys with the same lamellar grain size (300 μm) but two different lamellar interface spacings (120 nm and 450 nm). Initial stress of 280 MPa.

Beddoes et al. [626] proposed an explanation for the decrease in the minimum creep rate with decreasing lamellar interface spacing for a given stress. First, for two microstructures deforming in the glide-controlled creep regime (for example, for stresses higher than 190 MPa in Fig. 88), narrower spacings would increase the creep resistance since the mean free path for dislocations would significantly decrease. The lamellae interfaces would thus act as obstacles for gliding dislocations. In fact, dislocation pile-ups have been observed at interfaces in FL structures [628]. Second, there is a stress range (for example, stresses between 130 MPa and 190 MPa in Fig. 88) for which the rate-controlling creep mechanism in microstructures with very narrow lamellae would be dislocation climb (associated with lower strain-rates) whereas in others with wider lamellae it would be dislocation glide. This was attributed to the different backstresses originating at lamellae of different thickness. A larger Orowan stress is necessary to bow dislocations in narrow lamellae than in wider lamellae. Thus, an applied stress of 130–190 MPa would be high enough to cause dislocation bowing in the material with wider lamellae. Dislocation glide would be controlled by the interaction between dislocations and interfaces rather than climb. However, in narrower lamellae, the applied stress is not sufficient to cause dislocation bowing and dislocation movement is then controlled by climb.

Recently Viswanathan et al. [638,639] studied the creep properties of a FL Ti-48Al-2Cr-2Nb alloy at high stresses (207 MPa) and high temperatures (around

800°C), particularly examining the dislocation structures developed during deformation. They mainly observed unit a/2 [110] dislocations with jogs pinning the screw segments. They found that a distribution of lamellae spacings exists in a FL microstructure. A higher dislocation density was observed in the wider lamellae, suggesting, to these investigators, that wider lamellae contribute more to creep strain than thinner lamellae. Additionally, no subgrains were observed at the minimum creep-rate (at strains of about 1.5% to 2%). The absence of subgrain formation during secondary creep of single phase TiAl alloys under conditions where an activation energy similar to that of self-diffusion was observed (thus suggesting the predominance of dislocation climb), was also reported in [636,640]. In order to rationalize this apparent discrepancy between the behavior of single phase Ti-Al alloys and pure metals, Viswanathan et al. [639] proposed a modification of the jogged-screw creep model discussed in a previous chapter. The original model [641,642] suggests that the non-conservative motion of pinned jogs along screw dislocations is the rate-controlling process. Using this model, the "natural" stress exponent is derived. The conventional jogged-screw creep model predicts strain-rates that are several orders of magnitude higher than the measured values in TiAl [639]. The modification proposed by Viswanathan et al. [639] incorporates the presence of tall jogs instead of assuming that the jog height is equal to Burgers vector. Additionally, it is proposed that there should be an upper bound for the jog height, above which the jog becomes a source of dislocations. This maximum jog height, h_d, depends on the applied stress and can be approximated by the following expression,

$$h_d = \left(\frac{Gb}{\{8\pi(1-\nu)\tau\}} \right) \tag{125}$$

This is suggested to reasonably predict the strain rates in single phase TiAl alloys and could account for the absence of subgrain formation during secondary creep. At the same time, by introducing this additional stress dependence in the equation for the strain rate, the phenomenological stress exponent of 5 is obtained at intermediate stresses. This exponent increases with increasing stress. Viswanathan et al. [638] claim that the same model can be applied to creep of FL microstructures, where deformation mainly occurs within the wider γ-laths by jogged a/2 [110] unit dislocation slip.

Wang et al. [643] observed that, together with dislocation activity and some twinning, thinning and dissolution of α_2 lamellae and coarsening of γ-lamellae occurred during creep of two FL TiAl alloys at high stresses (e.g. > 200 MPa at 800°C and > 400 MPa at 650°C). They proposed a creep model based on the movement of ledges (or steps) at lamellar interfaces to rationalize these observations. Wang et al. [643] observed the presence of ledges at the lamellar interfaces already before deformation. Two such ledges of height h_L are illustrated in Figure 89 (from [643]). Growth of the γ phase at the expense of the α_2 phase could occur by ledge movement as a consequence of the applied stress. Ledge motion was suggested to involve glide of misfit dislocations and climb of misorientation dislocations. Ledge motion leading to the transformation from α_2 to γ may account for a significant amount of the creep deformation since, first, as mentioned above, it requires dislocation movement and, second, it also involves a volume change from α_2 to γ? At high stresses, multiple ledges (i.e., ledges that are several {111} planes in thickness) are suggested to be able to form and dissolve and, thus, deformation may occur. Diffusion of atoms is needed since the climb of misorientation dislocations is necessary for a ledge to move. Also, diffusion is also needed for the composition change associated with the phase transformation from α_2 to γ. Thus, lattice self-diffusion becomes rate-controlling, in agreement with previous observations of activation energies close to Q_{SD}.

Alloy additions are another factor that may influence the creep rate are. It is well known that additions of W greatly improve creep resistance [623]. It has been suggested that solute hardening by W occurs within the glide-controlled creep regime, whereas the addition of W may lower the diffusion rate, thus reducing the dislocation climb-rate in the climb-controlled creep regime. The effect of ternary or quaternary additions on the creep resistance may be more important than that of the lamellar interface spacing [626]. Additions of W, Si, C, N favor precipitation hardening [644–647], which may hinder dislocation motion and stabilize the lamellar microstructure. Other suggested hardening elements are Nb and Ta [623]. The

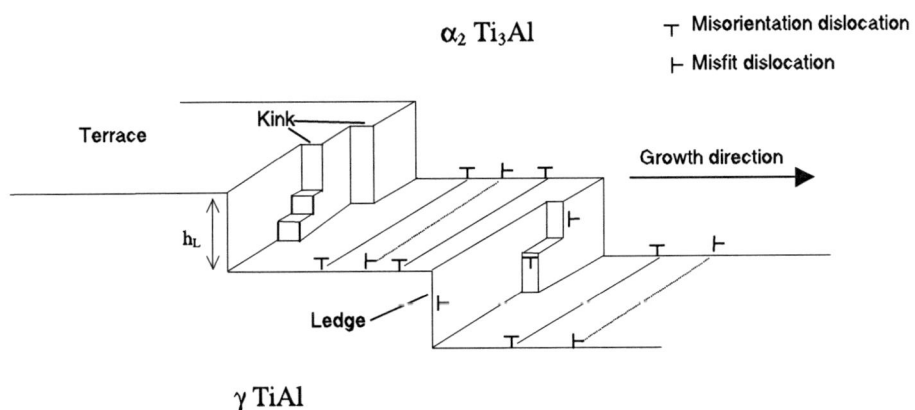

Figure 89. Interface separating γ and α_2 lamellae (from [643]). Ledge size is denoted by h_L.

addition of B does not seem to have any effect on the minimum strain-rate of FL microstructures [623].

The lamellar orientation also has a significant influence on the creep properties of FL TiAl alloys. Hard orientations (i.e., those in which the lamellae are parallel or perpendicular to the tensile axis) show improved creep resistance and low strain to failure soft orientations (those in which the lamellae form an angle of 30° to 60° with the tensile axis) are weaker but are more ductile [607]. The different behavior can be rationalized by considering changes in the Taylor factors and Hall-Petch strengthening [607]. Basically, in soft orientations the shear occurs parallel to the lamellar boundaries. In hard orientations, however, the resolved shear stress in the planes parallel to the lamellae is very low and therefore other systems are activated. Thus, the mean free path for dislocations is larger in soft orientations than in hard orientations [648–649].

b. Low Stress Regime. Hsiung and Nieh [629] investigated the rate-controlling creep mechanisms during "secondary creep" (or minimum creep-rate) at low stresses in a FL Ti-47Al-2Cr-2Nb alloy. In particular, they studied the stress/temperature range where stress exponents between 1 and 1.5 were observed. They reported an activation energy equal to 160 kJ/mol within this range, which is significantly lower than the activation energy for lattice diffusion of Ti in γ-TiAl (291 kJ/mole) [635] and much lower than the activation energy for lattice diffusion of Al in γ-TiAl. They suggested that dislocation climb is less important at low stresses. They also discarded grain boundary sliding as a possible deformation mechanism due to the presence of interlocking grain boundaries such as those shown in Fig. 85. These are boundaries in which there is not a unique boundary plane. Instead, the lamellae from adjacent (lamellar) grains are interpenetrating at the boundary, thus creating steps and preventing easy sliding [626]. TEM examination revealed both lattice dislocations (including those which are free within the γ-laths and threading dislocations which have their line ends within the lamellar interfaces) and interfacial (Shockley) dislocations, the density of the latter being much larger. They proposed that, due to the fine lamellar interface spacing (λ_l < 300 nm), the operation and multiplication of lattice dislocations at low stresses is very sluggish. Dislocations can only move small distances ($\approx \lambda_l$) and the critical stress to bow threading dislocation lines (which is inversely proportional to the lamellar interface spacing) is, on average, higher than the applied stress. Thus, Hsiung and Nieh [629] concluded that dislocation slip by threading dislocations could not rationalize the observed creep strain in alloys with thin laths. They proposed that the predominant deformation mechanism was interfacial sliding at lamellae interfaces caused by the viscous glide of interfacial (Shockley) dislocation arrays. These arrays might eventually be constituted by an odd number of partials, in which case a stacking fault is created at the interface.

Stacking faults are indeed observed by TEM [629]. Glide of pairs of partials would not create a stacking fault since the passage of the second partial along the fault created by the first one would regenerate it. Segregation of solute atoms may cause Suzuki locking. Thus, according to Hsiung and Nieh [629], the viscous glide of interfacial dislocations (dragged by solute atoms) is the rate-controlling mechanism. It was suggested that further reduction of the lamellar interface spacing (in the range of $\lambda_l \leq 300$ nm) would not significantly affect the creep rate once the γ-laths are thin enough for interfacial sliding to occur.

Zhang and Deevi [623] recently analyzed creep data of several TiAl alloys with Al concentrations ranging from 46 to 48 at.% compiled from numerous other studies. They proposed expressions relating the minimum creep-rate and the stress that could reasonably predict most of the data. They recognized that using the classical constitutive equations and power law models could be misleading, due to the large and gradual variations of the stress exponent with stress. Alternatively, they utilized,

$$\dot{\varepsilon}_{min} = \dot{\varepsilon} \sinh\left(\frac{\sigma}{\sigma_{int}}\right) \qquad (126)$$

where σ is the applied stress, and $\dot{\varepsilon}_0$ and σ_{int} are both temperature and materials dependent constants. Additionally,

$$\dot{\varepsilon}_0 \propto \rho_{sr} D_0 \exp = \left(\frac{-Q_{sd}}{kT}\right) \qquad (127)$$

where ρ_{sr} is the dislocation source density. The physical meaning of Eq. (127) is based on a viscous glide process. The modeling suggested [623] that σ_{int} and $\dot{\varepsilon}_0$ are independent of the lamellar interface spacing for FL microstructures with $\lambda_l > 0.3$ μm and that σ_{int} and $\dot{\varepsilon}_0$ increase with decreasing λ when $\lambda_l < 0.3$ μm $\dot{\varepsilon}_0$ is temperature dependent and Q_{sd} is about 375 kJ/mol for microstructures with different lamellar interface spacing. This value is slightly higher than the activation energy for diffusion of Ti in TiAl (291 kJ/mol). Zhang and Deevi [623] attributed this discrepancy to the fact that the dislocation source density, ρ_s, may not be constant as assumed in Eq. (127), and that the creep of TiAl may be controlled by diffusion of both Ti and Al in TiAl. Since the activation energy for self-diffusion of Al is higher than that of Ti, a combination of diffusion of both species could justify the higher Q values measured. In any case the activation energies have doubtful physical meaning since the stress exponent varies with stress.

However, the above equations do not fit the creep data of FL TiAl alloys obtained at both stresses lower than about 150 MPa and low temperatures. This was suggested to be due to grain boundary sliding being the dominant mechanism [629]. In this stress-temperature range, Zhang and Deevi found that most of the creep data could be described by:

$$\dot{\varepsilon}_{min}(GB) = 63.4 \exp(-2.18 \times 10^5/kT)g^{-2}\sigma^2 \qquad (128)$$

Despite being a thorough overview of creep results, this paper [623] works out well in bringing out trends but fails in the modeling aspects.

9.2.3 Primary Creep in FL Microstructures

γ-TiAl alloys are characterized by a pronounced primary creep regime. Depending on the temperature, the primary creep strain may exceed the acceptable limits for certain industrial applications. Thus, several investigations have focused on understanding the microstructural evolution during primary creep [626,631,637, 648,650,651].

Figure 90 (from [626]) illustrates the creep curves corresponding to a TiAl binary alloy deformed at 760°C and at an applied stress of 240 MPa. The creep curves correspond to a duplex (DP) microstructure and a fully lamellar (FL) microstructure. The FL microstructure shows lower strain-rates during primary creep. It has been suggested [637] that the pronounced primary creep regime in FL microstructures is due to the presence of a high density of interfaces and dislocations, since both may act as sources of dislocations. Careful TEM examination by Chen et al. [648] showed that dislocations formed loops that expand from the interface to the next lamellar interface. Other processes that may occur during primary creep

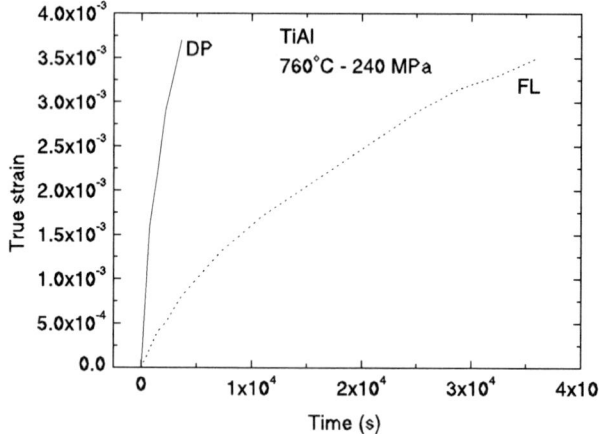

Figure 90. Primary creep behavior of a binary TiAl alloy deformed at 760°C at an applied stress of 240 MPa. The creep curves corresponding to duplex (DP) and fully lamellar (FL) microstructures are illustrated [626].

of TiAl alloys are twinning and stress-induced phase transformations (SIPT) ($\alpha_2 \to \gamma$ or $\gamma \to \alpha_2$). SIPT consists of the transformation of α_2 laths into γ laths (or vice versa). This transformation, which is aided by the applied stress, has been suggested to occur by the movement of ledge dislocations at the γ/α_2 interfaces, as illustrated in Fig. 89 [646]. The SIPT may be associated with a relatively large creep strain. The finding that the primary strain in microstructures with narrow lamellae (FLn) is not higher than the primary strain in alloys with wider lamellae (FLw) suggests that the contribution of interface boundary sliding by the motion of pre-existing interfacial dislocations [629] is less important [631]. It is possible that a sufficient number of interfacial dislocations need to be generated during primary creep before the onset of sliding.

Zhang and Deevi recently analyzed primary creep of TiAl based alloys [631] and concluded that the primary creep strain depends dramatically on stress. At stresses lower than a critical value σ_{cr}, the primary strain is low (about 0.1–0.2%), independent of the microstructure, temperature and composition. The relevant stresses anticipated for industrial applications are usually below σ_{cr} and therefore primary creep strain would be of less concern [631]. σ_{cr} seems to be mainly related to the critical stress to activate dislocation sources, twinning and stress-induced phase transformations. This value increases with W additions, with lamellar refinement and with precipitation of fine particles along lamellar interfaces [631]. For example, the value of σ_{cr} at 760°C for a FL Ti-47 at.%Al-2 at.%Nb-2 at.%Cr alloy with a lamellar spacing of 0.1 µm is close to 440 MPa, whereas the same alloy with a lamellar spacing larger than 0.3 µm has a σ_{cr} value of 180 MPa [631]. Primary creep strain increases significantly above the threshold stress. In order to investigate additional factors influencing the primary creep strain, Zhang and Deevi modeled primary creep of various TiAl alloys using the following expression, also utilized previously in other works [622,650],

$$\varepsilon_p = \varepsilon'_0 + A'(1 - \exp(-\alpha' t)) \qquad (129)$$

This expression reflects that primary creep strain consists of an "instantaneous" strain (ε'_0) that occurs immediately upon loading and a transient strain that is time dependent. Zhang and Deevi suggested that the influence of temperature on ε'_0, A', and α' could be modeled using the relation $X = X_0 \exp(-Q/RT)$, where X represents any of the three parameters. They obtained that $Q = 190$ kJ/mol for A', and $Q = 70$ kJ/mol for ε'_0 and α'. The physical basis for the temperature dependence is unclear. The effect of composition and microstructure on these parameters is also complex [631]. It is accepted that aging treatments before creep deformation have a beneficial effect in increasing the primary creep resistance [631,646,651]. Precipitation at lamellar interfaces has been suggested to hinder dislocation generation [631] and reduces the "instantaneous" strain and the strain hardening constant, A'. Additionally, the presence of fine precipitates may inhibit interface sliding and even twinning. Finally, the contribution of stress-induced phase transformation to the primary creep strain decreases in samples heat treated before creep deformation since metastable phases are eliminated [631].

9.2.4 Tertiary Creep in FL Microstructures

Several investigations have studied the effect of the microstructure on tertiary creep of FL TiAl alloys [622,626,648]. It has been suggested that tertiary creep is initiated due to strain incompatibilities between lamellar grains with soft and hard orientations leading to particularly elevated stresses [626]. These incompatibilities may lead to intergranular and interlamellar crack formation. Crack growth is prevented when lamellar grains are smaller than 200 μm (lamellar grain sizes are typically 500 μm in diameter), since cracks can be arrested by grain boundaries or by triple points. Thus, the creep life is improved due to an increase of the extent of tertiary creep. Grain boundary morphology also significantly influences tertiary creep behavior. In FL microstructures with wide lamellae, the latter bend close to grain boundaries and form a well interlocked lamellae network [626]. However, narrow lamellae are more planar. Grain boundaries in which lamellae are well interlocked offer greater resistance to cracking and allow larger strains to accummulate within the grains.

In summary, the creep behavior of FL TiAl alloys is influenced by many different microstructural features and it is difficult to formulate a model that incorporates all of the relevant variables. An optimum creep behavior may require [626]:

(a) a lamellar or colony grain size smaller than about 200 μm that helps to improve creep life by preventing early fracture.
(b) a narrow interlamellar spacing, that reduces the minimum strain rate during secondary creep.
(c) interlocked lamellar grain boundaries.
(d) stabilized microstructure.
(e) presence of alloying additions such as W, Nb, Mo, V (solution hardening) and C, B, N (precipitation hardening).

9.3 IRON ALUMINIDES

9.3.1 Introduction

Fe_3Al and FeAl based ordered intermetallic compounds have been extensively studied due to their excellent oxidation and corrosion resistance as well as other favorable properties such as low density, favorable wear resistance, and potentially lower cost than many other structural materials. Fe$_3$Al has a DO$_3$ structure. FeAl is a B2 ordered intermetallic phase with a simple cubic lattice with two atoms per lattice site, an Al atom at position (x, y, z) and a Fe atom at position $(x + 1/2, y + 1/2, z + 1/2)$. Several recent reviews summarizing the physical, mechanical and corrosion properties of these intermetallics are available [598,600,601,603,605–607,652–654]. Iron aluminides are especially attractive for applications at intermediate temperatures in the automotive and aerospace industry due to their high specific strength and stiffness. Additionally, they may replace stainless steels and nickel alloys to build long-lasting furnace coils and heat exchangers due to superior corrosion properties. The principal limitations of Fe-Al intermetallic compounds are low ambient temperature ductility (due mainly to the presence of weak grain boundaries and environmental embrittlement) and only moderate creep resistance at high temperatures [654]. Many efforts have been devoted in recent years to overcome these difficulties. The present review will discuss the strengthening mechanisms of Iron aluminides as well as other high temperature mechanical properties of these materials.

Figure 91 illustrates the Fe-Al phase diagram. The Fe$_3$Al phase, with a DO$_3$ ordered structure, corresponds to Al concentrations ranging from approximately 22 at.% and 35 at.%. A phase transformation to an imperfect B2 structure takes place above 550°C. The latter ultimately transforms to a disordered solid solution

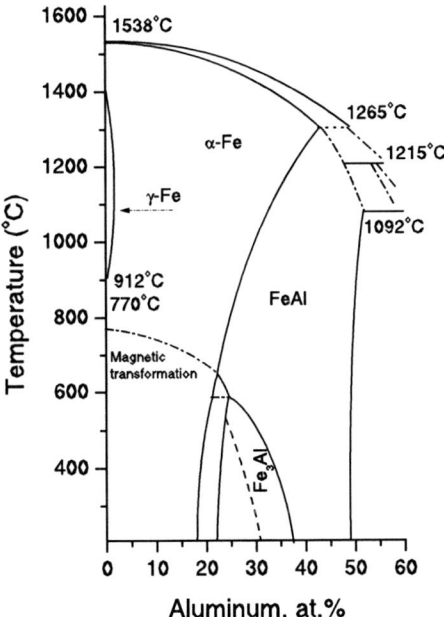

Figure 91. Fe-Al phase diagram [655].

with increasing temperature. This, in turn, leads to the degradation of creep and tensile resistance at high temperatures. The FeAl phase, which has a B2 lattice, is formed when the amount of Al in the alloy is between 35 at.% and 50 at.%.

9.3.2 Anomalous Yield Point Phenomenon

An anomalous peak in the variation of the yield stress with temperature has been observed in Fe-Al alloys with an Al concentration ranging from 25 at.% up to 45at.%. The peak appears usually at intermediate temperatures, between 400°C and 600°C [607,653–654,656–676]. This phenomenon is depicted in Figure 92 (from [661]), which illustrates the dependence of the yield strength with temperature for several large-grain Fe-Al alloys in tension at a strain rate of $10^{-4} s^{-1}$. Several mechanisms have been proposed to rationalize the yield-strength peak but the origin of this phenomenon is still not well understood. The different models are briefly described in the following. More comprehensive reviews on this topic as well as critical analyses of the validity of the different strengthening mechanisms are [653,656].

a. Transition from Superdislocations to Single Dislocations. Stoloff and Davies [662] suggested that the stress peak was related to the loss of order that occurs in Fe_3Al alloys at intermediate temperatures (transition between the DO_3 to the B2 structure). According to their model, at temperatures below the peak, superdislocations would lead to easy deformation, whereas single dislocations would, in turn, lead to easy deformation at high temperatures. At intermediate temperatures, both

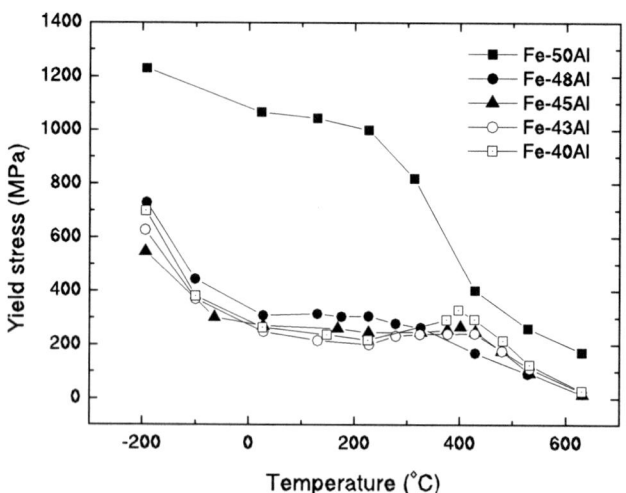

Figure 92. Variation of the yield strength with temperature for several large grain FeAl alloys strained in tension at a strain rate of 10^{-4} s^{-1} (from [661]).

superdislocations and single dislocations would move sluggishly, giving rise to the strengthening observed. It has been suggested, however, that this model cannot explain the stress peak observed in FeAl alloys, where no disordering occurs at intermediate temperatures and the B2 structure is retained over a large temperature interval. Recently, Morris et al. also questioned the validity of this model [659]. They observed that the stress peak occurred close to the disordering temperature at low strain-rates in two Fe_3Al alloys (Fe-28Al-5Cr-1Si-1Nb-2B and Fe-25Al, all atomic percent). For high strain-rates the stress peak occurred at higher temperatures than that corresponding to the transition. Thus, they concluded that the disordering temperature being about equal to the peak stress temperature at low strain rates was coincidental. Stein et al. [676] also did not find a correlation between these temperatures in several binary, ternary and quaternary DO_3-ordered Fe-26Al alloys.

b. Slip Plane Transitions {110} → {112}.   Umakoshi et al. [663] suggested that the origin of the stress peak was the cross slip of ⟨111⟩ superdislocations from {110} planes to {112} planes, where they become pinned. Experimental evidence supporting this observation was also reported by Hanada et al. [664] and Schroer et al. [665]. Since cross slip is thermally activated, dislocation pinning would be more pronounced with increasing temperatures, giving rise to an increase in the yield strength.

c. Decomposition of ⟨111⟩ Superdislocations: Climb Locking Mechanism.   The yield stress peak has often been associated with a change in the nature of dislocations responsible for deformation, from ⟨111⟩ superdislocations (dislocations formed by pairs of superpartial dislocations separated by an antiphase boundary) at low temperatures, to ⟨100⟩ ordinary dislocations at temperatures above the stress peak [666]. The ⟨100⟩ ordinary dislocations are sessile at temperatures below those corresponding to the stress peak and thus may act as pinning points for ⟨111⟩ superdislocations. The origin of the ⟨100⟩ dislocations has been attributed to the combination of two a/2[111] superdislocations or to the decomposition of ⟨111⟩ superdislocations on {110} planes into ⟨110⟩ and ⟨100⟩ segments on the same {110} planes. As the temperature increases, decomposition may take place more easily, and thus the amount of pinning points would increase leading to a stress peak. Experimental evidence consistent with this mechanism has been reported by Morris and Morris [667]. However, this mechanism was also later questioned by Morris et al. [659], where detailed TEM microstructural analysis suggested that anomalous strengthening is possible without the ⟨111⟩ to ⟨100⟩ transition in some Fe_3Al alloys.

d. Pinning of ⟨111⟩ Superdislocations by APB Order Relaxation.   An alternative mechanism for the appearance of the anomalous yield stress peak is the loss of order within APBs of mobile ⟨111⟩ superdislocations with increasing temperature [668]. This order relaxation may consist of structural changes as well as variations in the chemical composition. Thus, the trailing partial of the superdislocation would no longer be able to restore perfect order. A frictional force would therefore be created that will hinder superdislocation movement. With increasing temperature APB relaxation would be more favored and thus increasing superdislocation pinning would take place, leading to the observed stress peak.

e. Vacancy Hardening Mechanism. The concentration of vacancies in FeAl is in general very high, and it increases in Al-rich alloys. Constitutional vacancies are those required to maintain the B2 structure in Al-rich non-stoichiometric FeAl alloys. Thermal vacancies are those excess vacancies generated during annealing at high temperature and retained upon quenching. For example, the vacancy concentration is 40 times larger at 800°C than that corresponding to a conventional pure metal at the melting temperature. Constitutional vacancies may occupy up to 10% of the lattice sites (mainly located in the Fe sublattice) for Fe-Al compositions with high Al content (> 50% at.) [669]. The high vacancy concentration is due to the low value of the formation enthalpy of a vacancy as well as to the high value of the formation entropy (around 6 k) [670]. Vacancies have a substantial influence on the mechanical properties of iron aluminides [671].

It has been suggested that the anomalous stress peak is related to vacancy hardening in FeAl intermetallics [672]. According to this model, with rising temperatures a larger number of vacancies are created. These defects pin superdislocation movement and lead to an increase in yield strength. At temperatures higher than those corresponding to the stress peak the concentration of thermal vacancies is very large and vacancies are highly mobile. Thus, they may aid dislocation climb processes instead of acting as pinning obstacles for dislocations [673] and softening occurs.

The vacancy hardening model is consistent with many experimental observations. Recently Morris et al. [658] reported additional evidence for this mechanism in a Fe-40 at % Al alloy. First, they observed that some time is required at high temperature for strengthening to be achieved. This is consistent with the need of some time at high temperature to create the equilibrium concentration of vacancies required for hardening. Second, they noted that the stress peak is retained when the samples are quenched and tested at room temperature. They concluded that the point defects created after holding the specimen at high temperature for a given amount of time are responsible for the strengthening, both at high and low temperatures. Additionally, careful TEM examination suggested that vacancies were not present in the form of clusters. The small dislocation curvature observed, instead, suggested that single-vacancies were mostly present, which act as relatively weak obstacles to dislocation motion.

The concentration of thermal vacancies increases with increasing Al content and thus the effect of vacancy hardening would be substantially influenced by alloy composition. Additionally, the vacancy hardening mechanism implies that the hardening should be independent of the strain rate, since it only depends on the amount of point defects present. In a recent investigation, Morris et al. [659] analyzed the effect of strain rate on the flow stress of two Fe_3Al alloys with compositions Fe-28Al-5Cr-1Si-1Nb-2B and Fe-25Al-5Cr-1Si-1Nb-2B (at.%). The variation of the flow stress with temperature and strain rate (ranging from $4 \times 10^{-6} s^{-1}$ to $1\, s^{-1}$) for the Fe-28 at.%Al alloy is illustrated in Fig. 93. It can be observed that the "strengthening" part of the peak is rather insensitive to strain rate, consistent with the predictions of the vacancy hardening model. However, the investigators were skeptical regarding the effectiveness of this mechanism in Fe_3Al alloys, where the vacancy concentration is much lower than in FeAl alloys and, moreover, where the vacancy mobility is higher. Highly mobile vacancies are not as effective obstacles to dislocation motion. Another limitation of the vacancy model is that it fails to explain the orientation dependence of the stress anomaly as well as the tension-compression asymmetry in single crystals [657]. Thus, the explanation of the yield stress peak remains uncertain. On the other hand, it is evident in Fig. 93 that the softening part of the peak is indeed highly dependent

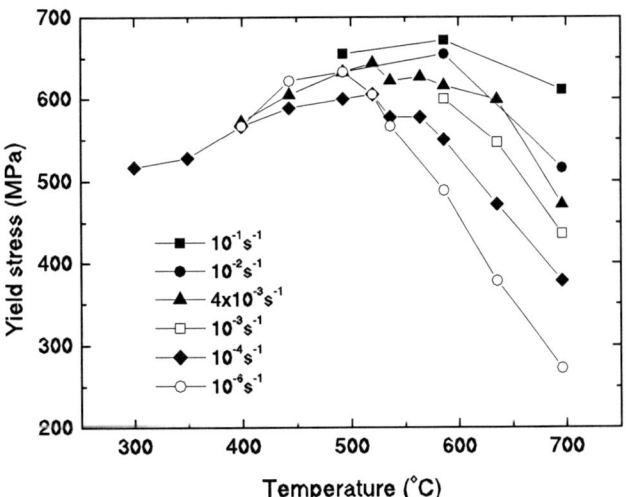

Figure 93. Variation of the yield stress with temperature and strain rate corresponding to the cast and homogeneized Fe-28 at%Al alloy (from [659]).

on strain rate. This is attributed to the onset of diffusional processes at high temperatures, where creep models may be applied and rate dependence may be more substantial [657,659].

9.3.3 Creep Mechanisms

The creep behavior of iron aluminides is still not well understood despite the large amount of creep data available on these materials [e.g. 654,677–689]. The values of the stress exponents and activation energies corresponding to several creep studies are summarized in Table 6 (based on [654] with added data). There are several factors that complicate the formulation of a general creep behavior of Fe-Al alloys. First, creep properties are significantly influenced by composition. Second, as discussed in [600,690], both the stress exponent and the activation energy have been observed to depend on temperature, in some cases. This suggests that several mechanisms may control creep of Fe-Al and Fe_3Al alloys. Other reasons may be the frequent absence of genuine steady-state conditions as well as the simultaneous occurrence of grain growth and discontinuous dynamic recrystallization. Nevertheless, in general terms, it can be inferred from Table 6 that diffusional creep or Harper-Dorn creep may be the predominant deformation mechanism at very low stresses and high temperatures. At intermediate temperatures and stresses, either diffusion controlled dislocation climb or viscous drag have been suggested [654,678,679,685].

a. Superplasticity in Iron Aluminides. Superplasticity has been observed in both FeAl and Fe_3Al with coarse grains ranging from 100 μm to 350 μm [691–695]. Elongations as high as 620% were achieved in a Fe-28 at%Al-2 at%Ti alloy deformed at 850°C and at a strain rate of $1.26 \times 10^{-3} s^{-1}$. The corresponding n value was equal to 2.5. Also, a maximum elongation of 297% was reported for a Fe-36.5 at.%Al-2 at.%Ti alloy with an n value close to 3. Moreover, Lin et al. [695] have reported an increasing number of boundaries misoriented between 3° and 6° with deformation. They suggest that these could be formed as a consequence of dislocation interaction, by a process of continuous recrystallization. The unusually large starting grain sizes as well as the values of the stress exponents (close to 3) would be consistent with a viscous drag deformation mechanism. However, significant grain refinement has been observed during deformation [694–695]. The correlation of grain refinement and large ductilities may, in turn, be indicative of the occurrence of grain boundary sliding, to some extent. Grain boundary sliding is favored when a large area fraction of grain boundaries is present and thus occurs readily in fine-grained microstructures.

Table 6. Stress exponents, activation energies, and suggested deformation mechanisms from various creep studies on iron aluminides [654,677–686].

Alloy	T (°C)	Q (kJ/mol)	n	Mechanism suggested	Ref.
Fe-19.4Al	500–600	305	4.6–6*	Diffusion controlled	[677]
Fe-27.8Al	550–615	276	—	Controlled by state of order	
	Higher T	418			
Fe-15/20Al	>500	260 to 305*		Diffusion controlled	[678]
	<500	σ dependent		Motion of jogged screw dislocations	
Fe-28Al	625	347	3.5 (low σ)	Viscous glide	
			7.7 (high σ)	Climb	
Fe-28Al-2Mo	650	335	1.4 (low σ)	Diffusional flow	[679]
			6.8 (high σ)	Climb	
Fe-28Al-1Nb-0.013Zr	650	335	1.8 (low σ)	Diffusional flow	
			19.0 (high σ)	Dispersion strengthening	
FA-180	593	627	7.9	Precipitation strengthening	[680]
Fe-28Al	600–675	—	3.4	Viscous glide	[554]
Fe-26Al-0.1C	600–675	305	3.0	Viscous glide	
	480–540	403	6.2	—	
Fe-28Al-2Cr	600–675	325	3.7	Viscous glide	
Fe-28Al-2Cr-0.04B	600–675	304	3.7	Viscous glide	
Fe-28Al-4Mn	600–675	302	2.6	Viscous glide	
FA-129	500–610	380–395	4–5.6		[681]
Fe-24Al-0.42 Mo-0.05B-0.09 C-0.1 Zr	650–750	—	5.5		[682]
FA-129	900–1200	335	4.81		[683]
Fe-39.7Al-0.05Z-50ppmB	500	260–300	11	Dispersion strengthening	[684]
	700	425–445	11	Climb	
Fe-27.6Al Fe-28.7Al-2.5Cr	425–625	375	2.7–3.4	Viscous glide	[685]
Fe-27.2Al-3.6Ti	425–625	325	3.5–3.8	Viscous glide	
	425–625	375	3.4–3.7	Viscous glide	

Continued

Table 6. Continued.

Alloy	T (°C)	Q (kJ/mol)	n	Mechanism suggested	Ref.
Fe-24Al-0.42Mo-0.1	800–1150	340–430	4–7	Diffusion-controlled (climb)	[686]
Zr-0.005B-0.11C-0.31O	1150 (Strain rate < $0.1 s^{-1}$)	365	3.3	Diffusion-controlled (superplasticity)	
Fe-30.2Al-3.9Cr-0.94Ti-1.9B-0.20Mn-0.16C	600–900	280	3.3	Viscous glide	[687]
Fe-47.5Al	827–1127 (g = 36 μm)	487	6.3–7.2		[688]
Fe-43.2Al	927–1127 g = 20 μm	368	5.6–9.7		[688]

*Dependent on Al concentration

9.3.4 Strengthening Mechanisms

The rather low creep strength of iron aluminides is an area that has received particular attention. Several strategies to increase the creep resistance have been suggested, that are reviewed in [690]. One possibility is the reduction of the high diffusion coefficient of the rate-controlling mechanism, which has been attempted by micro- and macroalloying with limited success. Another way to achieve strengthening is to add alloying elements that may hinder dislocation motion by forming solute atmospheres around dislocations or by modifying lattice order. For example, additions of Mn, Co, Ti and Cr moderately increase the creep resistance due to solid solution strengthening. Finally, the most promising strengthening mechanism seems to be the introduction of dispersions of second phases, such as carbides, intermetallic particles or oxide dispersions. Dislocation movement is hindered due to the attractive dislocation-particle interactions and thus the creep rate may be reduced. A sufficient volume fraction of precipitate phases (around 1–3%) must be present in order for this mechanism to be effective and the precipitates should be stable at the service temperatures. Alloying elements such as Zr, Hf, Nb, Ta and B have been effective in improving creep resistance of FeAl by precipitation hardening. Dispersoid particles also strengthen effectively iron aluminides [696–698].

Baligidad et al. [698–699] have reported that improved creep strength is obtained in a Fe-16 wt.%Al-0.5 wt.%C possibly due to combined carbon solid-solution strengthening and mechanical constraint from the $Fe_3AlC_{0.5}$ precipitates. They claim that creep is recovery controlled and that climb assists the recovery. Morris-Muñoz [700] analyzed the creep mechanisms in an oxide-dispersion-strengthened Fe-40 at.%Al intermetallic containing Y_2O_3 particles at 500°C and 700°C. The absence of substructure formation at either temperature suggested, to the investigators, constant-structure creep with a temperature-dependent threshold stress. Particle-dislocation interactions were also apparent. It was concluded that the threshold stress based on particle-dislocation interactions operates at 500°C (where dislocations are predominantly $\langle 111 \rangle$ superdislocations) and that climb-controlled processes occur at 700°C, where $\langle 100 \rangle$ dislocations are mainly present. The decrease in creep resistance observed between 500°C and 700°C was attributed to the rapid increase in diffusivity at high temperatures. Recently Sundar et al. [701] reported creep resistance values for two Fe-40 at.%Al alloys (with additions of Mo, Zr, and Ti for solute strengthening and additions of C and B for particle strengthening) that were comparable to, if not better than, those of many conventional Fe-based alloys. According to Sundar et al. [701], a combination of strengthening mechanisms is perhaps the best way to improve creep resistance of Iron aluminides.

9.4 NICKEL ALUMINIDES

9.4.1 Ni₃Al

The Ni-Al binary phase diagram is illustrated in Fig. 94. Ni$_3$Al forms at Al concentrations between 25 at.% and 27 at.%. This compound has a simple cubic Bravais lattice with four atoms per lattice site: one Al atom, located in the (x, y, z) position, and three Ni atoms, located, respectively, at the $(x + 1/2, y, z)$, $(x, y + 1/2, z)$, and $(x, y, z + 1/2)$ positions. (This intermetallic received substantial attention since it is the main strengthening phase in superalloys. Furthermore it has been considered to be a technologically important structural intermetallic alloy system especially after its successful ductilization by microalloying with boron [702]. Additionally, it exhibits a flow stress anomaly, i.e., the yield stress increases with increasing temperature over intermediate temperatures (230°C–530°C) as with iron aluminides. Thus, it has often been used as a model material for understanding intermetallic compounds in general. Commercialization of Ni$_3$Al for selected applications is underway [605,702].

The crystal structure of Ni$_3$Al is an ordered L1$_2$ (f.c.c.) structure having Al atoms at the unit cell corners and Ni atoms at the face centers. Similar to pure f.c.c. metals, the planes of easy glide are the octahedral planes {111}. Slip along {001} planes is

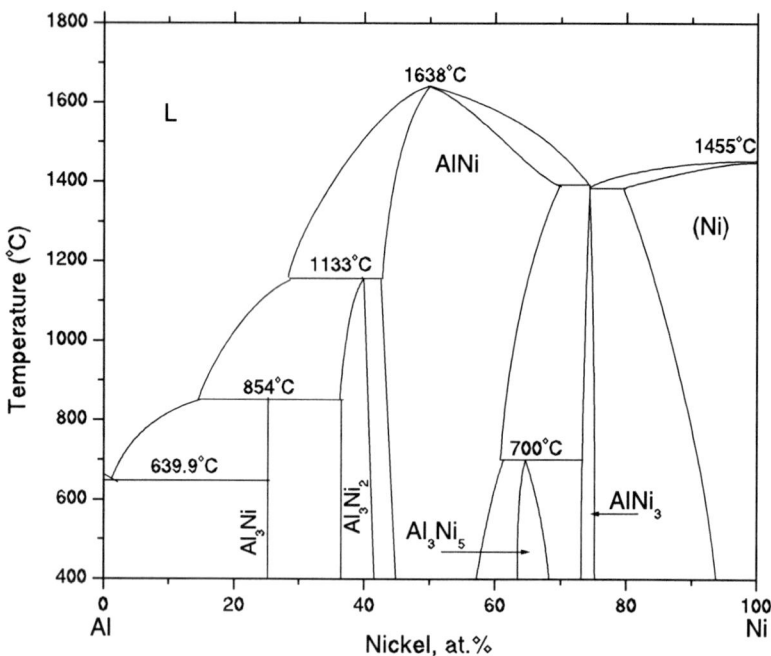

Figure 94. Binary Ni-Al phase diagram [655].

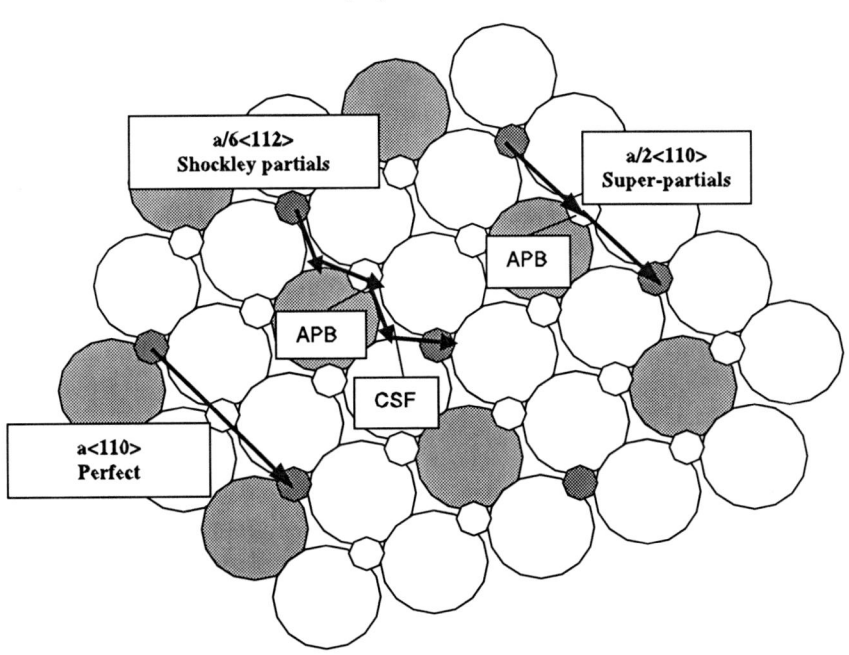

Figure 95. The octahedral {111} plane of the L1$_2$ crystal structure. The small circles are atoms one plane out (above) of the page. Unit dislocations, b = a⟨110⟩, can dissociate into superpartial dislocations with b = a/2⟨110⟩. The superpartials alter the neighboring lattice positions creating an antiphase boundary (APB). Superpartials can dissociate into Shockley partial dislocations, with b = a/6⟨112⟩, that are connected by a complex stacking fault (CSF) that includes both an APB and an ordinary stacking fault [703].

more difficult, since they are less compact, but it may occur by thermal activation [703]. Figure 95 illustrates an octahedral plane of this ordered structure. A perfect dislocation associated with primary octahedral glide has a Burgers vector, b = a ⟨110⟩, that is twice as large as that corresponding to a unit dislocation in the disordered f.c.c. lattice, and is termed "superdislocation." These dissociate into "super-partial" dislocations, with b = a/2 ⟨110⟩, and the latter may, in turn, dissociate into Shockley dislocations, with b = a/6 ⟨112⟩, as depicted in Fig. 95.

The creep behavior of Ni$_3$Al has drawn relatively little interest, perhaps due to inadequate creep strength as compared to that of commercial nickel-based superalloys and other intermetallic alloy systems such as NiAl and TiAl. The creep behavior of Ni$_3$Al will be briefly reviewed in the following.

a. Creep Curves. Creep tests have been performed on both single crystal [703–718] and polycrystalline [715–717,719–733] Ni$_3$Al alloys in tension and compression.

Most creep curves exhibit a normal shape, which consists of the conventional three stages. However, some [e.g. 719–721] show sigmoidal creep, where the creep-rate decreases quickly to a minimum and this is followed by a continuous increase in the creep-rate with strain. A steady state may or may not be achieved before reaching tertiary creep after Sigmoidal creep. This creep behavior is also frequently termed "inverse creep" among the intermetallics community. Primary creep is often limited to very small strains [612,708,718]. Figure 96 illustrates the creep curve of a Ni_3Al alloy (with 1 at.% Hf and 0.24 at.% B) deformed at 643°C at a constant stress of 745 MPa [703]. Initially, the creep rate decreases with increasing strain and normal primary creep occurs. This is followed by an extended region where the strain rate continually increases with strain. Steady state may or may not be reached afterwards, as will be explained later.

High-temperature creep refers to creep deformation at temperatures higher than T_p, the temperature at which the peak yield stress is observed. This temperature varies with alloy composition [732] and crystal orientation [704] but is typically observed from 0.5 to 0.6 T_m. Intermediate temperature creep usually refers to creep deformation at temperatures lower than (but close to) T_p. A steady-state regime is usually observed during high-temperature creep. The relationship between the strain rate and the stress in the high-temperature range usually follows a power-law relationship with a stress exponent of about 3. This may suggest that the viscous glide of dislocations is the rate-controlling mechanism [600]. However, at

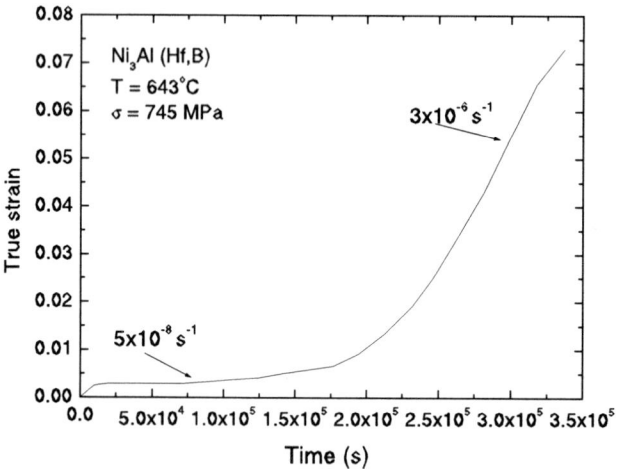

Figure 96. Sigmoidal creep curve corresponding to a Ni_3Al alloy (with 1 at.%Hf and 0.24 at.%B) deformed at 643°C at a constant stress of 745 MPa [703]. Normal primary creep is followed by a continuous increase in the strain rate. This creep behavior has been also termed "inverse creep".

intermediate temperatures, from about $0.3\,T_m$ to $0.6\,T_m$, sigmoidal creep may occur depending on both the temperature and the stress [720,721]. Nicholls and Rawlings [724] suggested that different creep mechanisms should operate below and above T_p.

b. Sigmoidal (or Inverse) Creep. Sigmoidal creep has been observed in both single- and polycrystalline Ni_3Al alloys [703,710,712,719–721], as well as in some other intermetallics, e.g. Ni_3Ga [735] and TiAl [721]. The onset of sigmoidal creep (i.e., the increase in the creep-rate after primary creep) usually takes place at very small strains. This strain-rate increase may extend over a large strain interval, leading directly to tertiary creep in the absence of a steady-state stage [703,710,719], as illustrated in Fig. 96, or it may occur only for a small strain previous to the steady state [705,720].

The conditions under which sigmoidal creep occurs are relatively narrow. Rong et al. [720] concluded that its occurrence depends on both temperature and stress. It is generally accepted that sigmoidal creep is more frequent and more pronounced at intermediate temperatures [721]. Smallman et al. [721] suggested that sigmoidal creep only occurs at temperatures below but very close to T_p and at stresses close to the yield stress.

Hemker et al. [703] and, previously, Nicholls and Rawling [724], observed a decrease of creep strength with increasing temperature in a single crystal alloy, with composition Ni-22.18 at.%Al-1 at.%Hf-0.24 at.%B [703], in the temperature regime where the yield strength is known to increase anomalously with temperature. This observation led them to infer that different dislocation mechanisms would be responsible for yielding (small strains) and for creep (large strains) and stimulated a detailed investigation of the deformation mechanisms. It is now well established that the yield stress anomaly in Ni_3Al is due to the operation of octahedral slip of $\langle 110 \rangle$ superdislocations that are retarded by cross-slip onto a cube plane at temperatures below T_p [598]. Slip on cube planes is thermally activated and is not favored at temperatures below T_p. However, superdislocations gliding on $\{111\}$ planes have a tendency to cross slip onto cube planes due to the presence of a torque caused by their anisotropic elastic fields. When the screw segment of a superdislocation is cross slipped onto a cube plane its mobility decreases significantly and this leads to an increase of the yield stress. Slip on cube planes becomes thermally activated and takes place easily at temperatures higher than T_p, leading to a decrease in the yield stress. Hemker et al. [703] proposed a model to explain the Sigmoidal creep of Ni_3Al based on careful microstructural examination. They suggested that octahedral slip during primary creep is exhausted by the formation of Kear-Wilsdorf (KW) locks, due to thermally activated cube cross-slip of the screw segments. Thus the strain rate is progressively reduced until the cross-slipped segments become thermally activated and are able to bow out and glide on the cube cross-slip plane. The KW locks act as Frank-Read type dislocation sources for glide on the {001} cube planes. The dislocation generation and subsequent glide on the cube planes leads to an increasing mobile dislocation density and thus to a larger strain rate and sigmoidal creep occurs. An alternative dislocation model for sigmoidal creep was proposed by Hazzledine and Schneibel [736]. They suggested that two highly stressed octahedral slip systems which share a common cube cross-slip plane, may interact "symbiotically" and unlock each other's superdislocations giving rise to an increasing number of ⟨001⟩ dislocations that are glissile on the cube plane. Thus, sigmoidal creep occurs.

Smallman et al. [721] pointed out that cube cross-slip is a necessary but not sufficient condition for sigmoidal creep. The operation of this mechanism, which leads to a strain rate increase under some conditions, is compensated by the strain rate decrease due to the exhaustion of dislocations on the octahedral slip systems. In fact, Smallman et al. [721] observed cube cross-slip in a polycrystalline Ni_3Al alloy creep deformed at 380°C ($T \ll T_p$), where sigmoidal creep was not apparent. Zhu et al. [704] also reported cube cross-slip in the absence of sigmoidal creep in single crystals of Ni_3Al with different orientations. In order to rationalize these observations, Rong et al. [720] suggested that sigmoidal creep would occur only when the length of a significant number of screw segments cross-slipped onto cube planes from octahedral planes, as suggested by Hemker et al. [703], is larger than a critical value. In this case, the density of mobile dislocations on the cube cross-slip planes would increase significantly, leading to an increase in the creep-rate. Rong et al. [720] also observed an anomalous temperature dependence of the creep strength in a polycrystalline Ni_3Al alloy, contrary to what Hemker et al. [703] had reported for their single crystal alloy. Rong et al. [720] found this anomalous dependence consistent with their TEM observations of a larger density of dislocations on cube cross-slip planes at the lower temperatures. They suggested that the average length of the screw segments on cube cross-slip planes would increase with decreasing temperature. Thus, at low temperatures, there would be more dislocations with lengths larger than the critical value or a higher mobile dislocation density and this would lead to a lower creep resistance.

The occurrence of sigmoidal creep has not only been found to depend on temperature and stress, but also on the prior deformation and processing history. For example, sigmoidal creep disappears in a Ni_3Al alloy prestrained 3% at ambient temperature [719]. A recent investigation on a single crystal of $Ni_3Al(0.5\%Ta)$ indicated that the temperature of pre-creep deformation also affects the subsequent creep behavior [714]. The implications of these observations in terms of the creep mechanisms occurring have not been discussed.

c. Steady State Creep. In Ni_3Al alloys, steady state creep can start very early and extend over a considerable strain range (up to 20%) at high [708] as well as at intermediate temperatures [704,718]. Occasionally this stage may be delayed or even absent at intermediate temperatures if Sigmoidal creep occurs, as described above. As in many other intermetallic systems, the minimum creep rate is used to calculate the stress exponent and the activation energy for creep, using the well-established power-law relations described elsewhere in this book, when clear steady state is not observed.

Table 7 summarizes some of the creep data obtained in various investigations on Ni_3Al-based alloys, mostly at high temperatures [706–711,715–718,722–725, 727–731,737–741]. This section will mainly focus on single phase alloys. The stress exponent, n, ranges mostly between 3.2 and 4.4 in both single crystal and polycrystalline alloys. A lower value of about 1 was reported at low stresses in a polycrystalline Ni_3Al(Hf, B) [725,728]. A few studies have reported higher values of 6.7 [718], 8 [730] and 9 [742]. The values of the activation energy Q_c for creep ranged from 263 to 530 kJ/mol, but are generally between 320 and 380 kJ/mol. It is not possible to normalize all the creep data of various Ni_3Al alloys in a single plot, such as in earlier chapters, due to the lack of diffusion coefficient and modulus of elasticity values at various temperatures over the large variety of compositions investigated.

The rate-controlling mechanism during creep of Ni_3Al intermetallics is still unclear. Based on the analysis of the stress exponents, several studies suggested dislocation glide ($n \approx 3$) [709–710,723,727,729], while others propose that dislocation climb predominates ($n \approx 4$–5) [710,714,722,731], and yet others point toward Coble or Nabarro-Herring creep ($n = 1$) [725,728]. However, others have questioned the predominance of a single mechanism, since the stress exponents vary from 3 to 5. Also, the Q_c values have been found to be stress-dependent, in some cases [717], and values are often much higher than the activation energy for diffusion (the activation energy for diffusion of Ni in Ni_3Al varies from 273 to 301 kJ/mol [743]). The diffusivity of Al in NiAl is believed to be higher but it has still not been measured directly due to the lack of suitable radioactive tracers [744].

Several TEM studies have been performed to investigate the microstructural evolution of Ni_3Al alloys during steady-state creep. In general, subgrains do not readily form. Wolfenstine et al. [710] observed randomly distributed, curved, dislocations in the $n = 3$ region between 810°C and 915°C, and a homogeneous dislocation distribution with some evidence for subgrain formation in the $n = 4.3$ region (lower stresses) but no evidence for subgrain formation in the $n = 3$ region (higher stresses) between 1015°C and 1115°C. Knobloch et al. [745] examined the microstructure of [001], [011], and [111] oriented Ni_3Al single crystals creep deformed at 850°C at a stress of 350 MPa. They observed an homogeneous dislocation distribution for all orientations and creep stages. Stress exponents and activation energies were not calculated. As mentioned above, the most common slip systems

Table 7. Creep data of Ni$_3$Al alloys.

Alloy	Structure	T (°C)	n	Q$_c$ (kJ/mol)	Ref.
Single-Phase					
Ni$_3$Al(10Fe)	P	871–1177	3.2	327	[723]
Ni$_3$Al(11Fe)	P	680–930	2.6	355	[724]
Ni$_3$Al(Zr, B)	P	860–965	4.4	406	[722]
Ni$_3$Al(Hf, B)	P	760	2-3 (HS)	–	[725]
			1 (LS)	–	
Ni$_3$Al(Zr, B)	P	760–860	2.9	339–346	[716]
Ni$_3$Al(8Cr, Zr, B)	P	760–860	3.3	391–400	[716]
Ni$_3$Al(5V)	P	850–950	2.89–3.37	–	[727]
Ni$_3$Al(Hf, B)	P	760–867	1 (LS)	313	[728]
Ni$_3$Al(Ta)	P	950–1100	3.3	383	[729]
Ni$_3$(Al, 4Ti)	P	750	8	–	[730]
Ni$_3$Al(8Cr, Hf, Ta, Mo...)	P	650–900	4.7	327	[731]
Ni$_3$Al(Hf, B)	S	924–1075	4.3	378	[708]
Ni-23.5Al	S	982	3.5	–	[707]
Ni$_3$Al(Cr, Ta, Ti, W, Co)	S	900–1000	3.5	380	[709]
Ni$_3$(Al, 4Ti)	S	852–902	3.3	282	[711]
Ni$_3$Al(Ta, B)	S	810–915	3.2	320	[710]
Ni$_3$Al(Ta, B)		1015–1115	3.2 (HS)	360	[710]
			4.3 (LS)	530	
Ni$_3$Al(Ta, B)	S	850–1000	3.5	420	[706]
Ni$_3$Al(X), X = Ti, Hf, Cr, Si	S	850–950	3.01–4.67	263–437	[715]
Ni$_3$Al(4Cr)	S	760–860	–	362–466	[717]
Ni$_3$Al(Ti, 2Ta)	S	850 1150	6.7 3.3	383	[718]
Multi-Phase (Precipitation Strengthed)					
Ni$_3$(Al, 4Ti)	γ/γ'	650, 750	31, 22	–	[730]
Ni-20.2Al-8.2Cr-2.44Fe	γ/γ'–α(Cr)	777–877	4.1	301	[737]
Oxide-Dispersion Strengthened Ni$_3$Al					
Ni$_3$Al(5Cr, B)	2 vol.% Y$_2$O$_3$	1000–1200	7.2, 7.8	650, 697	[738]
					[739]
Ni$_3$Al(5Cr, B)	2 vol.% Y$_2$O$_3$	649 732, 816 982	13.5 5.1 (LS) 22, 13 (HS) 9.1	– 239	[740] [741]

P = polycrystalline; S = single crystal; HS = high stress; LS = low stress.

operative during creep of Ni$_3$Al are the $\langle 110 \rangle \{111\}$ (octahedral slip) and $\langle 110 \rangle \{100\}$ (cube slip) [107,149], although dislocation glide on $\langle 100 \rangle \{100\}$ [747] and $\langle 110 \rangle \{110\}$ [745,748] systems has also been reported. This suggests that multiple slip takes place and that dislocation interactions may be important during creep [708,745].

d. Effect of Some Microstructural Parameters on Creep Behavior. Crystal orientation has a considerable influence on the creep behavior [704–708,718,746–748] and this influence is highly dependent on temperature. At high temperatures, [001] is the weakest orientation, showing the highest creep rate; [111] is the strongest orientation associated with the lowest creep rate, about 1/5 to 1/2 of that of the [001] orientation. Finally, the [011] and [123] orientations show an intermediate strength and the creep rate is about 1/3 to 1/2 of the creep rate of the [001] orientation [705, 707–708, 718,745]. At intermediate temperatures, the [111] orientation is softer than the [001], and the [123] has, again, an intermediate creep strength [704,708]. Thus, the orientation dependence of creep strength at intermediate temperatures is opposite to that at high temperatures. Models considering the operation of octahedral slip, cube slip, and multiple slip have been proposed to explain and predict the creep anisotropy at different temperatures [605,707,749]. However, it seems that the crystal orientation has no obvious influence on the stress exponent n [706–707,718] and on the activation energy Q_c for creep [706,718].

Only a few studies on the influence of grain size on the creep of Ni_3Al have been published. Schneibel et al. [725] observed a grain size dependence of the creep rate of a cast $Ni_3Al(Hf,B)$ alloy creep deformed at 760°C. They tested specimens with average grain sizes of 12 µm, 50 µm, and 120 µm. Figure 97 illustrates the strain rate vs. stress data from tests performed at high stresses in the samples with larger grain sizes. The stress exponent is 3 for small grain sizes (50 µm) and significantly higher for larger grain sizes (120 µm). The increase in the stress exponents is attributed to scatter of the experimental data corresponding to the lowest strain rate. Thus, the

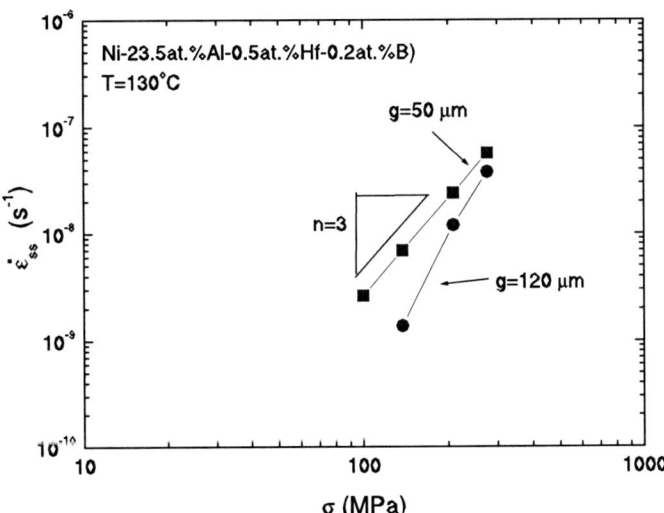

Figure 97. Stress dependence of the creep rate of Ni-23.5at.%Al-0.5 at.% Hf-0.2 at.%B at high stresses for two grain sizes (from [725]).

authors assume a stress exponent around 3 to be characteristic of the high stress regime, which would be consistent with viscous glide over the investigated grain sizes. The shear strain-rate in the samples with large grain sizes (> 50 μm) was found to be proportional to $g^{-1.9}$ at low stresses ($\sigma < 10$ MPa). This observation, together with the finding of a stress value equal to 1, may suggest Nabarro-Herring creep. For smaller grain sizes (12 μm), Coble creep may predominate, although conclusive evidence was not presented [725]. Hall-Petch strengthening was not discussed.

Hayashi et al. [727] and Miura et al. [715] investigated the effects of off-stoichiometry on the creep behavior of binary and ternary Ni_3Al alloys. They reported that, in both single- and polycrystalline alloys, the creep resistance increases with increasing Ni concentration on both sides of the stoichiometric composition and a discontinuity exists in the variation at the stoichiometric composition. The values of the activation energy Q_c for creep were also found to be strongly dependent on the Ni concentration and the alloying additions [715–716,727]. The characteristic variation in creep resistance with Ni concentration was explained by the strong concentration dependence of the activation energy for creep [727]. The n values (mostly about 3–4), however, appear nearly independent of the stoichiometric composition and the alloying additions [715,727].

Attempts have been made to improve the creep strength of Ni_3Al by adding various alloying elements [709,712,715,723,725,727,750–753]. Several solutes have been found to be beneficial, such as Hf, Cr, Zr and Ta [600]. In some cases the improvement in creep strength was accompanied by a non-desirable increase in density [600]. However, solid-solution strengthening has not been effective enough to increase creep resistance of Ni_3Al alloys above the typical values of Ni-based superalloys [709,752,754,755]. Therefore, research efforts have been directed to develop multiphase alloys based on Ni_3Al through precipitation strengthening [730,737] or dispersion strengthening (with addition of nonmetallic particles or fibers, e.g. oxides, borides, and carbides) [738–741].

A few investigations of creep in multiphase Ni_3Al alloys [730,737–741] are listed in Table 7. Three ranges of n and Q_c values were reported for multiphase alloys, i.e. $n = 4.1–5.1$ ($Q_c = 239$ kJ/mol), $n = 7.2–9.1$ ($Q_c = 301$ kJ/mol), and $n = 13–31$ ($Q_c = 650–697$ kJ/mol). The deformation mechanism governing creep of multiphase Ni_3Al alloys is still unclear. The steady state creep in a precipitation strengthened Ni_3Al alloy Ni–20.2 at.%Al–8.2 at.%Cr–2.44 at.%Fe [737], where $n = 4.1$ and $Q_c = 301$ kJ/mol were observed, was suggested to be controlled by diffusion-controlled climb of dislocation loops at Cr precipitate interfaces. In an oxide-dispersion strengthened (ODS) Ni_3Al alloy [738–741] the n values were shown to be strongly dependent on the temperature and the stress. The stress exponents increased from 5.1 (with $Q_c = 239$ kJ/mol) at low stresses to 13~22 at high stresses at temperatures of 732°C and 815°C [740,741], which are quite typical of ODS alloys as discussed earlier. At higher temperatures (from 1000 to 1273°C) [738,739] the stress exponents were 7.2 and 7.8 (with $Q_c = 650$ and 697 kJ/mol, respectively). It was suggested that the stress exponent of 5.1 in the ODS Ni$_3$Al should not be considered indicative of dislocation climb controlled creep as observed in pure metals and Class M alloys and as proposed for some single phase Ni$_3$Al alloys. Carreño et al. [738] emphasized that both Arzt and co-workers' detachment model (described in the chapter on second phase strengthening) and incorporating a threshold stress in any model [605] are not appropriate approaches to describe the creep behavior of ODS Ni$_3$Al at higher temperatures. There is relatively poor agreement between the data and the predictions by these models. Alternatively, they developed a "$\tilde{n}$-model" approach, which separates the contribution of the particles and that from the matrix. They assume the measured stress exponent is equal to the sum of the stress exponent corresponding to the matrix, termed $h_{\tilde{n}}$, and an additional stress exponent, $\tilde{n}$, that is necessary in order to account for the dislocation-particle interactions. The measured activation energy can be obtained by multiplying the activation energy corresponding to the matrix deformed under the same stress and temperature conditions by a factor equal to $(h_{\tilde{n}} + \tilde{n}/h)$. This approach satisfactorily models the data.

Figure 98 illustrates a comparison of the creep properties of an ODS Ni$_3$Al alloy, Ni-19 at.%Al-5 at.%Cr-0.1 at.%B with 2 vol.% of Y$_2$O$_3$ (filled circles) with a single crystal Ni$_3$Al alloy and two nickel-based superalloys, NASAIR 100 and MA6000.

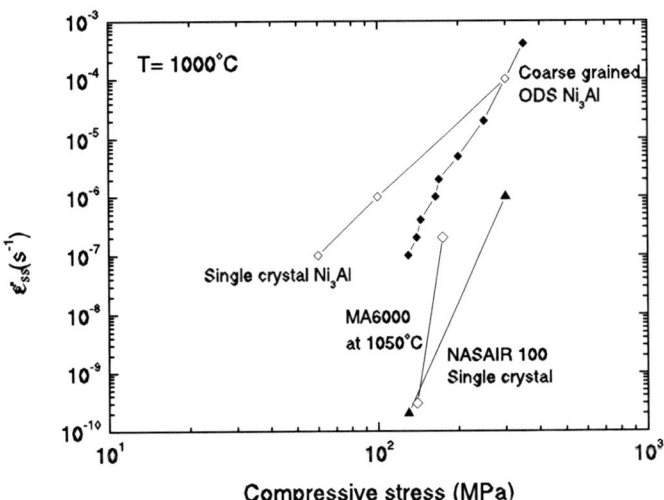

Figure 98. Comparison of the creep behavior corresponding to an ODS Ni$_3$Al alloy of composition Ni-19 at.%Al-5 at.%Cr-0.1 at.%B with 2 vol.% of Y$_2$O$_3$ (g~400 μm) [739] with a single crystal Ni$_3$Al alloy [708] and with the Ni superalloys NASAIR 100 [756], and MA6000 [757].

Although the introduction of an oxide dispersion contributed to strengthening of the Ni$_3$Al alloy with respect to the single crystal alloy, the creep performance of ODS Ni$_3$Al was still poorer than that of commercial nickel-based superalloys.

9.4.2 NiAl

a. Introduction. NiAl is a B2 ordered intermetallic phase with a simple cubic lattice with two atoms per lattice site, an Al atom at position (x, y, z) and a Ni atom at position $(x+1/2, y+1/2, z+1/2)$. This is a very stable structure that remains ordered until nearly the melting temperature. As illustrated in the phase diagram of Fig. 94, NiAl forms at Al concentrations ranging from 40 to about 55 at.%. Excellent reviews of the physical and mechanical properties of NiAl are available in [600,758,759].

NiAl alloys are attractive for many applications due to their favorable oxidation, carburization and nitridation resistance, as well as their high thermal and electrical conductivity. They are currently used to make electronic metallizations in advanced semiconductor heterostructures, surface catalysts, and high current vacuum circuit breakers [758]. Additionally, these alloys are attractive for aerospace structural applications due to their low density (5.98 g/cm^3) and high melting temperature [605]. However, two major limitations of single phase NiAl alloys are precluding their application as structural materials, namely poor creep strength at high temperatures and brittleness below about 400°C (brittle-ductile transition temperature). The following sections of this chapter will review the deformation mechanisms during creep of single-phase NiAl and the effects of different strengthening mechanisms.

b. Creep of Single-Phase NiAl. Most of the available creep data of NiAl were obtained from compression tests at constant strain rate or constant load [760–780]. Only limited data from tensile tests are available [768,781]. It is generally accepted that creep in single-phase NiAl is diffusion controlled. This has been inferred, first, from the analysis of the stress exponents and activation energies. The values for these parameters are listed in Table 8. In most cases, the stress exponents range from 4 to 7.5. Figure 99 illustrates the creep behavior of several binary NiAl alloys. Evidence for the predominance of dislocation climb was also inferred from the values of the activation energies which, in many investigations, are close to 291 kJ/mol, the value of the activation energy for bulk diffusion of Ni in NiAl. Additionally, subgrain formation during deformation has been observed [770], consistent with climb control.

However, occasionally stress exponents different from those mentioned above have been reported. For example, values as low as 3 were measured in NiAl single crystals by Forbes et al. [782] and Vanderwoort et al. [778], who suggested that both

Table 8. Creep parameters for NiAl (from [759] with additional data from recent publications).

Al, at.%	Grain size, μm	T (°C)	n	Q, kJ/mol	Ref.
48.25	5–9	727–1127	6.0–7.5	313	[774]
44–50.6	15–20	727–1127	5.75	314	[775]
50	12	927–1027	6	350	[776]
50	450	800–1045	10.2–4.6	283	[770]
50	500	900	4.7		[777]
50.4	1000	802–1474	7.0–3.3	230–290	[778]
50	Single crystal [720]	750–950	7.7–5.4		[779]
50	Single crystal	750–1055	4.0–4.5	293	[780]
49.8	39	727	5	260	[781]

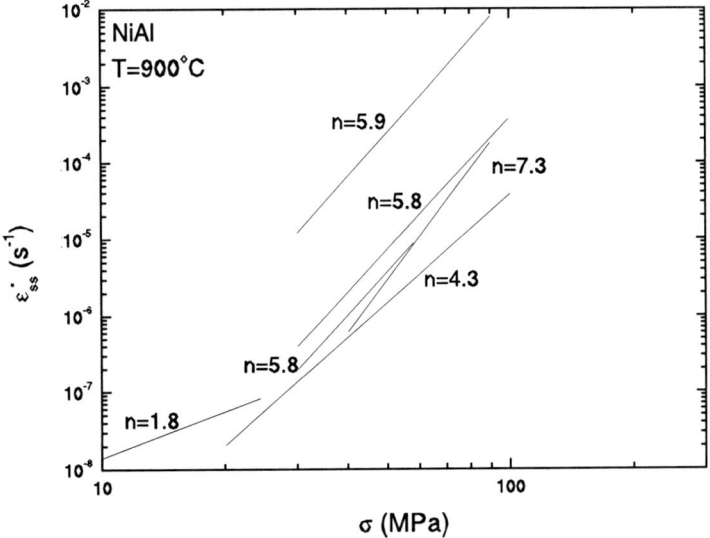

Figure 99. Creep behavior of several binary NiAl alloys [600].

viscous glide and dislocation climb would operate. The contributions of each mechanism would depend on texture, stress and temperature [778]. Recently, Raj [782] reported stress exponents as high as 13 in a Ni-50 at.%Al alloy tested in tension at 427°C, 627°C and 727°C and at constant stresses of 100–170 MPa, 40–80 MPa, and 35–65 MPa, respectively. Although no clear explanations for this high stress exponent value are provided by Raj [782], he notes that the creep behavior of NiAl in tension (his research) and compression (most of the previous works) is significantly different. For example, Raj observed that NiAl material creeps much faster in tension than in compression, especially at the lower temperatures.

Diffusional creep has been suggested to occur in NiAl when tested at low stresses ($\sigma < 30$ MPa) and high temperatures (T > 927°C) [783,784]. Stress exponents between 1 and 2 were reported under these conditions.

c. Strengthening Mechanisms. Several strengthening mechanisms have been utilized in order to improve the creep strength of NiAl alloys. Solid solution of Fe, Nb, Ta, Ti and Zr produced only limited strengthening. The presence of these solute atoms caused the lowering of the stress exponent. Therefore the solute strengthening effects are only significant at very high stresses [760,785]. Solute strengthening must be combined with other strengthening mechanisms in order to obtain improved creep strength. Precipitation hardening by additions of Nb, Ta or Ti renders NiAl more creep resistant than solid solution strengthening (i.e., alloys with the same composition and same alloying elements in smaller quantities) but still significant improvements are not achieved [761].

An alternative, more effective, strengthening method than solute or precipitation strengthening is dispersion strengthening. Artz and Grahle [764] mechanically alloyed dispersed particles in a NiAl matrix and obtained favorable creep strength up to 1427°C. Figure 100 (from [764]) compares the creep strengths of an ODS NiAl-Y_2O_3 alloy with a ferritic Superalloy (MA956) and a precipitation strengthened NiAl alloy in which the composition of the precipitates is AlN at 1200°C. The ODS Ni alloy is more creep resistant than the Ni Superalloy MA956 at these high

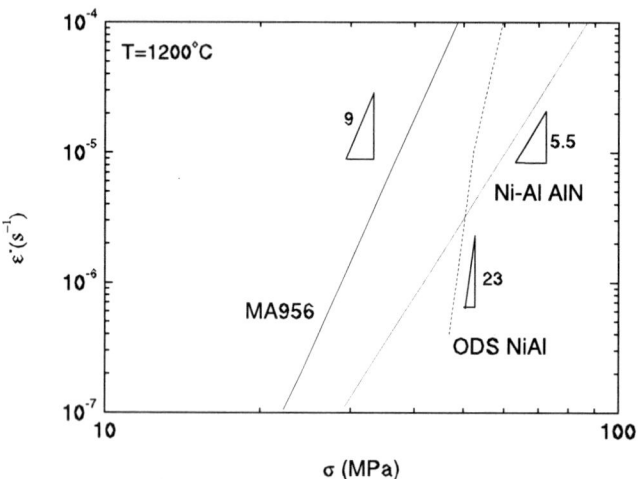

Figure 100. Comparison of the creep behavior of a coarse-grained ($g = 100$ μm) ODS NiAl alloy [764] with the ferritic ODS Superalloy MA956 and with a precipitation strengthened NiAl-AlN alloy produced by cryomilling [767].

temperatures. The ODS NiAl alloy is also more resistant than the precipitation strengthened alloy at low strain rates. Artz and Grahle [764] also observed that the creep behavior of ODS NiAl is significantly influenced by the grain size. In a coarse grain size ($g = 100\,\mu m$) ODS NiAl-Y_2O_3 alloy, the creep behavior showed the usual characteristics of dispersion strengthened systems, i.e., high stress exponents ($n = 17$) and activation energies much higher than that of lattice self-diffusion ($Q = 576\,kJ/mol$). However, intermediate stress exponents of 5 and very high activation energies ($Q = 659\,kJ/mol$) were measured in an ODS NiAl-Y_2O_3 alloy with a grain size of $0.9\,\mu m$. This creep data could not be easily modeled using established relationships for diffusional or for detachment-controlled dislocation creep [764]. Artz and Grahle [764] suggest that the presence of particles at grain boundaries partially suppresses the sink/source action of grain boundaries by pinning the grain boundary dislocations and, thus, hinders Coble creep. However, the low stress exponent indicates that Coble creep is not completely suppressed. Artz and Grahle [764] proposed a phenomenological model based on the coupling between Coble creep and the grain boundary dislocation-dispersoid interaction (controlled by thermally activated dislocation detachment from the particles, as indicated in an earlier chapter). The predictions of this model agree satisfactorily with the experimental data. HfC and HfB_2 can also provide significant particle strengthening [760].

Matrix reinforcement by larger particles such as TiB_2 and Al_2O_3 or whiskers also increases significantly the strength of NiAl alloys [760,765,786]. Xu and Arsenault [765] investigated the creep mechanisms of NiAl matrix composites with 20 vol.% of TiB_2 particles of 5 and 150 μm diameter, with 20 vol.% of Al_2O_3 particles of 5 and 75 μm diameter, and with 20 vol.% of Al_2O_3 whiskers. They measured stress exponents ranging from 7.6 to 8.4 and activation energies similar to that of lattice diffusion of Ni in NiAl. They concluded that the deformation mechanism is the same in unreinforced and reinforced materials, i.e., dislocation climb is rate-controlling in NiAl matrix composites during deformation at high temperature. Additionally, TEM examination revealed long screw dislocations with superjogs. Xu and Arsenault [765] suggested, based on computer simulation, that the jogged screw dislocation model, described previously in this book, can account for the creep behavior of NiAl metal matrix composites.

Chapter 10
Creep Fracture

10.1.	Background	215
10.2.	Cavity Nucleation	218
	10.2.1. Vacancy Accumulation	218
	10.2.2. Grain-Boundary Sliding	221
	10.2.3. Dislocation Pile-ups	222
	10.2.4. Location	224
10.3.	Growth	225
	10.3.1. Grain Boundary Diffusion-Controlled Growth	225
	10.3.2. Surface Diffusion-Controlled Growth	228
	10.3.3. Grain-Boundary Sliding	229
	10.3.4. Constrained Diffusional Cavity Growth	229
	10.3.5. Plasticity	234
	10.3.6. Coupled Diffusion and Plastic Growth	234
	10.3.7. Creep Crack Growth	237
10.4.	Other Considerations	239

Chapter 10
Creep Fracture

10.1 BACKGROUND

Creep plasticity can lead to tertiary or Stage III creep and failure. It has been suggested that creep fracture can occur by *w* or wedge-type cracking, illustrated in Figure 101(a), at grain-boundary triple points. Some have suggested that *w*-type cracks form most easily at higher stresses (lower temperatures) and larger grain sizes [787] when grain-boundary sliding is not accommodated. Some have suggested that the wedge-type cracks nucleate as a consequence of grain-boundary sliding. Another mode of fracture has been associated with *r*-type irregularities or cavities illustrated in Figure 102. The wedges may be brittle in origin or simply an accumulation of *r*-type voids [Figure 101(b)] [788]. These wedge cracks may propagate only by *r*-type void formation [789,790]. Inasmuch as *w*-type cracks are related to *r*-type voids, it is sensible to devote this short summary of creep fracture to cavitation.

There has been, in the past, a variety of reviews of creep fracture by Cocks and Ashby [791], Nix [792], and Needleman and Rice [793] and a series of articles in a single issue of a journal [793–799], chapter by Cadek [20] and particularly books by Riedel [800] and Evans [30], although most of these were published 15–20 years ago. This chapter will review these and, in particular, other more recent works.

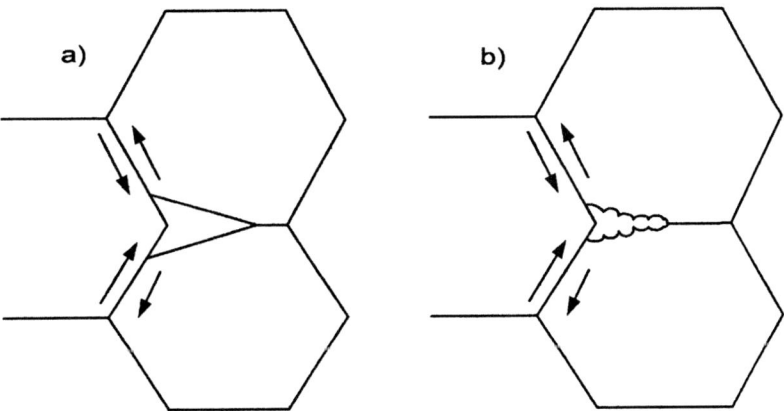

Figure 101. (a) Wedge (or *w*-type) crack formed at the triple junctions in association with grain-boundary sliding. (b) illustrates a Wedge crack as an accumulation of spherical cavities.

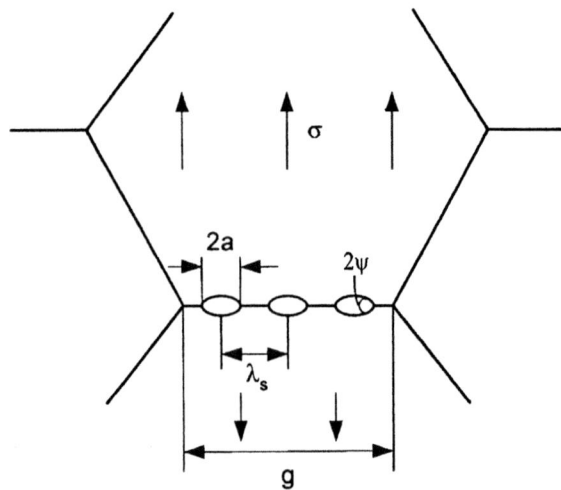

Figure 102. Cavitation (r-type) or voids at a transverse grain boundary. Often, ψ is assumed to be approximately 70°.

Some of these works are compiled in recent bibliographies [801] and are quite extensive, of course, and this chapter is intended as a balanced and brief summary. The above two books are considered particularly good references for further reading. This chapter will particularly reference those works published subsequent to these reviews.

Creep Fracture in uniaxial tension under constant stress has been described by the Monkman–Grant relationship [35], which states that the fracture of creep-deforming materials is controlled by the steady-state creep rate, $\dot{\varepsilon}_{ss}$, equation (4),

$$\dot{\varepsilon}_{ss}^{m''} t_f = k_{MG}$$

where k_{MG} is sometimes referred to as the Monkman–Grant constant and m'' is a constant typically about 1.0. Some data that illustrates the basis for this phenomenological relationship is in Figure 103, based on Refs. [30,802]. Although not extensively validated over the past 20 years, it has been shown recently to be valid for creep of dispersion-strengthened cast aluminum [803] where cavities nucleate at particles and not located at grain boundaries. Modifications have been suggested to this relationship based on fracture strain [804]. Although some more recent data on Cr–Mo steel suggests that equation (4) is valid [805], the same data has been interpreted to suggest the modified version. The Monkman–Grant (phenomenological) relationship, as will be discussed subsequently, places constraints on creep cavitation theories.

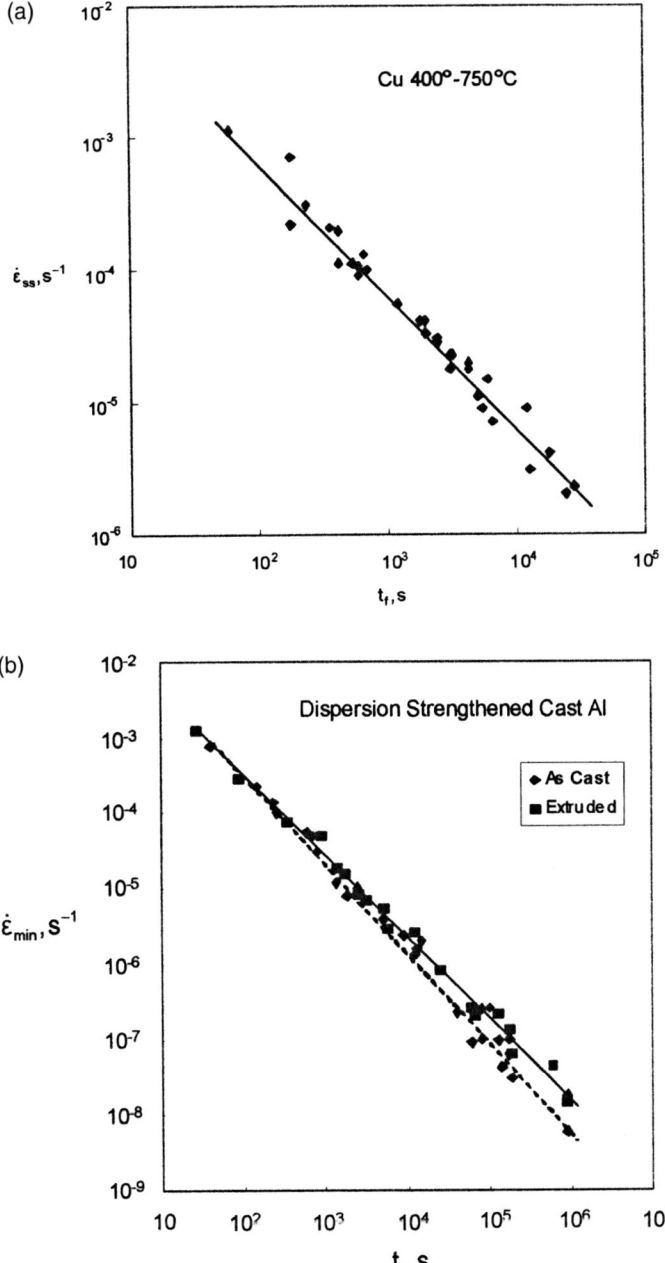

Figure 103. (a) The steady-state creep-rate (strain-rate) versus time-to-rupture for Cu deformed over a range of temperatures, adapted from Evans [30], and (b) dispersion-strengthened cast aluminum, adapted from Dunand et al. [803].

Another relationship to predict rupture time utilizes the Larson–Miller parameter [806] described by

$$\text{LM} = T[\log t_r + C_{\text{LM}}] \tag{130}$$

This equation is not derivable from the Monkman–Grant or any other relationship presented. The constant C_{LM} is phenomenologically determined as that value that permits LM to be uniquely described by the logarithm of the applied stress. This technique appears to be currently used for zirconium alloy failure time prediction [807]. C_{LM} is suggested to be about 20, independent of the material.

One difficulty with these equations is that the constants determined in a creep regime, with a given rate-controlling mechanism may not be used for extrapolation to the rupture times within another creep regime where the constants may change [807]. The Monkman–Grant relationship appears to be more popular.

The fracture mechanisms that will be discussed are those resulting from the nucleation of cavities followed by growth and interlinkage, leading to catastrophic failure. Figure 104 illustrates such creep cavitation in Cu, already apparent during steady-state (i.e., prior to Stage III or tertiary creep). It will be initially convenient to discuss fracture by cavitation as consisting of two steps, nucleation and subsequent growth.

10.2 CAVITY NUCLEATION

It is still not well established by what mechanism cavities nucleate. It has generally been observed that cavities frequently nucleate on grain boundaries, particularly on those transverse to a tensile stress, e.g. [788,808–812]. In commercial alloys, the cavities appear to be associated with second-phase particles. It appears that cavities do not generally form in some materials such as high-purity (99.999% pure) Al. Cavitation is observed in lower purity metal such as 99% Al [813] (in high-purity Al, boundaries are serrated and very mobile). The nucleation theories fall into several categories that are illustrated in Figure 105: (a) grain-boundary sliding leading to voids at the head (e.g., triple point) of a boundary or formation of voids by "tensile" GB ledges, (b) vacancy condensation, usually at grain boundaries at areas of high stress concentration, (c) the cavity formation at the head of a dislocation pile-up such as by a Zener–Stroh mechanism (or anti-Zener–Stroh mechanism [814]). These mechanisms can involve particles as well (d).

10.2.1 Vacancy Accumulation

Raj and Ashby [815] developed an earlier [816] idea that vacancies can agglomerate and form stable voids (nuclei) as in Figure 88(b). Basically, the free energy terms are

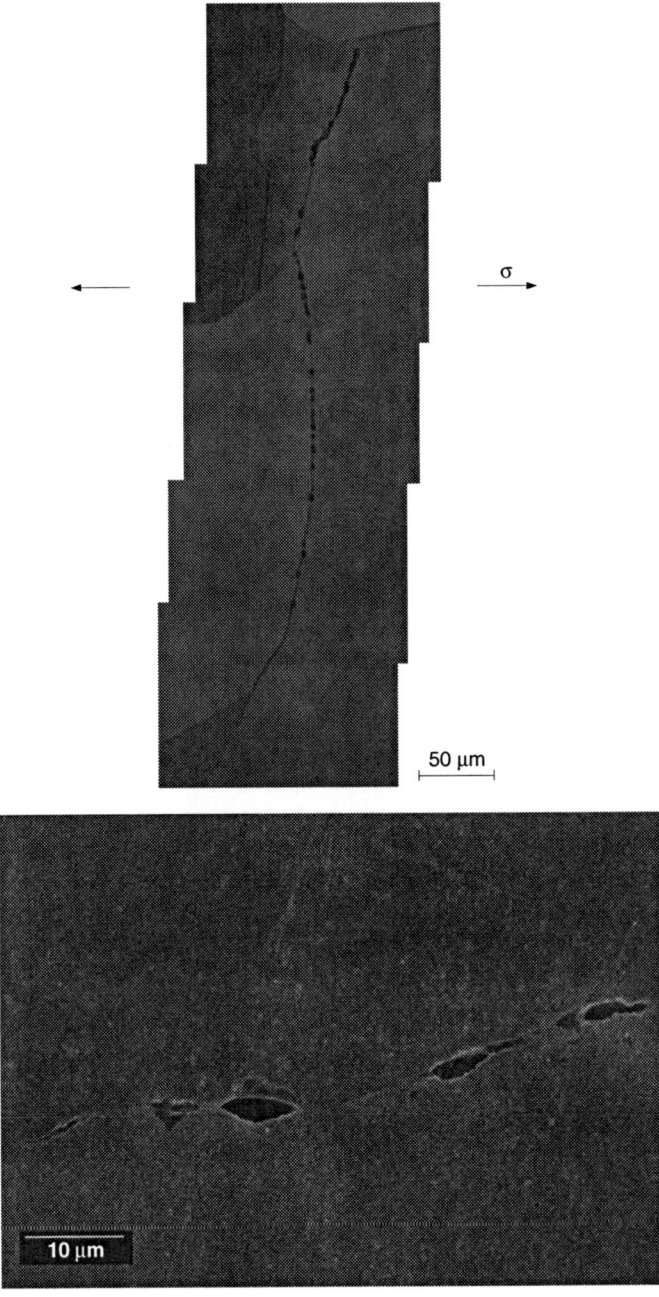

Figure 104. Micrograph of cavities in Cu deformed at 20 MPa and 550°C to a strain of about 0.04 (within stage II, or steady-state).

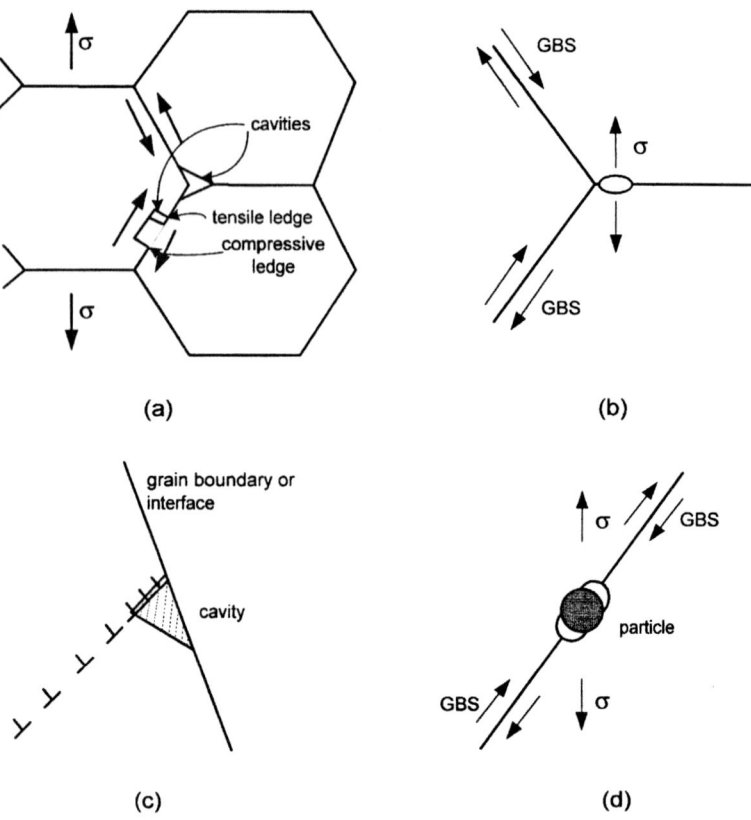

Figure 105. Cavity nucleation mechanism. (a) Sliding leading to cavitation from ledges (and triple points). (b) Cavity nucleation from vacancy condensation at a high-stress region. (c) Cavity nucleation from a Zener–Stroh mechanism. (d) The formation of a cavity from a particle-obstacle in conjunction with the mechanisms described in (a–c).

the work performed by the applied stress with cavity formation balanced by two surface energy terms. The change in total free energy is given by,

$$\Delta G_{\mathrm{T}} = -\sigma \Omega N + A_v \gamma_{\mathrm{m}} - A_{\mathrm{gb}} \gamma_{\mathrm{gb}} \tag{131}$$

where N is the number of vacancies, A_v and A_{gb} are the surface areas of the void and (displaced) area of grain boundary, and γ_{m} and γ_{gb} are surface and interfacial energy terms of the metal and grain boundary. (Note: all stresses and strain-rates are equivalent uniaxial and normal to the grain boundary in the equations in this chapter.)

This leads to a critical radius, a^*, and free energy, ΔG_T^*, for critical-sized cavities and a nucleation rate,

$$\dot{N} \cong n^* D_{gb} \qquad (132)$$

where $n^* = n_o \exp(-\Delta G_T^*/kT)$, D_{gb} is the diffusion coefficient at the grain boundary and n_0 is the density of potential nucleation sites. (The nucleation rate has the dimensions, $m^{-2} s^{-1}$.) (Some [24,800] have included a "Zeldovich" factor in equation (132) to account for "dissolution" of "supercritical" nuclei $a > a^*$.)

Some have suggested that vacancy supersaturation may be a driving force rather than the applied stress, but it has been argued that sufficient vacancy supersaturations are unlikely [800] in conventional deformation (in the absence of irradiation or Kirkendall effects).

This approach leads to expressions of nucleation rate as a function of stress (and the shape of the cavity). An effective threshold stress for nucleation is predicted. Argon et al. [817] and others [800] suggest that the cavity nucleation by vacancy accumulation (even with modifications to the Raj–Ashby nucleation analysis to include, among other things, a Zeldovich factor) requires large applied (threshold) stresses (e.g., 10^4 MPa), orders of magnitude larger than observed stresses leading to fracture, which can be lower than 10 MPa in pure metals [800].

Cavity nucleation by vacancy accumulation thus appears to require significant stress concentration. Of course, with elevated temperature plasticity, relaxation by creep plasticity and/or diffusional flow will accompany the elastic loading and relax the stress concentration. The other mechanisms illustrated in Figure 105 can involve Cavity nucleation by direct "decohesion" which, of course, also requires a stress concentration.

10.2.2 Grain-Boundary Sliding

Grain boundary sliding (GBS) can lead to stress concentrations at triple points and hard particles on the grain boundaries, although it is unclear whether the local stresses are sufficient to nucleate cavities [20,818]. These mechanisms are illustrated in Figures 88(a), (b) and (d). Another sliding mechanism includes (tensile) ledges [Figure 88(a)] where tensile stresses generated by GBS may be sufficient to cause Cavity nucleation [819], although some others [820] believed the stresses are insufficient. The formation of ledges may occur as a result of slip along planes intersecting the grain boundaries.

One difficulty with sliding mechanisms is that transverse boundaries (perpendicular to the principal tensile stress) appear to have a propensity to cavitate. Cavitation has been observed in bicrystals [821] where the boundary is perpendicular to the applied stress, such that there is no resolved shear and an absence of sliding. Hence, it appears that sliding is not a necessary condition for cavity nucleation. Others [795,808,822], however, still do not appear to rule out a relationship between GBS and cavitation along transverse boundaries. The ability to nucleate cavities via GBS has been demonstrated by prestraining copper bicrystals in an orientation favoring GBS, followed by subjecting the samples to a stress normal to the previously sliding grain boundary and comparing those results to tests on bicrystals that had not been subjected to GBS [819]. Extensive cavitation was observed in the former case while no cavitation was observed in the latter. Also, as will be discussed later, GBS (and concomitant cavitation) can lead to increased stress on transverse boundaries, thereby accelerating the cavitation at these locations. More recently, Ayensu and Langdon [808] found a relation between GBS and cavitation at transverse boundaries, but also note a relationship between GBS and strain. Hence, it is unclear whether GBS either nucleates or grows cavities in this case. Chen [809] suggested that transverse boundaries may slide due to compatibility requirements.

10.2.3 Dislocation Pile-ups

As transverse boundaries may not readily slide, perhaps the stress concentration associated with dislocation pile-ups against, particularly, hard second-phase particles at transverse grain boundaries, has received significant acceptance [799,824,825] as a mechanism by which vacancy accumulation can occur. Pile-ups against hard particles within the grain interiors may nucleate cavities, but these may grow relatively slowly without short-circuit diffusion through the grain boundary and may also be of lower (areal) density than at grain boundaries.

It is still not clear, however, whether vacancy accumulation is critical to the nucleation stage. Dyson [799] showed that tensile creep specimens that were prestrained at ambient temperature appeared to have a predisposition for creep cavitation. This suggested that the same process that nucleates voids at ambient temperature (that would not appear to include vacancy accumulation) may influence or induce void nucleation at elevated temperatures. This could include a Zener–Stroh mechanism [Figure 88(c)] against hard particles at grain boundaries. Dyson [799] showed that the nucleation process can be continuous throughout creep and that the growth and nucleation may occur together, a point also made by several other investigators [798,820,821,822]. This and the effect of prestrain are illustrated in Figure 106. The impact of cavitation rate on ductility is illustrated in Figure 107. Thus, the nucleation process may be controlled by the (e.g., steady-state) plasticity. The suggestion that cavity nucleation is associated with plastic deformation is consistent with the observation by Nieh and Nix [826], Watanabe et al. [828], Greenwood et al.

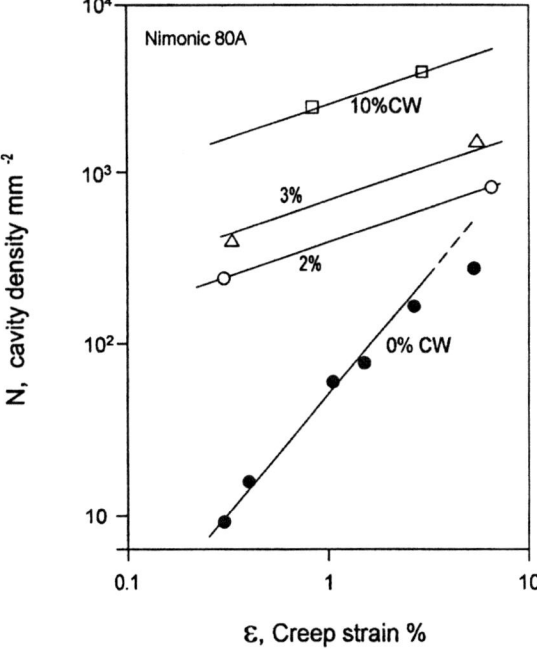

Figure 106. The variation of the cavity concentration versus creep strain in Nimonic 80A (Ni–Cr alloy with Ti and Al) for annealed and pre-strained (cold-worked) alloy. Adapted from Dyson [799]. Cavities were suggested to undergo unconstrained growth.

[829], and Dyson *et al.* [827] that the cavity spacing is consistent with regions of high dislocation activity (slip-band spacing). Goods and Nix [830] also showed that if bubbles are implanted, the ductility decreases. Davanas and Solomon [816] argue that if continuous nucleation occurs, modeling of the fracture process can lead to a Monkman–Grant relationship (diffusive and plastic coupling of cavity growth and cavity interaction considered). One consideration against the slip band explanations is that In situ straining experiments in the TEM by Dewald *et al.* [822] suggested that slip dislocations may easily pass through a boundary in a pure metal and the stress concentrations from slip may be limited. This may not preclude such a mechanism in combination with second-phase particles. Kassner *et al.* [2] performed creep fracture experiments on high-purity Ag at about $0.25\,T_m$. Cavities appeared to grow by (unstable) plasticity rather than diffusion. Nucleation was continuous, and it was noted that nucleation only occurred in the vicinity of high-angle boundaries where obstacles existed (regions of highly twinned metal surrounded by low twin-density metal). High-angle boundaries without barriers did not appear to cavitate. Thus,

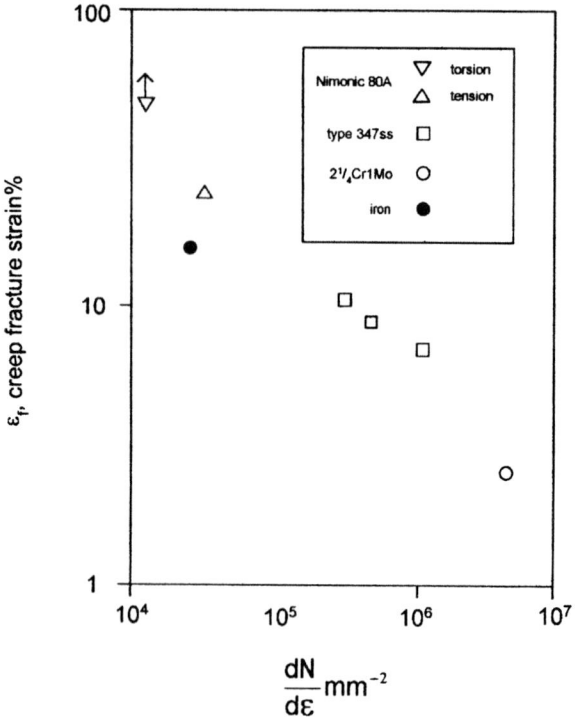

Figure 107. Creep ductility versus the "rate" of cavity production with strain. Adapted from Dyson [799] (various elevated temperatures and stresses).

nucleation (in at least transverse boundaries) appears to require obstacles and a Zener–Stroh or anti-Zener–Stroh appeared the most likely mechanism.

10.2.4 Location

It has long been suggested that (transverse) grain boundaries and second-phase particles are the common locations for cavities. Solute segregation at the boundaries may predispose boundaries to cavity nucleation [795]. This can occur due to the decrease in the surface and grain boundary energy terms.

Some of the more recent work that found cavitation associated with hard second-phase particles in metals and alloys includes [831–839]. Second-phase particles can result in stress concentrations upon application of a stress and increase cavity nucleation at a grain boundary through vacancy condensation by increasing the grain boundary free energy. Also, particles can be effective barriers to dislocation pile-ups.

The size of critical-sized nuclei is not well established but the predictions based on the previous equations is about 2–5 nm [20], which are difficult to detect. SEM under optimal conditions can observe (stable) creep cavities as small as 20 nm [840]. It has been suggested that the small angle neutron scattering can characterize cavity distributions from less than 10 nm to almost 1 μm) [20]. TEM has detected stable cavities (gas) at 3 nm [841]. Interestingly, observations of Cavity nucleation not only suggest continual cavitation but also no incubation time [842] and that strain rather than time is more closely associated with nucleation [20]. Figure 107 illustrates the effect of stress states on nucleation. torsion, for comparable equivalent uniaxial stresses in Nimonic 80, leads to fewer nucleated cavities and greater ductility than torsion. Finally, another nucleation site that may be important as damage progresses in a material is the stress concentration that arise around existing cavities. The initial (elastic) stress concentration at the cavity "tip" is a factor of three larger than the applied stress and, even after relaxation by diffusion, the stress may still be elevated [843] leading to increased local nucleation rates.

10.3 GROWTH

10.3.1 Grain Boundary Diffusion-Controlled Growth

The cavity growth process at grain boundaries at elevated temperature has long been suggested to involve vacancy diffusion. Diffusion occurs by cavity surface migration and subsequent transport along the grain boundary, with either diffusive mechanism having been suggested to be controlling depending on the specific conditions. This contrasts creep void growth at lower temperatures where cavity growth is accepted to occur by (e.g., dislocation glide-controlled) plasticity. A carefully analyzed case for this is described in Ref. [840].

Hull and Rimmer [844] were one of the first to propose a mechanism by which diffusion leads to cavity growth of an isolated cavity in a material under an applied external stress, σ. A stress concentration is established just ahead of the cavity. This leads to an initial "negative" stress gradient. However a "positive" stress gradient is suggested to be established due to relaxation by plasticity [30]. This implicit assumption in diffusion-controlled growth models appears to have been largely ignored in later discussions by other investigators, with rare exception (e.g. [30]). The equations that Hull and Rimmer and, later, others [800,815,845] subsequently derive for diffusion-controlled cavity growth are similar. Basically,

$$J_{gb} = -\frac{D_{gb}}{\Omega kT} \nabla f \qquad (133)$$

where J_{gb} is the flux, Ω the atomic volume, $f = -\sigma_{loc}\Omega$ and σ_{loc} is the local normal stress on the grain boundary. Also,

$$\nabla_f \sim \frac{\Omega}{\lambda_s}\left(\sigma - \frac{2\gamma_m}{a}\right) \tag{134}$$

where "a" is the cavity radius, σ the remote or applied normal stress to the grain boundary, and λ_s is the cavity separation. Below a certain stress $[\sigma_0 = (2\gamma_m/a)]$ the cavity will sinter. Equations (133) and (134) give a rate of growth,

$$\frac{da}{dt} \simeq \frac{D_{gb}\delta(\sigma - (2\gamma_m/a))\Omega}{2kT\lambda_s a} \tag{135}$$

where δ is the grain-boundary width. Figure 108 is a schematic that illustrates the basic concept of this approach.

By integrating between the critical radius (below which sintering occurs) and $a = \lambda_s/2$,

$$t_r \cong \frac{kT\lambda_s^3}{4D_{gb}\delta(\sigma - (2\gamma_m/a))\Omega} \tag{136}$$

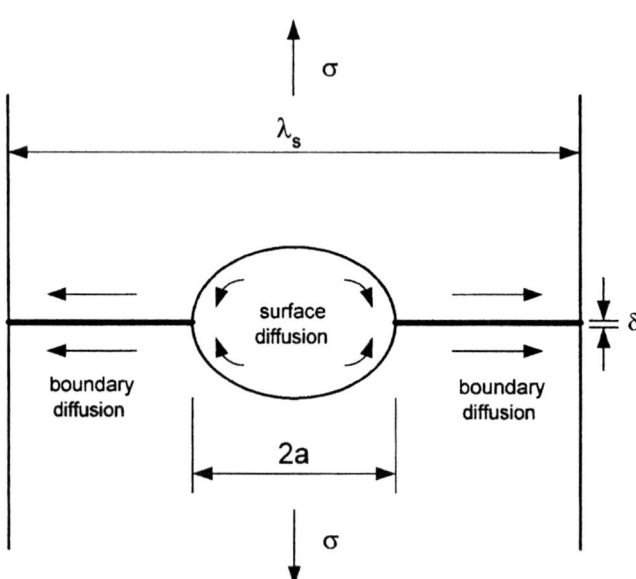

Figure 108. Cavity growth from diffusion across the cavity surface and through the grain boundaries due to a stress gradient.

This is the first relationship between stress and rupture time for (unconstrained) diffusive cavity growth. Raj and Ashby [815,846], Speight and Beere [844], Riedel [800], and Weertman [847] later suggested improved relationships between the cavity growth rate and stress of a similar form to that of Hull and Rimmer [equation (135)]. The subsequent improvements included modifications to the diffusion lengths (the entire grain boundary is a vacancy source), stress redistribution (the integration of the stress over the entire boundary should equal the applied stress), cavity geometry (cavities are not perfectly spherical) and the "jacking" effect (atoms deposited on the boundary causes displacement of the grains). Riedel, in view of these limitations, suggested that the equation for unconstrained cavity growth of widely spaced voids is, approximately,

$$\frac{da}{dt} = \frac{\Omega \delta D_{gb}[\sigma - \sigma_0'']}{1.22\, kT \ln(\lambda_s/4.24a)a^2} \tag{137}$$

where σ_0'' is the sintering stress. Again, integrating to determine the time for rupture shows that $t_r \propto 1/\sigma$. Despite these improvements, the basic description long suggested by Hull and Rimmer is largely representative of unconstrained cavity growth. An important point here is a predicted stress dependence of one and an activation energy of grain boundary diffusion for equations (135)–(137) for (unconstrained) cavity growth.

The predictions and stress dependence of these equations have been frequently tested [45,830,848–859]. Raj [856] examined Cu bicrystals and found the rupture time inversely proportional to stress, consistent with the diffusion-controlled cavity growth equations just presented. The fracture time for polycrystals increases orders of magnitude over bicrystals. Svensson and Dunlop [850] found that in α-brass, cavities grow linearly with stress. The fracture time appeared, however, consistent with Monkman–Grant and continuous nucleation was observed. Hanna and Greenwood [852] found that density change measurements in prestrained (i.e., prior Cavity nucleation) and with hydrogen bubbles were consistent with the stress dependency of the earlier equations. Continuous nucleation was not assumed. Cho et al. [853] and Needham and Gladman [858,859] measured the rupture times and/or cavity growth rate and found consistency with a stress to the first power dependency if continuous nucleation was assumed. Miller and Langdon [850] analyzed the density measurements on creep-deformed Cu based on the work of others and found that the cavity volume was proportional to σ^2 (for fixed T, t, and ε). If continuous nucleation occurs with strain, which is reasonable, and the variation of the nucleation rate is "properly" stress dependent (unverified), then consistency between the density trends and unconstrained cavity growth described by equations (135) and (137) can be realized.

Creep cavity growth experiments have also been performed on specimens with pre-existing cavities by Nix and coworkers [45,830,848,849]. Cavities, here, were created using water vapor bubbles formed from reacting dissolved hydrogen and oxygen. Cavities were uniformly "dispersed" (unconstrained growth). Curiously, the growth rate, da/dt, was found to be proportional to σ^3. This result appeared inconsistent with the theoretical predictions of diffusion-controlled cavity growth. The disparity is still not understood. Interestingly, when a dispersion of MgO particles was added to the Ag matrix, which decreased the Ag creep-rate, the growth rate of cavities was *unaffected*. This supports the suggestion that the controlling factor for cavity growth is diffusion rather than plasticity or GBS. Similar findings were reported by others [860]. Nix [792] appeared to rationalize the three-power observation by suggesting that only selected cavities participate in the fracture process. [As will be discussed later, it does not appear clear whether cavity growth in the Nix *et al.* experiments were generally unconstrained. That is, whether only diffusive flow of vacancies controls the cavity growth rate.]

10.3.2 Surface Diffusion-Controlled Growth

Chuang and Rice [861] and later Needleman and Rice [793] suggested that surface rather than grain-boundary diffusion may actually control cavity growth (which is not necessarily reasonable) and that these assumptions can give rise to a three-power stress relationship for cavity growth at low stresses [792,862].

$$\frac{da}{dt} \cong \frac{\Omega \delta D_s}{2kT\gamma_m^2} \sigma^3 \qquad (138)$$

At higher stresses, the growth rate varies as $\sigma^{3/2}$. The problem with this approach is that it is not clear in the experiments, for which three-power stress-dependent cavity growth is observed, that $D_s < D_{gb}$. Activation energy measurements by Nieh and Nix [848,849] for (assumed unconstrained) growth of cavities in Cu are inconclusive as to whether it better matches D_{gb} versus D_s. Also, the complication with all these growth relationships [equations (135)–(138)] is that they are inconsistent with the Monkman–Grant phenomenology. That is, for common Five-Power-Law Creep, the Monkman–Grant relationship suggests that the cavity growth rate ($1/t_f$) should be proportional to the stress to the fifth power rather than 1–3 power. This, of course, may emphasize the importance of nucleation in the rate-controlling process for creep cavitation failure, since cavity-nucleation may be controlled by the plastic strain (steady-state creep-rate). Of course, small nanometer-sized cavities (nuclei), by themselves, do not appear sufficient to cause cavitation failure. Dyson [798]

Creep Fracture

suggested that the Monkman–Grant relationship may reflect the importance of both (continuous) nucleation and growth events.

10.3.3 Grain-Boundary Sliding

Another mechanism that has been considered important for growth is grain-boundary sliding (GBS) [863]. This is illustrated in Figure 109. Here cavities are expected to grow predominantly in the plane of the boundary. This appears to have been observed in some temperature–stress regimes. Chen appears to have invoked GBS as part of the cavity growth process [795], also suggesting that transverse boundaries may slide due to compatibility requirements. A suggested consequence of this "crack sharpening" is that the tip velocity during growth becomes limited by surface diffusion. A stress to the third power, as in equation (138), is thereby rationalized. Chen suggests that this phenomenon may be more applicable to higher strain-rates and closely spaced cavities (later stages of creep) [823]. The observations that cavities are often more spherical rather than plate-like or lenticular, and that, of course, transverse boundaries may not slide, also suggest that cavity growth does not substantially involve sliding. Riedel [800] predicted that (constrained) diffusive cavity growth rates are expected to be a factor of $(\lambda_s/2a)^2$ larger than growth rates by (albeit, constrained) sliding. It has been suggested that sliding may affect growth in some recent work on creep cavitation of dual phase intermetallics [864].

10.3.4 Constrained Diffusional Cavity Growth

Cavity nucleation may be heterogeneous, inasmuch as regions of a material may be more cavitated than others. Adams [865] and Watanabe *et al.* [866] both suggested that different geometry (e.g., as determined by the variables necessary to characterize

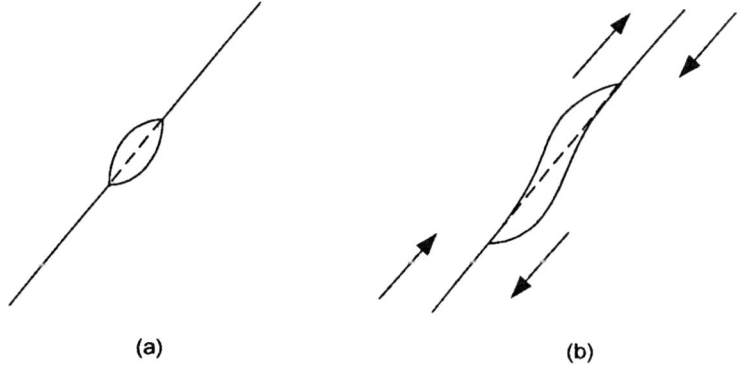

Figure 109. Cavity growth from a sliding boundary. From Ref. [795].

a planar boundary) high-angle grain boundaries have different tendencies to cavitate, although there was no agreement as to the nature of this tendency in terms of the structural factors. Also, of course, a given geometry boundary may have varying orientations to the applied stresses. Another important consideration is that the zone ahead of the cavity experiences local elongating with diffusional growth, and this may cause constraint in this region by those portions of the material that are unaffected by the diffusion (outside the cavity diffusion "zone"). This may cause a "shedding" of the load from the diffusion zone ahead of the cavity. Thus, cavitation is not expected to be homogeneous and uncavitated areas may constrain those areas that are elongating under the additional influence of cavitation. This is illustrated in Figure 110. Fracture could then be controlled by the plastic *creep-rate* in uncavitated regions that can also lead to cavity nucleation. This leads to consistency with the Monkman–Grant relationship [867,868].

Constrained diffusional growth was originally suggested by Dyson and further developed by others [791,808,823,861,868]. This constrained cavity growth rate has been described by the relationship [800]

$$\frac{da}{dt} \cong \frac{\sigma - (1-\omega)\sigma_0''}{(a^2 kT/\Omega \delta D_{gb}) + (\sigma_{ss}\pi^2(1+3/n)^{1/2}/\dot{\varepsilon}_{ss}\lambda_s^2 g)a^2} \quad (139)$$

where ω is the fraction of the grain boundary cavitated.

This is the growth rate for cavities expanding by diffusion. One notes that for higher strain-rates, where the increase in volume can be easily accommodated, the growth rate is primarily a function of the grain-boundary diffusion coefficient.

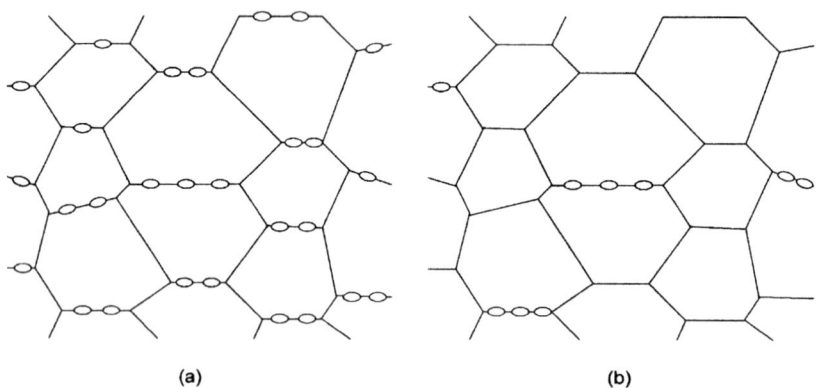

(a) (b)

Figure 110. Uniform (a) and heterogeneous (b) cavitation at (especially transverse) boundaries. The latter condition can particularly lead to constrained cavity growth [795].

If only certain grain-boundary facets cavitate, then the time for coalescence, t_c, on these facets can be calculated,

$$t_c \cong \frac{0.004\,kT\lambda^3}{\Omega\delta D_{gb}\sigma_{ss}} + \frac{0.24(1+3/n)^{1/2}\lambda}{\dot{\varepsilon}_{ss}g} \qquad (140)$$

where, again, n is the steady-state stress exponent, g is the grain size, and σ_{ss} and $\dot{\varepsilon}_{ss}$ are the steady-state stress and strain-rate, respectively, related by,

$$\dot{\varepsilon}_{ss} = A_0 \exp[-Q_c/kT](\sigma_{ss}/E)^n \qquad (141)$$

where $n = 5$ for classic five-power-law creep. However, it must be emphasized that failure is not expected by mere coalescence of cavities on isolated facets. Additional time may be required to join facet-size microcracks. The mechanism of joining facets may be rate-controlling. The advance of facets by local nucleation ahead of the "crack" may be important (creep-crack growth on a small scale). Interaction between facets and the nucleation rate of cavities away from the facet may also be important. It appears likely, however, that this model can explain the larger times for rupture (than expected based on unconstrained diffusive cavity growth). This likely also is the basis for the Monkman–Grant relationship if one assumes that the time for cavity coalescence, t_c, is most of the specimen lifetime, t_f, so that t_c is not appreciably less than t_f. Figure 111, adapted from Riedel, shows the cavity growth rate versus stress for constrained cavity growth as solid lines. Also plotted in this figure (as the dashed lines) is the equation for unconstrained cavity growth [equation (137)]. It is observed that the equation for unconstrained growth predicts much higher growth rates (lower t_f) than constrained growth rates. Also, the stress dependency of the growth rate for constrained growth leads to a time to fracture relationship that more closely matches that expected for steady-state creep as predicted by the Monkman–Grant relationship.

One must, in addition to considering constrained cases, also consider that cavities are continuously nucleated. For continuous nucleation and unconstrained diffusive cavity growth, Riedel suggests:

$$t_f = \left[\frac{kT}{5\Omega\delta D_{gb}\sigma}\right]^{2/5}\left(\frac{\omega_f}{\dot{N}}\right)^{3/5} \qquad (142)$$

where ω_f is the critical cavitated area fraction and, consistent with Figure 90 from Dyson [799]

$$\dot{N} = \alpha'\dot{\varepsilon} = \alpha'\beta\sigma^n \qquad (143)$$

with $\dot{\varepsilon}$ according to equation (141).

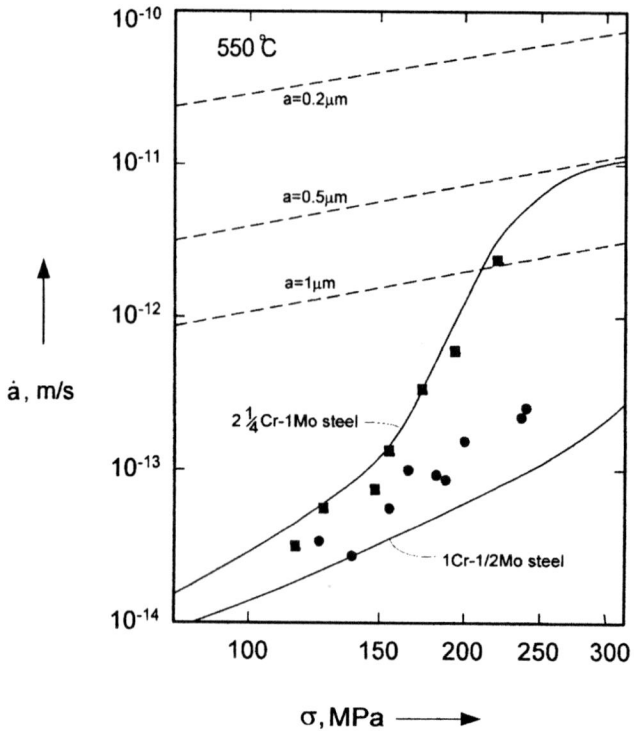

Figure 111. The cavity growth rate versus stress in steel. The dashed lines refer to unconstrained growth and solid lines to constrained growth. Based on Riedel [800].

Equation (142) can be approximated by

$$t_f \propto \frac{1}{\sigma^{(3n+2)/5}}$$

For continuous nucleation with the constrained case, the development of reliable equations is more difficult as discussed earlier and Riedel suggests that the time for coalescence on isolated facets is,

$$t_c = 0.38 \left[\frac{\pi(1 + (3/n))}{\dot{N}} \right]^{1/3} \frac{\omega_f}{[\dot{\varepsilon}g]^{2/3}} \quad (144)$$

which is similar to the version by Cho et al. [870]. Figure 112, also from Riedel, illustrates the realistic additional effects of continuous nucleation, which appear to match the observed rupture times in steel. The theoretical curves in Figure 112

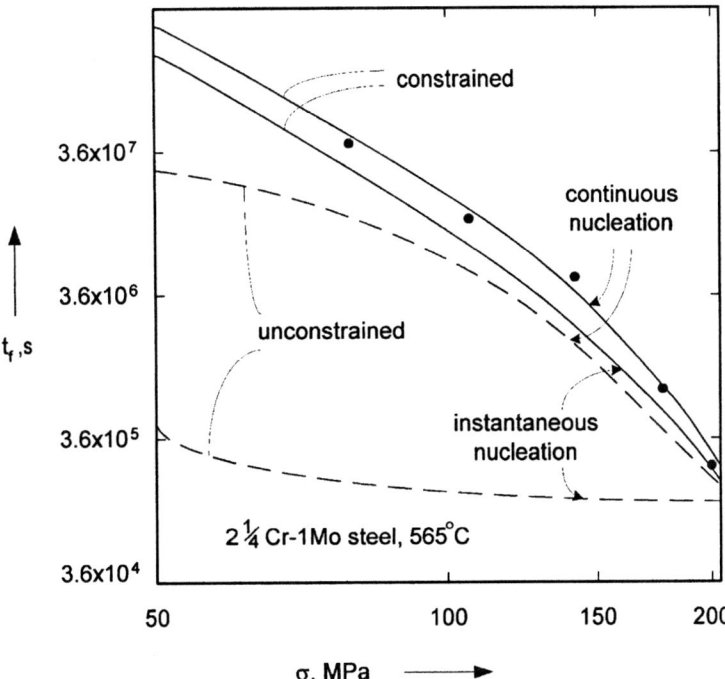

Figure 112. The time to rupture versus applied stress for (a) unconstrained (dashed lines) cavity growth with instantaneous or continuous nucleation. (b) Constrained cavity growth (t_c) with instantaneous and continuous nucleation. Dots refer to experimental t_f. Based on Riedel [800].

correspond to equations (136), (140), (142), and (144). One interesting aspect of this figure is that there is a very good agreement between t_c and t_f for constrained cavity growth. This data was based on the data of Cane [832] and Riedel [869], who determined the nucleation rate by apparently using an empirical value of α'. No adjustable parameters were used. Cho et al. [869] later, for NiCr steel, at 823 K, were able to reasonably predict rupture times assuming continuous nucleation and constrained cavity growth.

It should be mentioned that accommodated GBS can eliminate the constraint illustrated in Figure 93 (two-dimensional); however, in the three-dimensional case, sliding does not preclude constrained cavity growth, as shown by Anderson and Rice [871]. Nix et al. [872] and Yousefiani et al. [873] have used a calculation of the principal facet stress to predict the multiaxial creep rupture time from uniaxial stress data. Here, it is suggested that GBS is accommodated and the normal stresses on (transverse) boundaries are increased. Van der Giessen and Tvergaard [873] appear to analytically (3D) show that increased cavitation on inclined sliding boundaries may increase the normal stresses on transverse boundaries for constrained cavity growth. Thus, the Riedel solution may be non-conservative, in the sense that it overpredicts t_f. Dyson [875] suggested that within certain temperature and strain-rate regimes, there may be a transition from constrained to unconstrained cavity growth. For an aluminum alloy, it was suggested that decreased temperature and increased stress could lead to unconstrained growth. Interestingly, Dyson also pointed out that for constrained cavity growth, uncavitated regions would experience accelerated creep beyond that predicted by the decrease in load-carrying area resulting from cavitation.

10.3.5 Plasticity

Cavities can grow, of course, exclusively by plasticity. Hancock [876] initially proposed the creep-controlled cavity growth model based on the idea that cavity growth during creep should be analogous to McClintock's [833] model for a cavity growing in a plastic field. Cavity growth according to this model occurs as a result of creep deformation of the material surrounding the grain boundary cavities in the absence of a vacancy flux. This mechanism becomes important under high strain-rate conditions, where significant strain is realized. The cavity growth rate according to this model is given as

$$\frac{da}{dt} = a\dot{\varepsilon} - \frac{\gamma}{2G} \tag{145}$$

This is fairly similar to the relationship by Riedel [799] discussed earlier. It has been suggested, on occasion, that the observed creep cavity growth rates are consistent with plasticity growth (e.g., [849]) but it is not always obvious that constrained diffusional cavity growth is not occurring, which is also controlled by plastic deformation.

10.3.6 Coupled Diffusion and Plastic Growth

Cocks and Ashby [790], Beere and Speight [877], Needleman and Rice [793] and others [792,878–883] suggested that there may actually be a coupling of diffusive cavity growth of cavities with creep plasticity of the surrounding material from the far-field stress. It is suggested that as material from the cavity is deposited on the grain boundary via surface and grain-boundary diffusion, the length of the specimen increases due to the deposition of atoms over the diffusion length. This deposition distance is effectively increased (shortening the effective required diffusion-length) if there is creep plasticity in the region ahead of the diffusion zone. This was treated numerically by Needleman and Rice and

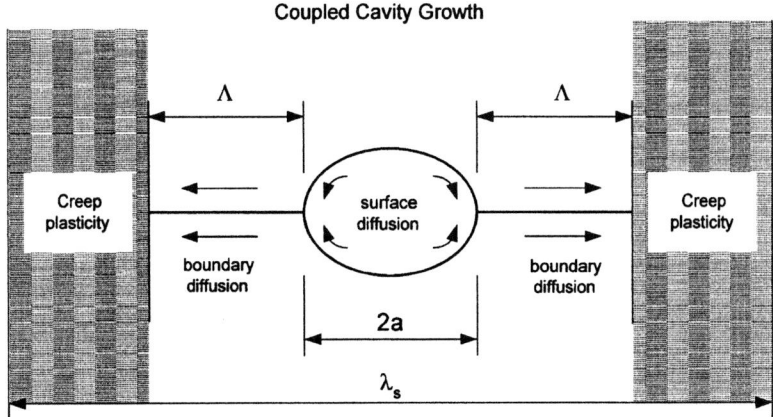

Figure 113. The model for coupled diffusive cavity growth with creep plasticity. The diffusion length is suggested to be reduced by plasticity ahead of the cavity. Based on Nix [792].

later by van der Giessen *et al.* [884]. Analytic descriptions were performed by Chen and Argon [879]. A schematic of this coupling is illustrated in Figure 113. The diffusion length is usually described as [793],

$$\Lambda = \left(\frac{D_{gb}\Omega\delta\sigma}{kT\dot{\varepsilon}}\right)^{1/3} \quad (146)$$

Chen and Argon [838] describe coupling by

$$\frac{dV}{dt} = \frac{\dot{\varepsilon}2\pi\Lambda^3}{[\ell n((a+\Lambda)/a) + (a/(a+\Lambda))^2 \times (1 - (1/4)(a/(a+\Lambda))^2) - (3/4)]} \quad (147)$$

as illustrated in Figure 114.

Similar analyses were performed by others with similar results [60,793,879,880]. It has been shown that when $\Lambda \ll a$ and λ [879,885], diffusion-controlled growth no longer applies. In the extreme, this occurs at low temperatures. Creep flow becomes important as a/Λ increases.

At small creep rates, but higher temperatures, Λ approaches $\lambda_s/2$, a/Λ is relatively small, and the growth rate can be controlled by diffusion-controlled cavity growth (DCCG). Coupling, leading to "enhanced" growth rates over the individual mechanisms, occurs at "intermediate" values of a/Λ as indicated in Figure 114. Of course, the important question is whether, under "typical creep" conditions, the addition of plasticity effects (or the coupling) is important. Needleman and Rice suggest that for $T > 0.5 T_m$, the plasticity effects are important only for $\sigma/G > 10^{-3}$

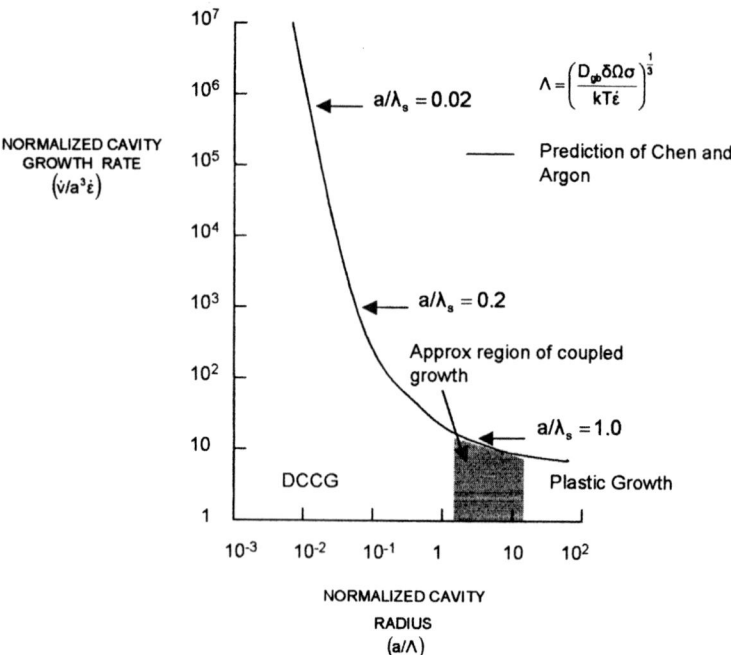

Figure 114. Prediction of growth rate for different ratios of cavity spacing λ and diffusion zone sizes Λ. From Ref. [795].

for pure metals (relatively high stress). Riedel suggests that, for pure metals, as well as creep-resistant materials, diffusive growth predominates over the whole range of creep testing. Figure 113 illustrates this coupling [792].

Even under the most relevant conditions, the cavity growth rate due to coupling is, at most, a factor of two different than the growth rate calculated by simply adding the growth rates due to creep and diffusion separately [885]. It has been suggested that favorable agreement between Chen and Argon's analytical treatments is fortuitous because of limitations to the analysis [881,882,885]. Of course, at lower temperatures, cavity growth occurs exclusively by plasticity [876]. It must be recognized that cavity growth by simply plasticity is not as well understood as widely perceived. In single-phase metals, for example, under uniaxial tension, a 50% increase in cavity size requires large strains, such as 50% [887]. Thus, a thousand-fold increase in size from the nucleated nanometer-sized cavities would not appear to be easily explained. Figure 103(b), interestingly, illustrates a case where plastic growth of cavities appears to be occurring. The cavities nucleate within grains at large particles in the dispersion-strengthened aluminum of this figure. Dunand et al. [888] suggest that this transgranular growth occurs by plasticity, as suggested by others

[791] for growth inside grains. Perhaps the interaction between cavities explains modest ductility. One case where plasticity in a pure metal is controlling is constrained thin silver films under axisymmetric loading where $\sigma_1/\sigma_2(=\sigma_3) \cong 0.82$ [2,840]. Here unstable cavity growth [840] occurs via steady-state deformation of silver. The activation energy and stress-sensitivity appear to match that of steady-state creep of silver at ambient temperature. Cavities nucleate at high-angle boundaries where obstacles are observed (high twin-density metal) by slip-plasticity. An SEM micrograph of these cavities is illustrated in Figure 115. The cavities in Figure 98 continuously nucleate and also appear to undergo plastic cavity growth. Interestingly, if a plastically deforming base metal is utilized (creep deformation of the constraining base metal of a few percent), the additional concomitant plastic strain (over that resulting from a perfectly elastic base metal) increases the nucleation rate and decreases the fracture time by several orders of magnitude, consistent with Figure 95. cavity growth can also be affected by segregation of impurities, as these may affect surface and grain-boundary diffusivity. Finally, creep fracture predictions must consider the scatter present in the data. This important, probabilistic, aspect recently has been carefully analyzed [890].

10.3.7 Creep Crack Growth

Cracks can occur in creeping metals from pre-existing flaws, fatigue, corrosion-related processes, and porosity [891,892]. In these cases, the cracks are imagined to develop relatively early in the lifetime of the metal. This contrasts the case where cracks can form in a uniformly strained (i.e., unconstrained) Cavity growth and uniform Cavity nucleation metal where interlinkage of cavities leading to crack formation is the final stage of the rupture life. Crack formation by cavity interlinkage in constrained cavity growth cases may be the rate-controlling step(s) for failure. Hence, the subject of Creep crack growth is quite relevant in the context of cavity formation. Figure 116 (from Ref. [893]) illustrates a Mode I crack. The stress/strain ahead of the crack leads to Cavity nucleation and growth. The growth can be considered to be a result of plasticity-induced expansion or diffusion-controlled cavity growth. Crack growth occurs by the coalescence of cavities with each other and the crack.

Nix et al. [893] showed that plastic growth of cavities ahead of the crack tip can lead to a "steady-state" crack growth rate. Nucleation was not included in the analysis. Nix et al. considered the load parameter to be the stress intensity factor for (elastic) metals with Mode I cracks, K_I.

$$v_c = \frac{k_g \lambda_s}{2(n-2)\ln(\lambda_s/2a)}\left(\frac{K_I}{n\sqrt{\lambda_s}}\right)^n \quad (148)$$

where k_g is a constant.

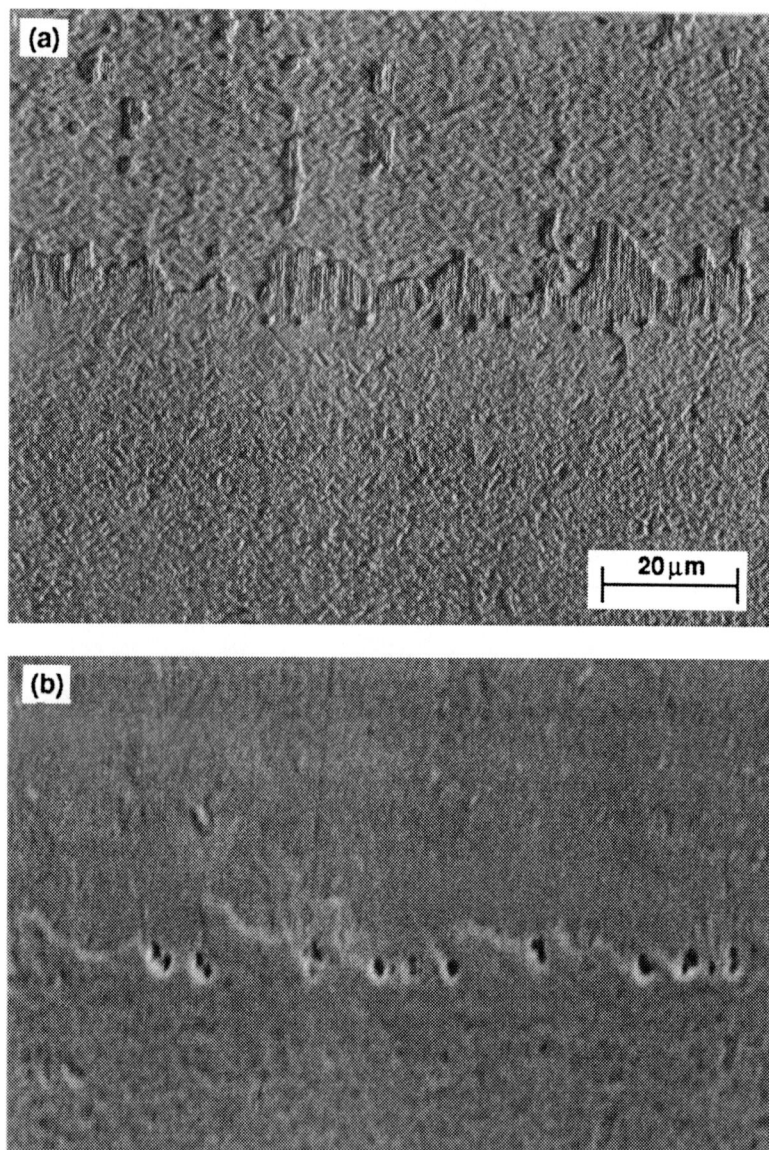

Figure 115. Creep cavitation in silver at ambient temperature. Cavities grow by unstable cavity growth, with the rate determined by steady-state creep of silver [2].

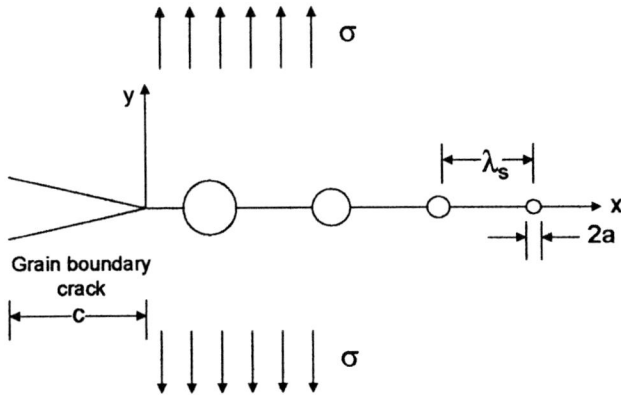

Figure 116. Grain boundary crack propagation controlled by the creep growth of cavities near a crack tip. From Ref. [893].

However, for cases of plasticity, and in the present case with time-dependent plasticity, the load parameters have been changed to J and C* [894], respectively. Much of the creep cavitation work since 1990 appears to have focused on creep cracks and analysis of the propagation in terms of C*. The C* term appears to be a reasonable loading parameter that correlates crack growth rates, although factors such as plane-stress/plane-strain (i.e., stress-state), crack branching, and extent of the damage zone from the crack tip may all be additionally important in predicting the growth rate [896–900].

Of course, another way that cracks can expand is by linking up with diffusionally growing cavities. This apppears to be the mechanism favored by Cadek [20] and Wilkinson and Vitek [900,901] and others [898]. Later, Miller and Pilkington [902] and Riedel [800] suggest that strain (plasticity) controlled growth models (with a critical strain criterion or with strain controlled nucleation) better correlate with existing crack growth data than diffusional growth models. However, Riedel indicates that the uncertainty associated with *strain*-controlled nucleation complicates the unambiguous selection of the rate-controlling growth process for cavities ahead of a crack. Figure 117 illustrates a correlation between the crack growth rate, ċ and the loading parameter C*. Riedel argued that the crack growth rate is best described by the plastic cavity growth relationship, based on a local critical strain criterion,

$$\dot{c} = k_9 \lambda^{1/n+1} (C^*)^{1/n+1} \left[\left(\frac{c - c_0}{\lambda_s} \right)^{1/n+1} - k_{10} \right] \quad (149)$$

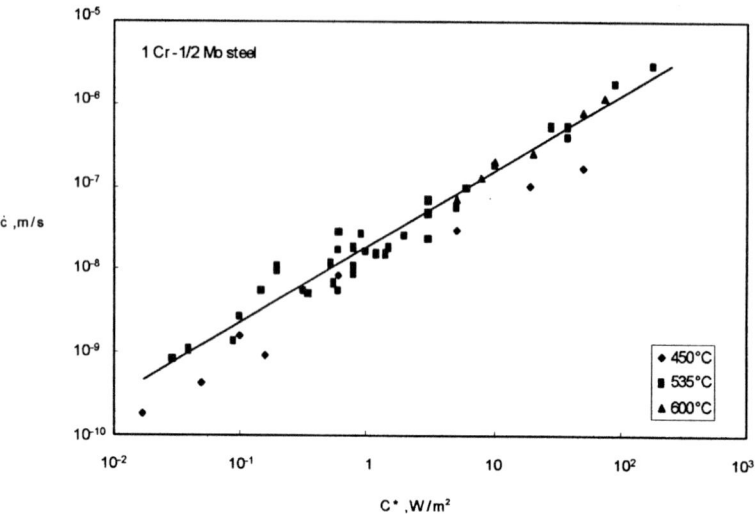

Figure 117. The crack growth rate versus loading parameter C^* for a steel. The line is represented by equation (144). From Ref. [810].

Riedel similarly argued that if Cavity nucleation occurs instantaneously, diffusional growth predicts,

$$\dot{c} = \frac{k_{11} D_b(\Omega \delta)}{2 \, kT \lambda_s^3} C^{*1/(n+1)} (c - c_0)^{n/(n+1)} \tag{150}$$

where c_0 is the initial crack length and c is the current crack length. These constants are combined (some material) constants from Riedel's original equation and the line in Figure 117 is based on equation (149) using some of these constants as adjustable parameters.

Note that equation (150) gives a strong temperature dependence (the "constants" of the equation are not strongly temperature dependent). Riedel also develops a relationship of strain-controlled cavity growth with strain-controlled nucleation, which also reasonably describes the data of Figure 117.

10.4 OTHER CONSIDERATIONS

As discussed earlier, Nix and coworkers [826,830,848,849] produced cavities by reacting with oxygen and hydrogen to produce water-vapor bubbles (cavities). Other (unintended) gas reactions can occur. These gases can include methane, hydrogen,

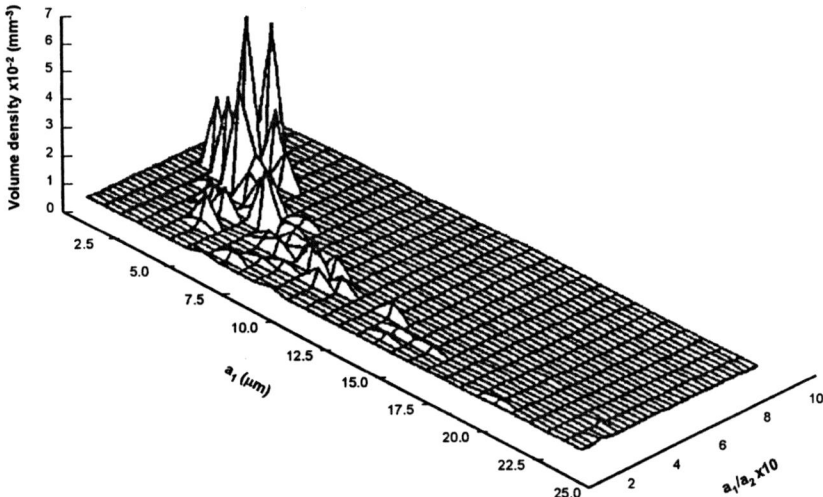

Figure 118. The cavity density versus size and aspect ratio of creep-deformed 304 stainless steel [882].

carbon dioxide. A brief review of environmental effects was discussed recently by Delph [885].

The randomness (or lack or periodicity) of the metal microstructure leads to randomness in cavitation and (e.g.) failure time. Figure 118 illustrates the cavity density versus major radius a_1 and aspect ratio a_1/a_2. This was based on metallography of creep-deformed AIS1304 stainless steel. A clear distribution in sizes is evident. Creep failure times may be strongly influenced by the random nature of grain-boundary cavitation.

References

1. Suri, S., Viswanathan, G.B., Neeraj, T., Hou, D.-H. & Mills, M.J. (1999) *Acta Mater.*, **47**, 1019.
2. Kassner, M.E., Rosen, R.S. & Henshall, G.A. (1990) *Metall. Trans.*, **21A**, 3085.
3. Basinski, Z.S. (1959) *Phil. Mag.*, **4**, 393.
4. Young, C.M., Robinson, S.L. & Sherby, O.D. (1975) *Acta Metall.*, **23**, 633.
5. Sherby, O.D. & Weertman, J. (1979) *Acta Metall.*, **27**, 387.
6. Sherby, O.D., Klundt, R.J. & Miller, A.K. (1977) *Metall. Trans. A*, **8A**, 843.
7. Kassner, M.E. (1986) *Res Mechanica*, **18**, 179.
8. Kassner, M.E. (1982) *Scripta Metall.*, **16**, 265.
9. Logan, R., Mukherjee, A.K. & Castro, R.G. (1983) *Scripta. Metall.*, **17**, 741.
10. Biberger, M. & Gibeling, J.C. (1995) *Acta Metall. et Mater.*, **43**, 3247.
11. Kassner, M.E., Pollard, J., Cerri, E. & Evangelista, E. (1994) *Acta Metall. et Mater.*, **42**, 3223.
12. Kassner, M.E. (1989) *Metall. Trans.*, **20A**, 2001.
13. Hughes, D.A. & Nix, W.D. (1988) *Metall. Trans.*, **19A**, 3013.
14. Adenstedt, H.K. (1949) *Metal Progress*, **65**, 658.
15. Tozera, T.A., Sherby, O.D. & Dorn, J.E. (1957) *Trans. ASM*, **49**, 173.
16. Sherby, O.D. & Burke, P.M. (1967) *Prog. Mater. Sci.*, **13**, 325.
17. Takeuchi, S. & Argon, A.S. (1975) *J. Mater. Sci.*, **11**, 1542.
18. Argon, A.S. (1996) *Physical Metallurgy*, Cahn, R.W. & Haasen, P. Eds., Elsevier, p. 1957.
19. Orlova, A. & Cadek, J. (1986) *Mater. Sci. and Eng.*, **77**, 1.
20. Cadek, J. (1988) *Creep in Metallic Materials*, Elsevier, Amsterdam.
21. Mukherjee, A.K. (1975) *Treatise on Materials Science and Technology*, Arsenault, R.J., Ed., Academic, New York, Vol. 6., p. 163.
22. Blum, W. (1993) Plastic Deformation and Fracture, in Cahn, R.W., Haasen, P. & Kramer, E.J. Eds. *Materials Science and Technology*, Mughrabi, H., Ed. VCH Publishers, Wienheim, Vol. 6, p. 339.
23. Nabarro, F.R.N. and de Villers, H.L. (1995). *The Physics of Creep*, Taylor and Francis, London.
24. Weertman, J. (1983) *Physical Metallurgy*, 3rd ed., Cahn, R.W. & Haasen, P., Eds. Elsevier, Amsterdam.
25. Weertman, J. (1999) *Mechanics and Materials Interlinkage*, John Wiley.
26. Nix, W.D. & Ilschner, B. (1980) *Strength of Metals and Alloys*, Haasen, P., Gerold, V. & Kostorz, G. Eds., Pergamon, Oxford, p. 1503.
27. Nix, W.D. & Gibeling, J.C. (1985) *Flow and Fracture at Elevated Temperatures*, Raj, R. Ed., ASM, Materials Park, OH, p. 1.
28. Evans, R.W. & Wilshire, B. (1985) *Creep of Metals and Alloys*, Inst. of Metals, London.
29. Kassner, M.E. and Pérez-Prado, M.-T. (2000) *Progress Mater. Sci.*, **45**, 1.
30. Evans, H.E. (1984) *Mechanisms of Creep Fracture*, Elsevier App. Science, London.
31. Shrzypek, J.J. (1993) *Plasticity and Creep*, CRC Press, Boca Raton.
32. Gittus, J. (1975) *Creep, Viscoelasticity and Creep Fracture in Solids*, Applied Science Pub., London.

33. Frost, H.J. & Ashby, M.E. (1982) *Deformation Mechanisms Maps, The Plasticity and Creep of Metals and Ceramics*, Pergamon Press.
34. Norton, E.H. (1929) *Creep of Steel at High Temperatures*, McGraw Hill, New York.
35. Monkman, F.C. & Grant, N.J. (1956) *Proc. ASTM*, **56**, 593.
36. Yavari, P. & Langdon, T.G. (1982) *Acta Metall.*, **30**, 2181.
37. McQueen, H.J., Solberg, J.K., Ryum, N. & Nes, E. (1989) *Phil. Mag. A*, **60**, 473.
38. Shrivastava, S.C., Jonas, J.J. & Canova, G.R. (1982) *J. Mech. Phys. Sol.*, **30**, 75.
39. Kassner, M.E., Nguyen, N.Q., Henshall, G.A. & McQueen, H.J. (1991) *Mater. Sci. Eng.*, **A132**, 97.
40. Campbell, A.N., Tao, S.S. & Turnbull, D. (1987) *Acta Metall.*, **35**, 2453.
41. Porrier, J.P. (1978) *Acta Metall.*, **26**, 629.
42. Morris, M.A. & Martin, J.L. (1984) *Acta Metall.*, **32**, 1609.
43. Morris, M.A. & Martin, J.L. (1984) *Acta Metall.*, **32**, 549.
44. Luthy, H., Miller, A.K. & Sherby, O.D. (1980) *Acta Metall.*, **28**, 169.
45. Goods, S.H. & Nix, W.D. (1978) *Acta Metall.*, **26**, 753.
46. Hughes, D.A., Liu, Q., Chrzan, D.C. & Hansen, N. (1997) *Acta Mater.*, **45**, 105.
47. Gil Sevillano, J., Van Houtte, R.V. and Aernoudt, E. (1980) *Prog. Mater. Sci.*, **25**, 69.
48. Garofalo, F. (1965) *Fundamentals of Creep Rupture in Metals*, Macmillan, New York.
49. Bird, J.E., Mukherjee, A.K. & Dorn, J.E. (1969) *Quantitative Relations Between Properties and Microstructure*, Brandon, D.G. & Rosen, A. Eds., Israel Univ. Press, Jerusalem, p. 255.
50. Harper, J. & Dorn, J.E. (1957) *Acta Metall.*, **5**, 654.
51. Herring, C. (1950) *J. Appl. Phys.*, **21**, 437.
52. Coble, R.L. (1963) *J. Appl. Phys.*, **34**, 1679.
53. Wu, M.Y. & Sherby, O.D. (1984) *Acta Metall.*, **32**, 1561.
54. Ardell, A.J. (1997) *Acta Mater.*, **45**, 2971.
55. Blum, W. & Maier, W. (1999) *Phys. Stat. Sol. A*, **171**, 467.
56. Owen, D.M. & Langdon, T.G. (1996) *Mater. Sci. and Eng.*, **A216**, 20.
57. Greenwood, G.W. (1994) *Scripta Metall. et Mater.*, **30**, 1527.
58. Bilde-Sorenson, J.B. and Smith, D.A. (1994) *Scripta Metall. et Mater.*, **30**, 1527.
59. Wolfenstine, J., Ruano, O.A., Wadsworth, J. & Sherby, O.D. (1993) *Scripta Metall. et Mater.*, **29**, 515.
60. Burton, B. & Reynolds, G.L. (1995) *Mater. Sci. and Eng.*, **A191**, 135.
61. Ruano, O.A., Wadsworth, J., Wolfenstine, J. & Sherby, O.D. (1993) *Mater. Sci. and Eng.*, **A165**, 133.
62. Konig, G. & Blum, W. (1980) *Strength of Metals and Alloys*, Haasen, P., Gerold, V. & Kostorz, G. Eds., Pergamon, Oxford, p. 363.
63. Shewmon, P. (1989) *Diffusion in Solids*, 2nd ed., TMS, Warrendale, PA.
64. Sherby, O.D., Lytton, J.L. & Dorn, J.E. (1957) *Acta Metall.*, **5**, 219.
65. Landon, P.R., Lytton, J.L., Shepard, L.A. & Dorn, J.E. (1959) *Trans. ASM*, **51**, 900.
66. Cuddy, L.J. (1970) *Metall. Trans.*, **1A**, 395.
67. Thompson, A.W. & Odegard, B. (1973) *Metall. Trans.*, **4**, 899.
68. Sherby, O.D. & Miller, A.K. (1979) *J. Eng. Mater., Technol.*, **101**, 387.
69. Spingarn, J.R., Barnett, D.M. & Nix, W.D. (1979) *Acta Metall.*, **27**, 1549.
70. Raj, S.V. & Langdon, T.G. (1989) *Acta Metall.*, **37**, 843.
71. Mecking, H. & Estrin, Y. (1980) *Scripta Metall.*, **14**, 815.
72. Mukherjee, A.K., Bird, J.E. & Dorn, J.E. (1969) *ASM Trans. Quart.*, **62**, 155.

References

73. Mohamed, F.A. & Langdon, T.G. (1974) *Acta Metall.*, **22**, 779.
74. Ardell, A.J. & Sherby, O.D. (1961) *Trans. AIME*, **239**, 1547.
75. Hood, G.M., Zou, H., Schultz, R.J., Bromley, E.H. & Jackman, J.A. (1994) *J. Nuc. Mater.*, **217**, 229.
76. Straub, S. & Blum, W. (1990) *Scripta Metall. et Mater.*, **24**, 1837.
77. Blum, W. (1991) *Hot Deformation of Aluminum Alloys*, Langdon, T.G., Merchant, H.D., Morris, J.G. & Zaidi, M.A. Eds., TMS, Warrendale, PA. p.181.
78. Kassner, M.E., M.-T. Perez-Prado, Long, M. & Vecchio, K.S. (2002) *Metall. and Mater. Trans.*, **33A**, 311.
79. Viswanathan, G.V., Karthikeyan, S., Hayes, R.W. & Mills, H.J. (2002) *Metall. and Mater. Trans.*, 33 in press.
80. Hayes, T.A., Kassner, M.E. & Rosen, R.S. (2002) *Metall. and Mater. Trans.*, 33 in press.
81. Blum, W. (1999) private communication.
82. Wilshire, B. (2002) *Metall. and Mater. Trans.*, 33.
83. Evans, R.W. & Wilshire, B. (1987) *Creep and Fracture of Structural Materials*, Wilshire, B. & Evans, R.W. Eds., Inst. Metals, London. p. 59.
84. McLean, D. (1968) *Trans. AIME*, **22**, 1193.
85. Ostrom, P. & Lagneborg, R. (1980) *Res Mechanica*, **1**, 59.
86. Ardell, A.J. & Przystupa, M.A. (1984) *Mech. Mater.*, **3**, 319.
87. Friedel, J., *Dislocations*, Pergamon Press, Oxford, 1964.
88. Hull, D. & Bacon, D.J. (1984) *Introduction to Dislocations*, 3rd edn, Butterworth, Heinemann, Oxford.
89. Ginter, T.J. & Mohamed, F.A. (1982) *J. Mater. Sci.*, **17**, 2007.
90. Straub, S. & Blum, W. (1996) *Hot Workability of Steels and Light Alloys*, McQueen, H.J., Konopleva, E.V. & Ryan, N.D. Eds., Canadian Inst. Mining, Metallurgy and Petroleum, Montreal, Quebec. p. 1889.
91. Hirsch, P.B., Howie, A., Nicholson, P.B., Pashley, D.W. & Whelan, M.J. (1977) *Electron Microscopy of Thin Crystals*, Kreiger, New York.
92. Blum, W., Absenger, A. & Feilhauer, R. (1980) *Strength of Metals and Alloys*, Haasen, P., Gerold, V. & Kostorz, G. Eds., Pergamon Press, Oxford, p. 265.
93. Hausselt, J. & Blum, W. (1976) *Acta Metall.*, **24**, 1027.
94. Randle, V. (1993) *The Measurement of Grain Boundary Geometry*, IOP, London.
95. Calliard, D. & Martin, J.L. (1982) *Acta Metall.*, **30**, 791.
96. Cailliard, D. & Martin, J.L. (1982) *Acta Metall.*, **30**, 437.
97. Cailliard, D. & Martin, J.L. (1983) *Acta Metall.*, **31**, 813.
98. Kassner, M.E., Elmer, J.W. & Echer, C.J. (1986) *Metall. Trans. A*, **17**, 2093.
99. Kassner, M.E. (1984) *Mater. Lett.*, **5B**, 451.
100. Raj, S.V. & Pharr, G.M. (1986) *Mater. Sci. and Eng.*, **81**, 217.
101. Ardell, A.J. & Lee, S.S. (1986) *Acta Metall.*, **34**, 2411.
102. Lin, P., Lee, S.S. & Ardell, A.J. (1989) *Acta Metall.*, **37**, 739.
103. Challenger, K.D. & Moteff, J. (1973) *Metall. Trans.*, **4**, 749.
104. Parker, J.D. & Wilshire, B. (1980) *Phil. Mag.*, **41A**, 665.
105. Evans, E. & Knowles, G. (1977) *Acta Metall.*, **25**, 963.
106. Ajaja, O. (1990) *Scripta Metall.*, **24**, 1435.
107. Kassner, M.E. (1990) *J. Mater. Sci.*, **25**, 1997.
108. Shi, L.Q. & Northwood, D.O. (1995) *Phys Stat. Sol. A*, **149**, 213.

109. Burton, B. (1982) *Phil. Mag.*, **45**, 657.
110. Kassner, M.E., Miller, A.K. & Sherby, O.D. (1982) *Metall. Trans. A*, **13**, 1977.
111. Kassner, M.E., Ziaai-Moayyed, A.A. and Miller, A.K. (1985) *Metall. Trans A*, **16**, 1069.
112. Karashima, S., Iikubo, T., Watanabe, T. & Oikawa, H. (1971) *Trans. Jpn. Inst. Metals*, **12**, 369.
113. Hofmann, U. & Blum, W. (1993) *7th Inter. Symp. on Aspects of High Temperature Deformation and Fracture of Crystalline Materials*, Jap. Inst. Metals, Sendai, p. 625.
114. Gibbs, G.B. (1966) *Phil. Mag.*, **13**, 317.
115. Ahlquist, C.N. & Nix, W.D. (1969) *Scripta Metall.*, **3**, 679.
116. Ahlquist, C.N. & Nix, W.D. (1971) *Acta Metall.*, **19**, 373.
117. Gibeling, J.C. & Nix, W.D. (1980) *Acta Metall.*, **28**, 1743.
118. Mills, M.J., Gibeling, J.C. & Nix, W.D. (1985) *Acta Metall.*, **33**, 1503.
119. Gibeling, J.C. & Nix, W.D. (1980) *Mater. Sci. and Eng.*, **45**, 123.
120. Nakayama, G.S. & Gibeling, J.C. (1990) *Acta Metall.*, **38**, 2023.
121. Weckert, E. & Blum, W. (1979) *Strength of Metals and Alloys*, McQueen, H.J., J.-P. Bailon, Dickson, J.I., Jonas, J.J. & Akben, M.G. Eds., Pergamon Press, Oxford. p. 773.
122. Blum, W., Cegielska, A., Rosen, A. & Martin, J.L. (1989) *Acta Metall.*, **37**, 2439.
123. Blum, W. & Finkel, A. (1982) *Acta Metall.*, **30**, 1705.
124. Blum, W. & Weckert, E. (1987) *Mater. Sci. and Eng.*, **86**, 147.
125. Muller, W., Biberger, M. & Blum, W. (1992) *Phil. Mag.*, **A66**, 717.
126. Evans, R.W., Roach, W.J.F. & Wilshire, B. (1985) *Mater. Sci. and Eng.*, **73**, L5.
127. Ferreira, P.I. & Stang, R.G. (1983) *Acta Metall.*, **31**, 585.
128. Mohamed, F.A., Soliman, M.S. & Mostofa, M.S. (1985) *Phil Mag.*, **51**, 1837.
129. Huang, Y. & Humphreys, F.J. (1997) *Acta Mater.*, **45**, 4491.
130. Carrard, M. & Martin, J.L. (1985) *Strength of Metals and Alloy*, McQueen, H.J., Bailon, J.-P., Dickson, J.I., Jonas, J.J. & Akben, M.G. Eds., Pergamon Press, Oxford. p. 665.
131. Northwood, D.O. & Smith, I.O. (1989) *Phys. Stat. Sol. (a)*, **115**, 1495.
132. Kassner, M.E. & Mukherjee, A.K. (1983) *Scripta Metall.*, **17**, 741.
133. Kocks, U.F., Argon, A.S. & Ashby, M.F. (1975) *Prog. Mater. Sci.*, **19**, 1.
134. Parker, J.D. & Wilshire, B. (1976) *Phil. Mag.*, **34**, 485.
135. Blum, W., Hausselt, J. and König, G. (1976) *Acta Metall.*, **24**, 239.
136. Straub, S., Blum, W., Maier, H.J., Ungar, T., Borberly, A. & Renner, H. (1996) *Acta Mater.*, **44**, 4337.
137. Argon, A.S. & Takeuchi, S. (1981) *Acta Metall.*, **29**, 1877.
138. Mughrabi, H. (1983) *Acta Metall.*, **31**, 1367.
139. Nix, W.D., Gibeling, J.C. & Fuchs, K.P. (1982) *Mechanical Testing for Deformation Model Development*, ASTM STP 765, Rhode, R.W. & Swearengin, J.C. Eds., ASTM, p. 301
140. Peralta, P., Llanes, P., Bassani, J. & Laird, C. (1994) *Phil. Mag.*, **70A**, 219.
141. Hughes, D.A. & Godfrey, A. (1998) *Hot Deformation of Aluminum Alloys*, Bieler, T.A., Lalli, L.A. & MacEwen, S.R. Eds., TMS, Warrendale, PA, p. 23.
142. Kassner, M.E. (1989) *Metall. Trans.*, **20A**, 2182.
143. Thiesen, K.E., Kassner, M.E., Pollard, J., Hiatt, D.R. & Bristow, B. (1993) *Titanium '92*, Froes, F.H. & Caplan, I.L. Eds., TMS, Warrendale, PA, p. 1717.
144. Widersich, H. (1964) *J. Met.*, **16**, 423.
145. Kocks, U.F. (1980) *Strength of Metals and Alloys*, Haasen, P., Gerold, V. & Kostorz, G. Eds., Pergamon Press, Oxford. p. 1661.

146. Kassner, M.E. & McMahon, M.E. (1987) *Metall. Trans.*, **18A**, 835.
147. Kassner, M.E. & Li, X. (1991) *Scripta. Metall et Mater.*, **25**, 2833.
148. Kassner, M.E. (1993) *Mater. Sci. Eng.*, **166**, 81.
149. Kassner, M.E. & Kyle, K. (2003) *Acta Mater.*, in press.
150. Bailey, J.E. & Hirsch, P.B. (1960) *Phil. Mag.*, **5**, 485.
151. Jones, R.L. & Conrad, H. (1969) *TMS*, AIME, **245**, 779.
152. Ajaja, O. & Ardell, A.J. (1977) *Scripta Metall.*, **11**, 1089.
153. Ajaja, O. & Ardell, A.J. (1979) *Phil. Mag.*, **39**, 65.
154. Shi, L. & Northwood, D.O. (1993) *Phys. Stat. Sol.* (*a*), **137**, 75.
155. Shi, L. & Northwood, D.O. (1993) *Phys. Stat. Sol.* (*a*), **140**, 87.
156. Henshall, G.A., Kassner, M.E. & McQueen, H.J. (1992) *Metall. Trans.*, **23A**, 881.
157. Mills, M.J., unpublished research, presented at the Conference for Creep and Fracture of Structural Materials, Swansea, UK, April 1987.
158. Da, E.N., Andrade, C. (1910) *Proc. Royal Soc.*, **A84**, 1.
159. Cottrell, A.H. & Aytekin, V. (1947) *Nature*, **160**, 328.
160. Conway, J.B. (1967) *Numerical Methods for Creep and Rupture Analyses*, Gordon and Breach, New York.
161. Garofalo, F., Richmond, C., Domis, W.F. and von Gemmingen, F. (1963) *Proc. Joint. Int. Conf. on Creep*, Inst. Mech. Eng., London, p. 1.
162. Barrett, C.R., Nix, N.D. & Sherby, O.D. (1966) *Trans. ASM*, **59**, 3.
163. Sikka, V.K., Nahm, H. & Moteff, J. (1975) *Mater. Sci. Eng.*, **20**, 55.
164. Orlova, A., Pahutova, M. & Cadek, J. (1972) *Phil. Mag.*, **25**, 865.
165. Daily, S. & Ahlquist, C.N. (1972) *Scripta Metall.*, **6**, 95.
166. Orlova, A., Tobolova, Z. & Cadek, J. (1972) *Philos. Mag.*, **26A**, 1263.
167. Suh, S.H., Cohen, J.B. & Weertman, J. (1981) *Metall. Trans.*, **12**, 361.
168. Petry, F. & Pchenitzka, F.P. (1984) *Mater. Sci. Eng.*, **68**, L7.
169. Kassner, M.E., *Acta Mater.*, 2003, in press.
170. Langdon, T.G., Vastava, R.D. & Yavari, P. (1980) *Strength of Metals and Alloys*, Haasen, P., Gerold, V. & Kostorz, G. Eds., Pergamon, Oxford. p. 271.
171. Evans, B.W., Roach, W.J.F. & Wilshire, B. (1985) *Scripta Metall.*, **19**, 999.
172. Delos-Reyes, M.M.S. (1996) Thesis, Department of Mechanical Engineering, Oregon State University, Corvallis, OR.
173. Sleeswyk, A.W., James, M.R., Plantinga, D.H. & Maathuis, W.S.T. (1978) *Acta Metall.*, **126**, 1265.
174. Orowan, E. (1959) *Internal Stress and Fatigue in Metals*, General Motors Symposium, Elsevier, Amsterdam. p. 59.
175. Hasegawa, T., Ikeuchi, Y. & Karashima, S. (1972) *Metal. Sci.*, **6**, 78.
176. Mughrabi, H. (1981) *Mater. Sci. Eng. A*, **85**, 15.
177. Derby, B. & Ashby, M.F. (1987) *Acta Metall*, **35**, 1349.
178. Vogler, S. & Blum, W. (1990) *Creep and Fracture of Engineering Materials and Structures*, Wilshire, B. & Evans, R.W. Eds., Inst. Metals, London. p. 65.
179. Lepinoux, J. & Kubin, L.P. (1985) *Phil. Mag., A*, **57**, 675.
180. Kassner, M.E., Delos-Reyes, M.A. and Wall, M.A. (1997) *Metall. and Mater. Trans.*, **28A**, 595.
181. Borbély, A., Hoffmann, G., Aernoudt, E. & Ungar, T. (1997) *Acta Mater.*, **45**, 89.
182. Kassner, M.E., Weber, F.J., Koike, J. & Rosen, R.S. (1996) *J. Mater. Sci.*, **31**, 2291.
183. Levine, L.E., private communication, March 1998.

184. Gaal, I. (1984) *Proc. 5th Int. Riso Symposium*, Hersel Andersen, N., Eldrup, M., Hansen, N., Juul Jensen, D., Leffers, T., Lilkolt, H., Pedersen, O.B. & Singh, B.N. Eds., Roskilde, Denmark, p. 249.
185. Kassner, M.E., M.T. Perez-Prado, Vecchio, K.S. & Wall, M.A. (2000) *Acta Mater.*, **48**, 4247.
186. Kassner, M.E., M.-T. Pérez-Prado and Vecchio, K.S., *Mater. Sci. and Eng.*, in press.
187. Weertman, J. (1975) *Rate Process in Plastic Deformation of Metals*, Li, J.C.M. & Mukherjee, A.K. (Eds.), ASM, Materials Park, OH, p. 315.
188. Weertman, J. (1968) *Trans. Quart. ASM*, **61**, 680.
189. Weertman, J. (1955) *J. Appl. Phys.*, **26**, 1213.
190. Barrett, C.R. & Nix, W.D. (1965) *Acta Metall.*, **12**, 1247.
191. Watanabe, T. & Karashima, S. (1970) *Trans. JIM*, **11**, 159.
192. Ivanov, L.I. & Yanushkevich, V.A. (1964) *Fiz. Metal. Metal.*, **17**, 112.
193. Blum, W. (1971) *Phys. Status. Solidi*, **45**, 561.
194. Weertman, J. (1984) *Creep and Fracture of Engineering Materials*, B. Wilshire, et al. Eds., Pineridge, Swansea. p. 1.
195. Weertman, J. & Weertman, J.R. (1987) *Constitutive Relations and Their Physical Basis*, Andersen, S.E., Bilde-Sorensen, J.B., Hansen, N., Leffers, T., Lilholt, H., Pedersen, O.B. & Ralph, B. Eds., *Proc. of 8th Risø Int. Symp. on Materials Science*, Risø, Denmark, p. 191.
196. Mitra, S.K. & McLean, D. (1967) *Metal. Sci.*, **1**, 192.
197. Lagneborg, R. (1972) *Metal Sci. J.*, **6**, 127.
198. Mott, N.F. (1956) *Conference on Creep and Fracture of Metals at High Temperatures*, H.M. Stationery Office, London.
199. Raymond, L. & Dorn, J. (1964) *Trans. Met. Soc. AIME*, **230**, 560.
200. Gittus, J. (1974) *Acta Metall.*, **22**, 789.
201. Nes, E., Blum, W. & Eisenlohr, P. (2002) *Metall. Mater. Trans. A*, **33**, 305.
202. Maruyama, K., Karashima, S. & Oikawa, H. (1983) *Res Mechanica*, **7**, 21.
203. Orowan, E. (1934) *Z. Physik*, **89**, 614.
204. Bailey, R.W. (1926) *Inst. Metals*, **35**, 27.
205. Daehn, G.S., Brehm, H. & Lim, B.S., in press.
206. Brehm, H. & Daehn, G.S. (2002) *Metall. Mater. Trans.*, **33A**, 363.
207. Bendersky, L., Rosen, A. & Mukherjee, A.K. (1985) *Int. Met. Rev.*, **30**, 1.
208. Parker, J.D. & Wilshire, B. (1978) *Metal. Sci.*, **12**, 453.
209. Li, J.C.M. (1960) *Acta Metall.*, **8**, 296.
210. Henderson-Brown, M. & Hale, K.F. (1974) *HVEM*, Swann, P.R., Humphreys, C.J. & Goringe, M.J. Eds., Academic Press, London, p. 206.
211. Klundt, R.J. (1978) PhD Thesis, Department of Materials Science and Engineering, Stanford University, Stanford, CA.
212. Parker, J.D. & Wilshire, B. (1980) *Mater. Sci. and Eng.*, **43**, 271.
213. Levinstein, H.J. & Robinson, W.H. (1963) *The Relations between Structure and the Mechanical Properties of Metal*, Symp. at the National Physical Lab., January 1963 (Her Majesty's Stationery Office), p. 180. From Weertman, J. & Weertman, J.L. (1983) *Physical Metallurgy*, Cahn, R.W. & Hassen, P. Eds., Elsevier, p. 1259.
214. Blum, W., private communication, 2002.
215. Nes, E. (1998) *Prog. in Mater. Sci.*, **41**, 129.
216. Gorman, J.A., Wood, D.S. & Vreeland, T. (1969) *J. App. Phys.*, **40**, 833.

217. Kocks, U.F., Stout, M.G. & Rollett, A.D. (1989) *Strength of Metals and Alloys*, V. 1, Kettunen, P.O., Lepistö, T.K. & Lehtoness, M.E. Eds., Pergamon, Oxford, p. 25.
218. Kassner, M.E., Myshlyaev, M.M. & McQueen, H.J. (1989) *Mater. Sci. and Eng.*, **108A**, 45.
219. McQueen, H.J., Knustad, O., Ryum, N. & Solberg, J.K. (1985) *Scripta Metall.*, **19**, 73.
220. Myshlyaev, M.M., Senkov, O.N. & Likhachev, V.A. (1985) in McQueen, H.J., Bailon, J.-P., Dickson, J.I., Jonas, J.J. & Akben, M.G. *Strength of Metals and Alloys*, Pergamon, Oxford. p. 841.
221. Schmidt, C.G., Young, C.M., Walser, B., Klundt, R.H. & Sherby, O.D. (1982) *Metall. Trans.*, **13A**, 447.
222. Doherty, R.D., Hughes, D.A., Humphreys, F.J., Jonas, J.J., Juul-Jensen, D., Kassner, M.E., King, W.E., McNelley, T.R., McQueen, H.J. & Rollett, A.D. (1997) *Mater. Sci. and Eng.*, **A238**, 219.
223. Matlock, D.K. & Nix, W.D. (1974) *Metall Trans.*, **5A**, 961.
224. Petkovic, R.A., Luton, M.J. & Jonas, J.J. (1979) *Metal Sci.*, 569.
225. Conrad, H. (1963) *Acta. Metall.*, **11**, 75.
226. Hansen, D. & Wheeler, M.A. (1931) *J. Inst. Met.*, **45**, 229.
227. Weinberg, F.A. (1958) *Trans. AIME*, **212**, 808.
228. Barrett, C.R., Lytton, J.L. & Sherby, O.D. (1967) *Trans. TMS-AIME*, **239**, 170.
229. Garofalo, F., Domis, W.F. and von Gemmingen, R. (1968) *Trans. TMS-AIME*, **240**, 1460.
230. Kikuchi, S. & Yamaguchi, A. (1985) *Strength of Metals and Alloys*, McQueen, H.J., Bailon, J.-P., Dickson, J.I., Jonas, J.J. & Akben, M.G. Eds., Pergammon Press, Oxford. p. 899.
231. Kassner, M.E. (1995) *Grain Size and Mechanical Properties: Fundamentals and Applications*, Otooni, M.A., Armstrong, R.W., Grant, N.J. & Rshizaki, K. Eds., *MRS*, p. 157.
232. Al-Haidary, J.T., Petch, M.J. and Delos-Rios, E.R. (1983) *Phil. Mag.*, **47A**, 863.
233. Perdrix, C., Perrin, Y.M. & Montheillet, F. (1981) *Mem. Sci. Rev. Metall.*, **78**, 309.
234. Evans, W.J. & Wilshire, B. (1974) *Scripta Metall.*, **8**, 497.
235. Evans, W.J. & Wilshire, B. (1970) *Metall. Trans.*, **1**, 2133.
236. Warda, R.D., Fidleris, V. & Teghtsoonian, E. (1973) *Metall. Trans.*, **4**, 1201.
237. Nabarro, F.R.N. (1948) *Rept. of Conf. on the Solids*, The Physical Society, London, p. 75.
238. Greenwood, G.W. (1992) *Proc. R. Soc. Lond. A*, **436**, 187.
239. Fiala, J. & Cadek, J. (1985) *Mater. Sci. Eng.*, **75**, 117.
240. Fiala, J., Novotny, J. & Cadek, J. (1983) *Mater. Sci. Eng.*, **60**, 195.
241. Crossland, I.G. & Jones, R.B. (1977) *Met. Sci.*, **11**, 504.
242. Sritharan, T. & Jones, H. (1979) *Acta Metall.*, **27**, 1293.
243. Mishra, R.S., Jones, H. & Greenwood, G.W. (1989) *Phil. Mag. A*, **60**(6), 581.
244. Pines, B. Ya. and Sirenko, A.F. (1959) *Fiz. Met. Metall.*, **7**, 766.
245. Jones, R.B. (1965) *Nature*, **207**, 70.
246. Burton, B. & Greenwood, G.W. (1970) *Acta Metall.*, **18**, 1237.
247. Harris, K.E. & King, A.H. (1998) *Acta Mater.*, **46**(17), 6195.
248. Coble, R.L. (1965) *High Strength Materials*, John Wiley, New York.
249. Burton, B. & Reynolds, G.L. (1975) *Physical Metallurgy of Reactor Fuel Elements*, Metals Soc., London, p. 87.

250. Folweiler, R. (1950) *J. Appl. Phys.*, **21**, 437.
251. Warshau, S.I. & Norton, F.H. (1962) *J. Am. Ceram. Soc.*, **45**, 479.
252. Chang, R. (1959) *J. Nucl. Mater.*, **2**, 174.
253. Bernstein, I.M. (1967) *Trans. AIME*, **239**, 1518.
254. Burton, B. & Greenwood, G.W. (1970) *Met. Sci. J.*, **4**, 215.
255. Crossland, I.G. (1975) *Physical Metallurgy of Reactor Fuel Elements*, Metals Soc., London, p. 66.
256. Crossland, I.G., Burton, B. & Bastow, B.D. (1975) *Metal Sci.*, **9**, 327.
257. Towle, D.J. & Jones, H. (1975) *Acta Metall.*, **24**, 399.
258. Passmore, E.M., Duff, R.H. & Vasilos, T.S. (1966) *J. Am. Ceram. Soc.*, **49**, 594.
259. Langdon, T.G. & Pask, J.A. (1970) *Acta Metall.*, **18**, 505.
260. Knorr, D.B., Cannon, R.M. & Coble, R.L. (1989) *Acta Metall.*, **37**(8), 2103.
261. Langdon, T.G. (1996) *Scripta Mater.*, **35**(6), 733.
262. Harris, J.E. (1973) *Met. Sci. J.*, **7**, 1.
263. Squires, R.L., Weiner, R.T. & Phillips, M. (1963) *J. Nucl. Mat.*, **8**(1), 77.
264. *Proc. Int. Conf. Creep of Advanced Materials for the 21st Century*, San Diego, 1999, TMS, Warrendale, PA, in press.
265. Ruano, O.A., Sherby, O.D., Wadsworth, J. & Wolfenstine, J. (1998) *Scripta Mater.*, **38**(8), 1307.
266. Wadsworth, J., Ruano, O.A. & Sherby, O.D. (2002) *Metall. Mater. Trans.*, **33A**, 219.
267. Ruano, O.A., Wadsworth, J. and Sherby, O.D. (1988) *Scripta Metall.*, **22**, 1907.
268. Ruano, O.A., Sherby, O.D., Wadsworth, J. & Wolfenstine, J. (1996) *Mater. Sci. Eng. A*, **211**, 66.
269. Barrett, C.R., Muehleisen, E.C. & Nix, W.D. (1972) *Mater. Sci. Eng. A*, **10**, 33.
270. Wang, J.N. (1994) *J. Mater. Sci.*, **29**, 6139.
271. Kloc, L. (1996) *Scripta Mater.*, **35**(6), 733.
272. Fiala, J. & Langdon, T.G. (1992) *Mater. Sci. Eng. A*, **151**, 147.
273. Greenfield, P., Smith, C.C. & Taylor, A.M. (1961) *Trans. AIME*, **221**, 1065.
274. McNee, K.R., Greenwood, G.W. & Jones, H. (2002) *Scripta Mater.*, **46**, 437.
275. McNee, K.R., Greenwood, G.W. & Jones, H. (2001) *Scripta Mater.*, **44**, 351.
276. Nabarro, F.R.N. (2002) *Metall. Mater. Trans. A*, **33**, 213.
277. Nabarro, F.R.N. (1999) private communication, San Diego, CA.
278. Lifshitz, I.M. (1963) *Sov. Phys.* (JETP), **17**, 909.
279. Mori, T., Onaka, S. & Wakashima, K. (1998) *J. Appl. Phys.*, **83**(12), 7547.
280. Onaka, S., Huang, J.H., Wakashima, K. & Mori, T. (1998) *Acta Mater.*, **46**(11), 3821.
281. Onaka, S., Madgwick, A. & Mori, T. (2001) *Acta Mater.*, **49**, 2161.
282. Sahay, S.S. & Murty, G.S. (2001) *Scripta Mater.*, **44**, 841.
283. Burton, B. (1977) *Diffusional Creep of Polycrystalline Materials*, Trans. Tech. Publications, Zurich, Switzerland, p. 61.
284. Stevens, R.N. (1971) *Phil. Mag.*, **23**, 265.
285. Aigeltinger, A.E. & Gifkins, R.C. (1975) *J. Mater. Sci.*, **10**, 1889.
286. Cannon, W.R. (1972) *Phil. Mag.*, **25**, 1489.
287. Gates, R.S. (1975) *Phil. Mag.*, **31**, 367.
288. Arieli, A., Gurewitz, G. & Mukherjee, A.K. (1981) *Metals Forum*, **4**, 24.
289. Gibbs, G.B. (1980) *Mater. Sci. Eng. A*, **2**, 262.
290. Raj, R. & Ashby, M.F. (1971) *Metall. Trans.*, **2**, 1113.
291. Beere, W.B. (1976) *Met. Sci.*, **10**, 133.

292. Speight, M.V. (1975) *Acta Metall.*, **23**, 779.
293. Mott, N.F. (1953) *Proc. Roy. Soc. A*, **220**, 1.
294. Mohamed, F.A., Murty, K.L. & Morris, J.W. (1973) *Metall. Trans.*, **4**, 935.
295. Yavari, P., Miller, D.A. & Langdon, T.G. (1982) *Acta Metall.*, **30**, 871.
296. Mohamed, F.A. & Ginter, T.J. (1982) *Acta Metall.*, **30**, 1869.
297. Malakondaiah, G. and Rama Rao, P. (1981) *Acta Metall.*, **29**, 1263.
298. Novotny, J., Fiala, J. & Cadek, J. (1985) *Acta Metall.*, **33**, 905.
299. Malakondaiah, G. and Rama Rao, P. (1982) *Mater. Sci. Eng.*, **52**, 207.
300. Dixon-Stubbs, P.G. and Wilshire, B. (1982) *Phil. Mag. A*, **45**, 519.
301. Langdon, T.G. (1983) *Phil. Mag. A*, **47**, L29.
302. Ruano, O.A., Wolfenstine, J., Wadsworth, J. & Sherby, O.D. (1991) *Acta Metall. Mater.*, **39**, 661.
303. Ruano, O.A., Wolfenstine, J., Wadsworth, J. & Sherby, O.D. (1992) *J. Am. Ceram. Soc.*, **75**(7), 1737.
304. Ramesh, K.S., Yasuda, E.Y. & Kimura, S. (1986) *J. Mater. Sci.*, **21**, 3147.
305. Wang, J.N. (1994) *Scripta Metall. Mater.*, **30**, 859.
306. Wang, J.N. (1994) *Phil. Mag. Lett.*, **70**(2), 81.
307. Wang, J.N., Shimamoto, T. & Toriumi, M. (1994) *J. Mater. Sci. Lett.*, **13**, 1451.
308. Wang, J.N. (1994) *J. Am. Ceram. Soc.*, **77**(11), 3036.
309. Banerdt, W.B. & Sammis, C.G. (1985) *Phys. Earth Planet. Int.*, **41**, 108.
310. Wolfenstine, J., Ruano, O.A., Wadsworth, J. & Sherby, O.D. (1991) *Scripta Metall. Mater.*, **25**, 2065.
311. Poirier, J.P., Peyronneau, J., Gesland, J.K. & Brebec, G. (1983) *Phys. Earth Planet. Int.*, **32**, 273.
312. Beauchesne, S. & Poirier, J.P. (1990) *Phys. Earth Planet. Int.*, **61**, 1982.
313. Wang, J.N., Hobbs, B.E., Ord, A., Shimamoto, T. & Toriumi, M. (1994) *Science*, **265**, 1204.
314. Wang, J.N. (1994) *Mater. Sci. Eng. A*, **183**, 267.
315. Wang, J.N. (1994) *Mater. Sci. Eng. A*, **187**, 97.
316. Berbon, M.Z. & Langdon, T.G. (1996) *J. Mater. Sci. Lett.*, **15**, 1664.
317. Mohamed, F.A. (1978) *Mater. Sci. Eng. A*, **32**, 37.
318. Langdon, T.G. & Yavari, P. (1982) *Acta Metall.*, **30**, 881.
319. Wang, J.N. & Langdon, T.G. (1994) *Acta Metall. Mater.*, **42**(7), 2487.
320. Weertman, J. & Blacic, J. (1984) *Geophys. Res. Lett.*, **11**, 117.
321. Nabarro, F.R.N. (1989) *Acta Metall.*, **37**(8), 2217.
322. Ruano, O.A., Wadsworth, J. & Sherby, O.D. (1988) *Acta Metall.*, **36**(4), 1117.
323. Wang, J.N. (1993) *Scripta Metall. Mater.*, **29**, 1267.
324. Wang, J.N. & Nieh, T.G. (1995) *Acta Metall.*, **43**(4), 1415.
325. Wang, J.N. (1995) *Phil. Mag.*, **71A**, 105.
326. Wang, J.N., Wu, J.S. & Ding, D.Y. (2002) *Mater. Sci. Eng. A*, **334**, 275.
327. Wang, J.N. (1995) *Phil. Mag. A*, **71**(1), 105.
328. Wang, J.N. (1996) *Acta Metall.*, **44**(3), 855.
329. Przystupa, M.A. & Ardell, A.J. (2002) *Metall. Mater. Trans. A*, **33**, 231.
330. Nabarro, F.R. (2000) *Phys. Stat. Sol.*, **182**, 627.
331. Blum, W., Eisenlohr, P. & Breutinger, F. (2002) *Metall. Mater. Trans. A*, **33**, 291.
332. Nes, E., Pettersen, T. & Marthinsen, K. (2000) *Scripta Mater.*, **43**, 55.
333. Marthinsen, K. & Ness, E. (2001) *Mater. Sci. Tech.*, **17**, 376.

334. Mohamed, F.A. (2002) *Metall. Mater. Trans. A*, **33**, 261.
335. Ginter, T.J., Chaudhury, P.K. & Mohamed, F.A. (2001) *Acta Mater.*, **49**, 263.
336. Ginter, T.J. & Mohamed, F.A. (2002) *Mater. Sci. Eng. A*, **322**, 148.
337. Langdon, T.G. (2002) *Metall. Mater. Trans. A*, **33**, 249.
338. Yavari, P., Mohamed, F.A. & Langdon, T.G. (1981) *Acta Metall.*, **29**, 1495.
339. Mohamed, F.A. (1998) *Mater. Sci. Eng. A*, **245**, 242.
340. Langdon, T.G. (1996) *Mater. Trans. JIM*, **37**(3), 359.
341. Oikawa, H., Sugawara, K. & Karashima, S. (1978) *Mater. Trans. JIM*, **19**, 611.
342. Endo, T., Shimada, T. & Langdon, T.G. (1984) *Acta Metall.*, **32**(11), 1991.
343. Sherby, O.D., Anderson, R.A. & Dorn, J.E. (1951) *Trans. AIME*, **191**, 643.
344. Weertman, J. (1960) *Trans. AIME*, **218**, 207.
345. Weertman, J. (1957) *J. Appl. Phys.*, **28**(10), 1185.
346. Lothe, J. (1962) *J. Appl. Phys.*, **33**(6), 2116.
347. Hirth, J.P. & Lothe, J. (1968) *Theory of Dislocations*, McGraw-Hill, NY, p. 584.
348. Oikawa, H., Matsuno, N. & Karashima, S. (1975) *Met. Sci.*, **9**, 209.
349. Horiuchi, R. & Otsuka, M. (1972) *Trans. JIM*, **13**, 284.
350. Cottrell, A.H. & Jaswon, M.A. (1949) *Proc. Roy. Soc. London A*, **199**, 104.
351. Fisher, J.C. (1954) *Acta Metall.*, **2**, 9.
352. Suzuki, H. (1957) *Sci. Rep. Res. Inst. Tohoku University A*, **4**, 455.
353. Shoeck, G. (1961) *Mechanical Behavior of Materials at Elevated Temperature*, Dorn, J.E. Ed., McGraw-Hill, New York, p. 77.
354. Snoek, J. (1942) *Physica*, **9**, 862.
355. Takeuchi, S. & Argon, A.S. (1976) *Acta Metall.*, **24**, 883.
356. Oikawa, H., Honda, K. & Ito, S. (1984) *Mater. Sci. Eng.*, **64**, 237.
357. Mohamed, F.A. & Langdon, T.G. (1974) *Acta Metall.*, **22**, 779.
358. Darken, L.S. (1948) *Trans. Am. Inst. Min. Engrs.*, **175**, 184.
359. Mohamed, F.A. (1983) *Mater. Sci. Eng.*, **61**, 149.
360. Soliman, M.S. & Mohamed, F.A. (1984) *Metall. Trans. A*, **15**, 1893.
361. Fuentes-Samaniego, R. and Nix, W.D. (1981) *Scripta Metall.*, **15**, 15.
362. Cannon, W.R. & Sherby, O.D. (1970) *Metall. Trans.*, **1**, 1030.
363. King, H.W. (1966) *J. Mater. Sci.*, **1**, 79.
364. Kassner, M.E., McQueen, H.J. & Evangelista, E. (1993) *Mater. Sci. Forum*, **113–115**, 151.
365. Laks, H., Wiseman, C.D., Sherby, O.D. & Dorn, J.E. (1957) *J. Appl. Mech.*, **24**, 207.
366. Sherby, O.D., Trozera, T.A. & Dorn, J.E. (1956) *Proc. Am. Soc. Test. Mat.*, **56**, 789.
367. Sherby, O.D. & Taleff, E.M. (2002) *Mater. Sci. Eng. A*, **322**, 89.
368. Oikawa, H., Kariya, J. & Karashima, S. (1974) *Met. Sci.*, **8**, 106.
369. Murty, K.L. (1973) *Scripta Metall.*, **7**, 899.
370. Murty, K.L. (1974) *Phil. Mag.*, **29**, 429.
371. Orlova, A. & Cadek, J. (1974) *Z. Metall.*, **65**, 200.
372. Mohamed, F.A. (1979) *Mater. Sci. Eng.*, **38**, 73.
373. Blum, W. (2001) *Mater. Sci. Eng. A*, **319–321**, 8.
374. Vagarali, S.S. & Langdon, T.G. (1982) *Acta Metall.*, **30**, 1157.
375. Jones, B.L. & Sellars, C.M. (1970) *Met. Sci. J.*, **4**, 96.
376. Oikawa, H., Maeda, M. & Karashima, S. (1973) *J JIM*, **37**, 599.
377. Matsuno, N., Oikawa, H. & Karashima, S. (1974) *J JIM*, **38**, 1071.
378. McNelley, T.R., Michel, D.J. & Salama, A. (1989) *Scripta Metall.*, **23**, 1657.

379. McQueen, H.J. & Kassner, M.E. (1990) *Superplasticity in Aerospace II*, McNelley, T.R. & Heikkenen, C. Eds., TMS, Warrendale, PA, p. 77.
380. Burke, M.A. & Nix, W.D. (1975) *Acta Metall.*, **23**, 793.
381. Hart, E.W. (1967) *Acta Metall.*, **15**, 351.
382. Taleff, E.M., Henshall, G.A., Nieh, T.G., Lesuer, D.R. & Wadsworth, J. (1998) *J. Metall. Trans.*, **29**, 1081.
383. Taleff, E.M., Lesuer, D.R. & Wadsworth, J. (1996) *J. Metall. Trans.*, **27A**, 343.
384. Taleff, E.M. & Nevland, P.J. (1999) *JOM*, **51**, 34.
385. Taleff, E.M. & Nevland, P.J. (2001) *Metall. Trans.*, **32**, 1119.
386. Watanabe, H., Tsutsui, H., Mukai, T., Kohzu, M., Tanabe, S. & Higashi, K. (2001) *Int. J. Plasticity*, **17**, 387.
387. Murty, K.L., Mohamed, F.A. & Dorn, J.E. (1972) *Acta Metall.*, **20**, 1009.
388. Chaudhury, P.K. & Mohamed, F.A. (1988) *Mater. Sci. Eng. A*, **101**, 13.
389. Nortman, A. and Neuhäuser, H. (1998) *Phys. Stat. Sol. (a)*, **168**, 87.
390. Blandin, J.J., Giunchi, D., Suéry, M. and Evangelista, E. (2002) *Mater. Sci. Tech.*, **18**, 333.
391. Li, Y. & Langdon, T.G. (1999) *Metall. Trans.*, **30**, 2059.
392. Tan, J.C. & Tan, M.J. (2003) *Mater. Sci. Eng. A*, **339**, 81.
393. Wadsworth, J., Dougherty, S.E., Kramer, P.A. & Nieh, T.G. (1992) *Scripta Metall. Mater.*, **27**.
394. Hayes, R.W. & Soboyejo, W.O. (2001) *Mater. Sci. Eng. A*, **319-321**, 827.
395. Mukhopadhyay, J., Kaschner, G.C. & Mukherjee, A.K. (1990) *Superplasticity in Aerospace II*, McNelley, T.R. & Heikkenen, C. Eds., *TMS*, Warrendale, PA, p. 33.
396. Nieh, T.G. & Oliver, W.C. (1989) *Scripta Metall.*, **23**, 851.
397. Hsiung, L.M. & Nieh, T.G. (1999) *Intermetallics*, **7**, 821.
398. Wolfenstine, J. (1990) *J. Mater. Sci. Lett.*, **9**, 1091.
399. Lin, D., Shan, A. & Li, D. (1994) *Scripta Metall. Mater.*, **31**(11), 1455.
400. Sundar, R.S., Kutty, T.R.G. & Sastry, D.H. (2000) *Intermetallics*, **8**, 427.
401. Li, D., Shan, A., Liu, Y. & Lin, D. (1995) *Scripta Metall. Mater.*, **33**, 681.
402. Yaney, D.L. & Nix, W.D. (1988) *J. Mater. Sci.*, **23**, 3088.
403. Yang, H.S., Lee, W.B. & Mukherjee, A.K. (1993) *Structural Intermetallics*, Darolia, R., Lewandowski, J.J., Liu, C.T., Martin, P.L., Miracle, D.B. & Nathal, M.V. Eds., *TMS*, Warrendale, PA. p. 69.
404. Li, Y. & Langdon, T.G. (1999) *Metall. Trans.*, **30**, 315.
405. González-Doncel, G. and Sherby, O.D. (1993) *Acta Metall.*, **41**, 2797.
406. Langdon, T.G. & Wadsworth, J. (1991) *Proc. Int. Conf. on Superplasticity in Advanced Materials*, Hori, S., Tokizane, M. & Furushiro, N. Eds., The Japan Society of Research on Superplasticity, Osaka, Japan. p. 847.
407. Bengough, G.D. (1912) *J. Inst. Metals*, **7**, 123.
408. Sherby, O.D., Nieh, T.G. & Wadsworth, J. (1994) *Materials Science Forum*, **13**, 170.
409. Pearson, C.E. (1934) *J. Inst. Metals*, **54**, 111.
410. Valiev, R.Z., Korznikov, A.V. & Mulyukov, R.R. (1993) *Mater. Sci. Eng. A*, **141**, 168.
411. Wakai, F. Sakaguchi, S. and Matsuno, Y. (1986) *Adv. Cerm. Mater.*, **1**, 259.
412. Langdon, T.G. (1993) *Ceramics International*, **19**, 279.
413. Wu, M.Y., Wadsworth, J. & Sherby, O.D. (1984) *Scripta Metall.*, **18**, 773.
414. Sikka, V.K., Liu, C.T. & Loria, E.A. (1987) *Processing of Structural Metals by Rapid Solidification*, Froes, F.H. & Savage, S.J. Eds., *ASM*, Metals Park, Ohio, p. 417.
415. Sherby, O.D. & Wadsworth, J. (1989) *Prog. Mater. Sci.*, **33**, 169.

416. Kaibyshev, O.A. (1992) *Superplasticity of Alloys, Intermetallics and Ceramics*, Springer-Verlag, Berlin, Germany.
417. Chokshi, A.H., Mukherjee, A.K. & Langdon, T.G. (1993) *Mater. Sci. Eng. R*, **10**, 237.
418. Nieh, T.G., Wadsworth, J. & Sherby, O.D. (1997) *Superplasticity in Metals and Ceramics*, Clarke, D.R., Suresh, S. & Ward, I.M. Eds., Cambridge University Press, Cambridge, UK.
419. Nieh, T.G., Henshall, C.A. & Wadsworth, J. (1984) *Scripta Metall.*, **18**, 1405.
420. Higashi, K., Mabuchi, M. & Langdon, T.G. (1996) *ISIJ International*, **36**(12), 1423.
421. Valiev, R.Z., Salimonenko, D.A., Tsenev, N.K., Berbon, P.B. & Langdon, T.G. (1997) *Scripta Mater.*, **37**(12), 1945.
422. Komura, S., Berbon, P.B., Furukawa, M., Horita, Z., Nemoto, M. & Langdon, T.G. (1998) *Scripta Mater.*, **38**(12), 1851.
423. Berbon, P.B., Furukawa, M., Horita, Z., Nemoto, M., Tsenev, N.K., Valiev, R.Z. & Langdon, T.G. (1998) *Phil. Mag. Lett.*, **78**(4), 313.
424. Langdon, T.G., Furukawa, M., Horita, Z. & Nemoto, M. (1998) *J. Metals*, **50**(6), 41.
425. Cline, H.E. & Alden, T.H. (1967) *Trans. TMS-AIME*, **239**, 710.
426. Zehr, S.W. & Backofen, W.A. (1968) *Trans. ASM*, **61**, 300.
427. Alden, T.H. (1968) *Trans. ASM*, **61**, 559.
428. Alden, T.H. (1967) *Acta Met.*, **15**, 469.
429. Holt, D.L. & Backofen, W.A. (1966) *Trans. ASM*, **59**, 755.
430. Lee, D. (1969) *Acta Met.*, **17**(8), 1057.
431. Hart, E.W. (1967) *Acta Met.*, **15**, 1545.
432. Alden, T.H. (1969) *Acta Met.*, **17**(12), 1435.
433. Langdon, T.G. (1994) *Acta Metall. Mater.*, **42**(7), 2437.
434. Raj, R. & Ashby, M.F. (1971) *Metall. Trans.*, **2**, 1113.
435. Zelin, M.G. & Mukherjee, A.K. (1996) *Mater. Sci. Eng. A*, **208**, 210.
436. Ashby, M.F. & Verral, R.A. (1973) *Acta Metall.*, **21**, 149.
437. Spingarn, J.R. & Nix, W.D. (1978) *Acta Metall.*, **26**, 1389.
438. Padmanabhan, K.A. (1979) *Mater. Sci. Eng.*, **40**, 285.
439. Langdon, T.G. (1991) *Mater. Sci. Eng.*, **137**, 1.
440. Taplin, D.M.R., Dunlop, G.L. & Langdon, T.G. (1979) *Annu. Rev. Mater. Sci.*, **9**, 151.
441. Weertman, J. (1978) *Philos. Trans. Roy. Soc. London A*, **288**, 9.
442. Ruano, O.A., Wadsworth, J. & Sherby, O.D. (1986) *Mater. Sci. Eng.*, **84**, L1.
443. Ball, A. & Hutchison, M.M. (1969) *Met. Sci. J.*, **3**, 1.
444. Langdon, T.G. (1970) *Phil. Mag.*, **26**, 945.
445. Mukherjee, A.K. (1971) *Mater. Sci. Eng. A*, **8**, 83.
446. Arieli, A. & Mukherjee, A.K. (1980) *Mater. Sci. Eng. A*, **45**, 61.
447. Langdon, T.G. (1994) *Acta Mater.*, **42**, 2437.
448. Gifkins, R.C. (1976) *Metall. Trans. A*, **7**, 1225.
449. Fukuyo, H., Tsai, H.C., Oyama, T. & Sherby, O.D. (1991) *ISIJ International*, **31**, 76.
450. Valiev, R.Z. & Langdon, T.G. (1993) *Mater. Sci. Eng. A*, **137**, 949.
451. Langdon, T.G. (1994) *Mater. Sci. Eng. A*, **174**, 225.
452. Pérez-Prado, M.T. (1998) PhD dissertation, Universidad Complutense de Madrid.
453. Cutler, C.P., Edington, J.W., Kallend, J.S. & Melton, K.N. (1974) *Acta Metall.*, **22**, 665.
454. Melton, K.N. & Edington, J.W. (1974) *Scripta Metall.*, **8**, 1141.
455. Bricknell, R.H. & Edington, J.W. (1979) *Acta Metall.*, **27**, 1303.
456. Bricknell, R.H. & Edington, J.W. (1979) *Acta Metall.*, **27**, 1313.

457. Pérez-Prado, M.T., Cristina, M.C., Ruano, O.A. and González-Doncel, G. (1998) *Mater. Sci. Eng.*, **244**, 216.
458. Pérez-Prado, M.T., McNelley, T.R., Ruano, O.A. and González-Doncel, G. (1998) *Metall. Trans. A*, **29**, 485.
459. Tsuzaki, K., Matsuyama, H., Nagao, M. and Maki, T. (1990) *Mater. Trans. JIM*, **31**, 983.
460. Qing, L., Xiaoxu, H., Mei, Y. & Jinfeng, Y. (1992) *Acta Mater.*, **40**, 1753.
461. Liu, J. & Chakrabarti, D.J. (1996) *Acta Mater.*, **44**, 4647.
462. Blackwell, P.L. & Bate, P.S. (1993) *Metall. Trans. A.*, **24**, 1085.
463. Bate, P.S. (1992) *Metall. Trans. A*, **23**, 1467.
464. Blackwell, P.L. & Bate, P.S. (1994) *Superplasticity: 60 Years after Pearson*, Ridley, N. Ed., The Institute of Materials, Manchester, UK, p. 183.
465. Johnson, R.H., Parker, C.M., Anderson, L. & Sherby, O.D. (1968) *Phil. Mag.*, **18**, 1309.
466. Naziri, N. & Pearce, R. (1970) *J. Inst. Met.*, **98**, 71.
467. McDarmaid, D.S., Bowen, A.W. & Partridge, P.G. (1984) *Mater. Sci. Eng. A*, **64**, 105.
468. Ruano, O.A. & Sherby, O.D. (1988) *Revue Phys. Appl.*, **23**, 625.
469. Sherby, O.D. & Ruano, O.A. (1982) *Proc. Int. Conf. Superplastic Forming of Structural Alloys*, Paton, N.E. & Hamilton, C.H. Eds., TMS, Warrendale, PA, p. 241.
470. Watts, B.M., Stowell, M.J., Baike, B.L. and Owen, D.G.E. (1976) *Met. Sci. J.*, **10**, 189.
471. Grimes, R. (1988) *Advances and Future Directions in Superplastic Materials*, NATO-AGARD Lecture Series, No. 168, p. 8.1.
472. Wert, J.A., Paton, N.E., Hamilton, C.H. & Mahoney, M.W. (1981) *Metall. Trans. A*, **12**, 1267.
473. Pérez-Prado, M.T., McMahon, M.E. & McNelley, T.R. (1998) *Modeling the Mechanical Response of Structural Materials*, Taleff, E.M. & Mahidhara, R.K. Eds., TMS, Warrendale, PA, p. 181.
474. Melton, K.N., Edington, J.W., Kallend, J.S. & Cutler, C.P. (1974) *Acta Metall.*, **22**, 165.
475. Randle, V. (1992) *Microtexture Determination and Its Applications*, The Institute of Metals.
476. Kocks, U.F., Tomé, C.N. & Wenk, H.R. (1998) *Texture and Anisotropy*, Cambridge Univ. Press.
477. Hosford, W.F. (1993) *The Mechanics of Crystals and Textured Polycrystals*, Oxford Univ. Press.
478. Mabuchi, M. & Higashi, K. (1998) *JOM*, **50**(6), 34.
479. Hosokawa, H. & Higashi, K. (Eds.) (2000) *Materials Science Research International*, Vol. 6, (no.3), Fourth International Symposium on Microstructure and Mechanical Properties of Engineering Materials (IMMM'99), Beijing, China, September 20-23, 1999, Soc. Mater. Sci. Japan, 153.
480. Langdon, T.G., Furukawa, M., Horita, Z. & Nemoto, M. (1999) *JOM*, **50**(6), 41.
481. Mishra, R.S. (2001) *Journal of Metals*, **53**(3), 23.
482. Mabuchi, M., Koike, J., Iwasaki, H., Higashi, K. & Langdon, T.G. (1994) *Materials Science Forum*, **170-172**, 503.
483. Kim, H.Y. & Hong, S.H. (1994) *Scripta Metall. Mater.*, **30**, 297.
484. Huang, X., Liu, Q., Yao, C. & Yao, M. (1991) *J. Mater. Sci. Lett.*, **10**, 964.
485. Matsuki, K., Matsumoto, H., Tokizawa, M., Takatsuji, N., Isogai, M., Murakami, S. & Murakami, Y. (1991) *Science and Engineering of Light Metals*, Hirano, K., Oikawa, H. & Ikeda, K., Eds., The Japan Institute of Light Metals, Tokyo.

486. Nieh, T.G., Imai, T., Wadsworth, J. & Kojima, S. (1994) *Scripta Metall. Mater.*, **31**, 1685.
487. Mabuchi, M. & Imai, T. (1990) *J. Mater. Sci. Lett.*, **9**, 763.
488. Imai, T., Mabuchi, M., Tozawa, Y. & Yamada, M. (1990) *Journal of Materials Science Letters*, **9**, 255.
489. Mabuchi, M., Higashi, K., Inoue, K. & Tanimura, S. (1992) *Scripta Metall. Mater.*, **26**,. 1839.
490. Imai, T., L'Esperance, G. & Hong, B.D. (1991) *Scripta Metall. Mater.*, **25**, 2503.
491. Nieh, T.G. & Wadsworth, J. (1991) *Mater. Sci. Eng.*, **147**, 129.
492. Nieh, T.G., Wadsworth, J. & Imai, T. (1992) *Scripta Metall. Mater.*, **26**, 703.
493. Mabuchi, M., Higashi, K. & Langdon, T.G. (1994) *Acta Metall. Mater.*, **42**(5), 1739.
494. Mabuchi, M. & Higashi, K. (1994) *Phil. Mag. Lett.*, **70**(1), 1.
495. Koike, J., Mabuchi, M. & Higashi, K. (1994) *Acta Metall. Mater.*, **43**(1), 199.
496. Mabuchi, M., Koike, J., Iwasaki, H., Higashi, K. & Langdon, T.G. (1994) *Mater. Sci. Forum*, **170-172**, 503.
497. Mabuchi, M. & Higashi, K. (1994) *Mater. Trans. JIM*, **35**(6), 399.
498. Higashi, K., Nieh, T.G., Mabuchi, M. & Wadsworth, J. (1995) *Scripta Metall. Mater.*, **32**(7), 1079.
499. Mishra, R.S. & Mukherjee, A.K. (1991) *Scripta Metall. Mater.*, **25**, 271.
500. Mishra, R.S., Bieler, T.R. & Mukherjee, A.K. (1995) *Acta Metall. Mater.*, **43**(3), 877.
501. Bieler, T.R., Mishra, R.S. & Mukherjee, A.K. (1996) *JOM*, **48**, 52.
502. Mishra, R.S. Bieler, T.R. & Mukherjee, A.K. (1997) *Acta Mater.*, **45**(2), 561.
503. Artz, E., Ashby, M.F. & Verrall, R.A. (1983) *Acta Metall.*, **31**, 1977.
504. Mabuchi, M. & Higashi, K. (1995) *Mater. Trans. JIM*, **36**(3), 420.
505. Mabuchi, M. & Higashi, K. (1996) *Scripta Mater.*, **34**(12), 1893.
506. Li, Y. & Langdon, T.G. (1998) *Acta Mater.*, **46**(11), 3937.
507. Langdon, T.G. (1999) *Mater. Sci. Forum*, **304–306**, 13.
508. Li, Y. & Langdon, T.G. (1999) *Acta Mater.*, **47**(12), 3395.
509. Mabuchi, M. & Higashi, K. (1996) *Phil. Mag. A.*, **74**, 887.
510. Park, K.T. & Mohamed, F.A. (1995) *Metall. Mater. Trans.*, **26**, 3119.
511. Nieh, T.G., Gilman, P.S. & Wadsworth, J. (1985) *Scripta Metall.*, **19**, 1375.
512. Bieler, T.R., Nieh, T.G., Wadsworth, J. & Mukherjee, A.K. (1988) *Scripta Metall. Mater.*, **22**, 81.
513. Bieler, T.R. & Mukherjee, A.K. (1990) *Mater. Sci. Eng. A*, **128**, 171.
514. Higashi, K., Okada, T., Mukai, T. & Tanimura, S. (1992) *Mater. Sci. Eng.*, **159**, L1.
515. Higashi, K. (1993) *Mater. Sci. Eng.*, **166**, 109.
516. Gregory, J.K., Gibeling, J.C. & Nix, W.D. (1985) *Metall. Trans. A*, **16**.
517. Higashi, K., Nieh, T.G. & Wadsworth, J. (1995) *Acta Metall. Mater.*, **43**(9), 3275.
518. Watanabe, H., Mukai, T., Mabuchi, M. & Higashi, K. (1999) *Scripta Metall.*, **41**(2), 209.
519. Hsiao, I.C. & Huang, J.C. (1999) *Scripta Mater.*, **40**(6), 697.
520. Watanabe, H., Mukai, T., Ishikawa, K., Mabuchi, M. & Higashi, K. (2001) *Mater. Sci. Eng.*, **307**, 119.
521. Sergueeva, A.V., Stolyarov, V.V., Valiev, R.Z. & Mukherjee, A.K. (2000) *Scripta Mater.*, **43**, 819.
522. Horita, Z., Furukawa, M., Nemoto, M., Barnes, A.J. & Langdon, T.G. (2000) *Acta Mater.*, **48**, 3633.
523. Bhattacharya, S.S., Betz, U. & Hahn, H. (2001) *Scripta Mater.*, **44**, 1553.

524. Komura, S., Horita, Z., Furukawa, M., Nemoto, M. & Langdon, T.G. (2001) *Metall. Trans. A*, **32**, 707.
525. Mohamed, F.A. & Li, Y. (2001) *Mater. Sci. Eng. A*, **298**, 1.
526. Mishra, R.S., Valiev, R.Z., McFadden, S.X., Islamgaliev, R.K. & Mukherjee, A.K. (2001) *Mater. Sci. Eng.*, **81**, 37.
527. McFadden, S.X., Mishra, R.S., Valiev, R.Z., Zhilyaev, A.P. & Mukherjee, A.K. (1999) *Nature*, **398**, 684.
528. Mishra, R.S., McFadden, S.X. & Mukherjee, A.K. (1999) *Mater. Sci. Forum*, **304–306**, 31.
529. Islamgaliev, R.K., Valiev, R.Z., Mishra, R.S. & Mukherjee, A.K. (2001) *Mater. Sci. Eng.*, **304–306**, 206.
530. Bohn, R., Klassen, T. & Bormann, R. (2001) *Intermetallics*, **9**, 559.
531. Mishra, R.S. & Mukherjee, A.K. (1998) *Int. Conf. On Superplasticity and Superplastic Forming*, Ghosh, A.K. & Bieler, T.R. Eds., The Minerals, Metals & Materials Society. p. 109.
532. Valiev, R.Z., Gayanov, R.M., Yang, H.S. & Mukherjee, A.K. (1991) *Scripta Metall. Mater.*, **25**, 1945.
533. Valiev, R.Z., Song, C., McFadden, S.X., Mukherjee, A.K. & Mishra, R.S. (2001) *Phil. Mag. A*, **81**(1), 25.
534. Yamakov, V., Wolf, D., Salazar, M., Phillpot, S.R. & Gleiter, H. (2001) *Acta Mater.*, **49**, 2713.
535. Turnbull, D. & Fisher, J.C. (1949) *J. Chem. Phys.*, **17**, 71.
536. Christian, J.W. (1975) *The Theory of Transformations in Metals and Alloys*, 2nd edn., Pergamon Press, Oxford.
537. Humphreys, F.J. & Hatherly, M. (1995) *Recrystallization and Related Annealing Phenomena*, Pergamon Press, Oxford.
538. Cahn, R.W. (1949) *J. Inst. Met.*, **76**, 121.
539. Jonas, J.J. (1994) *Mater. Sci. and Eng.*, **A184**, 155.
540. Sakai, T. & Jonas, J.J. (1984) *Acta Metall.*, **32**, 189.
541. Barrabes, S., Daraio, C., Kassner, M.E., Hayes, T.A. & Wang, M.-Z. (2002). *Light Metals*, Lewis, T., Ed., CIM, Montreal, p. 825.
542. Hornbogen, E. & Koster, U. (1978) *Recrystallization of Metallic Materials*, Haessner, F. and Dr. Riederer, Eds., Verlag, Berlin, p. 159.
543. Blum, W. & Reppich, B. (1985) *Creep Behavior of Crystalline Solids*, Wilshire, B. & Evans, R.W. Eds., Pineridge Press, Swansea, UK. p. 83.
544. Reppich, B. (1993) *Materials Science and Technology*, Cahn, R.W., Haasen, P. & Kramer, E.J. Eds., VCH, Weinheim, Vol. 6, p. 312.
545. Reppich, B. (2002) *Zeit. Metallk.*, **93**, 605.
546. Arzt, E. (1991) *Res Mechanica*, **31**, 399.
547. Arzt, E., Dehm, G., Gumbsch, P., Kraft, O. and Weiss, D. (2001) *Prog. Mater. Sci.*, **46**, 283.
548. Ansell, G.S. (1968) *Oxide Dispersion Strengthening*, Ansell, G.S., Cooper, T.D. & Lenel, F.V. Eds., Gordon and Breach, New York. p. 61.
549. Ashby, M.F. (1970) *Strength of Metals and Alloys*, ASM, Materials Park, OH. p. 507.
550. Brown, L.M. (1982) *Fatigue and Creep of Composite Materials, Proc. 3rd. Riso Int. Symp. on Metallurgy and Materials Science*, Linholt, H. & Talreja, R. Eds., p. 1.

551. Bilde-Sorensen, J.B. (1983) *Deformation of Multi-Phase and Particle Containing Materials*, 4th Riso Int. Symp. on Metallurgy and Materials Science, Bilde-Sorensen, J.B. Ed., p. 1.
552. Sellars, C.M. and Petkovic-Luton, R. (1980) *Mater. Sci. and Eng.*, **46**, 75.
553. Lin, J. & Sherby, O.D. (1981) *Res Mech.*, **2**, 251.
554. Martin, J.W. (1980) *Michromechanics in Particle-Hardened Alloys*, Cambridge.
555. Ashby, M.F. (1966) *Acta Metall.*, **14**, 679.
556. Zhu, Q., Blum, W. and McQueen, H.J. (1996) *Materials Sci. Forum*, **217–222**, 1169.
557. Lund, R.W. & Nix, W.D. (1976) *Acta Metall.*, **24**, 469.
558. Pharr, G.M. & Nix, W.D. (1976) *Scripta Metall.*, **10**, 1007.
559. Ansell, G.S. & Weertman, J. (1959) *Trans AIME*, **215**, 838.
560. Arzt, E. & Ashby, M.F. (1982) *Scripta Metall.*, **16**, 1282.
561. Brown, L.M. & Ham, R.K. (1971) *Strengthening Methods in Crystals*, Kelly, A. & Nicholson, R.B. Eds., Applied Science, London. p. 9.
562. Arzt, E. & Wilkinson, D.S. (1986) *Acta Metall.*, **34**, 1893.
563. Haussett, J.H. & Nix, W.D. (1977) *Acta Metall.*, **25**, 1491.
564. Stevens, R.A. & Flewett, P.E.J. (1981) *Acta Metall.*, **29**, 867.
565. Evans, H.E. & Knowles, G. (1980) *Metal Sci.*, **14**, 262.
566. Bacon, D.J., Kocks, U.F. and Scattergood, R.O. (1973) *Phil. Mag.*, **8**, 1241.
567. Lagneborg, R. (1973) *Scripta Metall.*, **7**, 605.
568. Nardone, V.C. & Tien, J.K. (1983) *Scripta Metall.*, **17**, 467.
569. Schroder, J.H. & Arzt, E. (1985) *Scripta Metall.*, **19**, 1129.
570. Reppich, B. (1997) *Acta Mater.*, **46**, 61.
571. Srolovitz, D., Luton, M.J., Petkovic-Luton, R., Barnett, D.M. and Nix, W.D. (1984) *Acta Metall.*, **32**, 1079.
572. Behr, R., Mayer, J. and Arzt, E. (1999) *Intermetall.*, **17**, 423.
573. Rosler, J. & Arzt, E. (1988) *Acta Metall.*, **36**, 1043.
574. Arzt, E. & Rosler, J. (1988) *Acta Metall.*, **36**, 1053.
575. Rosler, J. & Arzt, E. (1990) *Acta Metall.*, **38**, 671.
576. Arzt, E. & Gohring, E. (1998) *Acta Metall.*, **46**, 6584.
577. Reppich, B., Brungs, F., Hummer, G. & Schmidt, H. (1990) *Creep and Fracture of Eng. Mater. and Structures*, Wilshire, B. & Evans, R.W. Eds., Inst. of Metals, London, p. 141.
578. Ajaja, O., Towson, T.E., Purushothaman, S. and Tien, J.K. (1980) *Mater. Sci. and Eng.*, **44**, 165.
579. Reppich, B., Bugler, H., Leistner, R. & Schutze, M. (1984) *Creep and Fracture of Engineering Materials and Structures*, Wilshire, B. & Owen, D.R.J. Eds., Pineridge, p. 279.
580. Lagneborg, R. & Bergman, B. (1976) *Met. Sci.*, **10**, 20.
581. Heilmaier, M. & Reppich, B. (1999) *Creep Behavior of Advanced Materials for the 21st Century*, Mishra, R.S., Mukherjee, A.K. and Linga Murty, K. Eds., TMS, Warrendale. p. 267.
582. Arzt, E. & Rosler, J. (1988) *Dispersion Strengthened Aluminum Alloys*, Kim, Y.-W. & Griffith, W.M. Eds., TMS, Warrendale. p. 31.
583. Kucharova, K., Zhu, S.J. and Cadek, J. (2003). *Mater. Sci. and Eng.*, **355**, 267.
584. Cadek, J., Oikawa, H. and Sustek, V. (1995) *Mater. Sci. and Eng.*, **A190**, 9.
585. Clauer, A.H. & Hansen, N. (1984) *Acta Metall.*, **32**, 269.
586. Timmins, R. & Arzt, E. (1988) *Scripta Metall.*, **22**, 1353.

587. Spigarelli, S. (2002) *Mater. Sci. and Eng.*, **A337**, 306.
588. Stephens, J.J. & Nix, W.D. (1985) *Metall. Trans.*, **16A**, 1307.
589. Dunand, D.C. & Jansen, A.M. (1997) *Acta Mater.*, **45**, 4569.
590. Jansen, A.M. & Dunand, D.C. (1997) *Acta Mater.*, **45**, 4583.
591. Blum, W. & Portella, P.D. (1983) *Deformation of Multi-Phase and Particle Containing Materials*, J.B. Bilde-Sorenson et al. Eds., Riso National Lab., Roskilde. p. 493.
592. Blum, W., private communication, 2002.
593. Sauthoff, G. & Peterseim, J. (1986) *Steel Res.*, **57**, 19.
594. Gregory, J.K., Gibeling, J.C. and Nix, W.D. (1985) *Metall. Trans.*, **16A**, 777.
595. Singer, R.F. & Arzt, E. (1986) *High Temperature Alloys for Gas Turbines and Other Applications*, W. Betz et al. Eds., p. 97.
596. Nix, W.D. (1984) *Proceedings Superplastic Forming Symp.*, ASM, Materials Park, OH.
597. Seidman, D.N., Marquis, E.A. and Dunand, D.C. (2002) *Acta Mater.*, **50**, 4021.
598. Sauthoff, G. (1995) *Intermetallics*, VCH, New York.
599. Hemker, K.J. and Nix, W.D. (1997) *Structural Intermetallics*, TMS, Warrendale, PA, p. 21.
600. Stoloff, N.S. and Sikka, V.K., Eds. (1996) *Physical Metallurgy and Processing of Intermetallic Compounds*, Chapman & Hall, New York.
601. Westbrook, J.H. and Fleischer, R.L., Eds. (2000) *Structural Applications of Intermetallic Compounds*, Wiley, New York.
602. Cahn, R.W. (2002) *Mater. Sci. Eng. A*, **324**, 1.
603. Izumi, O. (1989) *Mater Trans JIM*, **30**, 627.
604. Yamaguchi, M. and Umakoshi, Y. (1990) *Prog. Mater. Sci.*, **34**, 1.
605. Deevi, S.C., Sikka, V.K. and Liu, C.T. (1997) *Prog. Mater. Sci.*, **42**, 177.
606. Westbrook, J.H. and Fleischer, R.L., Eds. (2000) *Basic Mechanical Properties and Lattice Defects of Intermetallic Compounds*, Wiley, New York.
607. Veyssière, P. (2001) *Mater. Sci. Eng. A*, **44**, 309–310.
608. Yamaguchi, M., Inui, H. and Ito, K. (2000) *Acta Mater.*, **48**, 307.
609. Oikawa, H. and Langdon, T.G. (1985) *Creep Behavior of Crystalline Solids*, Wilshire, B., Evans, R.W., Eds., Pineridge, Swansea, UK, p. 33.
610. Nabarro, F.R.N. and de Villiers, H.L. (1995) *The Physics of Creep*, Taylor & Francis, Bristol.
611. Shah, D., Lee, E. and Westbrook, J.H., (2002) *Intermetallic Compounds*, Fleischer, R.L., Ed., Vol. 3, John Wiley & Sons, Chichester, UK, p. 297.
612. Schneibel, J.H. and Hazzledine, P.M. (1992) *Applied Sciences*, **213**, 565.
613. Smallman, R.E., Rong, T.S. and Jones, I.P. (1996) *The Johannes Weertman Symposium*, TMS, Warrendale, PA, p. 11.
614. Sauthoff, G. (1993) *Structural Intermetallics*, TMS, Warrendale, PA, p. 845.
615. Sauthoff, G. (1993) *Diffusion in Ordered Alloys*, Fultz, B., Chan, R.W. and Gupta, D., Eds., TMS, Warrendale, PA, p. 205.
616. Oikawa, H., (1992) *The Processing, Properties and Applications of Metallic and Ceramic Materials*, Warley, UK, p. 383.
617. Kumpfert, J. (2001) *Adv. Eng. Mater.*, **3**, 851.
618. Mishra, R.S., Nandy, T.K., Sagar, P.K., Gogia, A.K. and Banerjee, D. (1996) *Trans. India Inst. Met.* **49**, 331.
619. Stoloff, N.S., Alven, D.A. and McKamey, C.G. (1997) *Nickel and Iron Aluminides: Processing, Properties, and Applications*, ASM International, Materials Park, OH, p. 65.

620. Kim, Y.W. (1994) *J. Metals*, **46**, 31.
621. Kim, Y.W. (1989) *J. Metals*, **41**, 24.
622. Beddoes, J., Wallace, W. and Zhao, L. (1995) *Int. Mater. Reviews* **40**, 197.
623. Zhang, W.J. and Deevi, S.C. (2001) *Structural Intermetallics 2001*, Hemker, K.J., Dimiduk, D.M., Clemens, H., Darolia, R., Inui, H., Larsen, J.M., Sikka, V.K., Thomas, M. and Whittenberger, J.D., Eds., TMS, Warrendale, PA, p. 699.
624. Hall, E.L. and Huang, S.C. (1990) *Acta Mater.*, **38**, 539.
625. Huang, S.C. and Hall, E.L. (1991) *Metall. Trans. A* **22**, 427.
626. Beddoes, J., Zhao, L., Au, P., Dudzinsky, D. and Triantafillou, J. (1997) *Structural Intermetallics 1997*, Nathal, M.V., Darolia, R., Liu, C.T., Martin, P.L., Miracle, D.B., Wagner, R. and Yamaguchi, M., Eds., TMS, Warrendale, PA, p. 109.
627. Es-Souni, M., Bartels, A. and Wagner, R. (1995) *Mater. Sci. Eng. A*, **192/193**, 698.
628. Worth, D.B., Jones, J.W., and Allison, J.E. (1995) *Metall. Trans. A*, **26**, 2947.
629. Hsiung, L.M. and Nieh, T.G. (1999) *Intermetallics*, **7**, 821.
630. Zhang, W.J. and Deevi, S.C. (2002) *Intermetallics*, **10**, 603.
631. Zhang, W.J. and Deevi, S.C. (2003) *Intermetallics*, **11**, 177.
632. Seo, D.Y., Saari, H., Beddoes, J. and Zhao, L. (2001) *Structural Intermetallics 2001*, Hemker, K.J., Dimiduk, D.M., Clemens, H., Darolia, R., Inui, H., Larsen, J.M., Sikka, V.K., Thomas, M. and Whittenberger, J.D., Eds., TMS, Warrendale, PA, p. 653.
633. Weller, M., Chatterjee, A., Haneczok, G., Wanner, A., Appel, F. and Clemens, H. (2001) *Structural Intermetallics 2001*, Hemker, K.J., Dimiduk, D.M., Clemens, H., Darolia, R., Inui, H., Larsen, J.M., Sikka, V.K., Thomas, M. and Whittenberger, J.D., Eds., TMS, Warrendale, PA., p. 465.
634. Wolfenstine, J. and González-Doncel, G. (1994) *Materials Letters*, **18**, 286.
635. Kroll, S., Mehrer, H., Stolwijk, N., Herzig, C., Rosenkranz, R. and Frommeyer, G. (1992) *Z. Metallkde.*, **83**, 591.
636. Lu, M. and Hemker, K. (1997) *Acta Mater.*, **45**, 3573.
637. Es-Souni, M., Bartels, A. and R. Wagner. (1995) *Acta Metal. Mater.*, **43**, 153.
638. Viswanathan, G.B., Kartikeyan, S., Mills, M.J. and Vasudevan, V.K. (2001) *Mater. Sci. Eng. A*, **833**, 319–321.
639. Viswanathan, G.B., Vasudevan, V.K. and Mills, M.J. (1999) *Acta Mater.*, **47**, 1399.
640. Ishikawa, Y. and Oikawa, H. (1994) *Mater. Trans. JIM.*, **35**, 336.
641. Hirsch, P.B. and Warrington, D.H. (1961) *Phil. Mag. A*, **6**, 715.
642. Friedel, J. (1956) *Phil. Mag. A*, **46**, 1169.
643. Wang, J.N. and Nieh, T.G. (1998) *Acta Mater.*, **46**, 1887.
644. Gouma, P.I., Subramanian, K., Kim, Y.W. and Mills, M.J. (1998) *Intermetallics*, **6**, 689.
645. Chen, W.R., Zhao, L. and Beddoes, J. (1999) *Scripta Mater.*, **41**, 597.
646. Seo, D.Y., Beddoes, J., Zhao, L. and Botton, G.A. (2002) *Mater. Sci. Eng. A*, **810**, 329–331.
647. Chen, W.R., Beddoes, J. and Zhao, L. (2002) *Mater. Sci. Eng. A*, **323**, 306.
648. Chen, W.R., Triantafillou, J., Beddoes, J. and Zhao, L. (1999) *Intermetallics*, **7**, 171.
649. Parthasarathy, T.A., Subramanian, P.R., Mendiratta, M.G. and Dimiduk, D.M. (2000) *Acta Mater.*, **48**, 541.
650. Bieler, T.R., Seo, D.Y., Everard, T.R. and McQuay, P.A. (1999) *Creep Behavior of Materials for the 21st century*, Mishra, R.S., Mukherjee, A.K. and Murty, K.L., Eds., TMS, Warrendale, PA, p. 181.

651. Beddoes, J., Seo, D.Y., Chen, W.R. and Zhao, L. (2001) *Intermetallics*, **9**, 915.
652. Liu, C.T., Stringer, J., Mundy, J.N., Horton, L.L. and Angelini, P. (1997) *Intermetallics*, **5**, 579.
653. Baker, I. and Munroe, P.R. (1997) *Int. Mater. Reviews*, **42**, 181.
654. Stoloff, N.S. (1998) *Mater. Sci. Eng. A*, **258**, 1.
655. Massalski, T.B. (1986) *Binary Alloy Phase Diagrams*, ASM, Metals Park, OH, p.112.
656. Morris, D.G. and Morris, M.A. (1997) *Mater. Sci. Eng. A*, **23**, 239–240.
657. George, E.P. and Baker, I. (1998) *Intermetallics*, **6**, 759.
658. Morris, D.G., Liu, C.T. and George, E.P. (1999) *Intermetallics*, **7**, 1059.
659. Morris, D.G., Zhao, P. and Morris-Muñoz, M.A. (2001) *Mater. Sci. Eng. A*, **297**, 256.
660. Park, J.W. (2002) *Intermetallics*, **10**, 683.
661. Baker, I., Xiao, H., Klein, O., Nelson, C. and Whittenberger, J.D. (1995) *Acta Metal. Mater.*, **30**, 863.
662. Stoloff, N.S. and Davies, R.G. (1966) *Prog. Mater. Sci.*, **13**, 1.
663. Umakoshi, Y., Yamaguchi, M., Namba, Y. and Murakami, M. (1976) *Acta Metal.* **24**, 89.
664. Hanada, S., Watanabe, S., Sato, T. and Izumi, O. (1981) *Scripta metall.*, **15**, 1345.
665. Schroer, W., Hartig, C. and Mecking, H. (1993) *Z. Metallkde.*, **84**, 294.
666. Yoshimi, K., Hanada, S. and Yoo, M.H. (1995) *Acta Metall. Mater.*, **44**, 4141.
667. Morris, D.G. and Morris, M.A. (1997) *Intermetallics*, **5**, 245.
668. Morris, D.G. (1992) *Ordered Intermetallics-Physical Metallurgy and Mechanical Behavior*, Liu, C.T., Cahn, R.W. and Sauthoff, G., Eds., Kluwer, Dordrecht, p. 123.
669. Paris, D. and Lesbats, P. (1978) *J. Nucl. Mater.*, **69/70**, 628.
670. Wurschum, R., Grupp, C. and Schaefer, H. (1995) *Phys. Rev. Lett.*, **75**, 97.
671. Jordan, J.L. and Deevi, S.C. (2003) *Intermetallics*, **11**, 507.
672. Carleton, R., George, E.P. and Zee, R.H. (1995) *Intermetallics*, **3**, 433.
673. George, E.P. and Baker, I. (1998) *Phil. Mag.*, **77**, 737.
674. Park, J.W. (2002) *Intermetallics*, **10**, 683.
675. Pesicka, J. and Schmitz, G. (2002) *Intermetallics*, **10**, 717.
676. Stein, F., Schneider, A. and Frommeyer, G. (2003) *Intermetallics*, **11**, 71.
677. Ziegler, N. (1932) *Trans. AIME*, **100**, 267.
678. Davies, R.G. (1963) *Trans. AIME*, **227**, 22.
679. McKamey, C.G., Masiasz, P.J. and Jones, J.W. (1992) *J. Mater. Res.*, **7**, 2089.
680. Sikka, V.K., Gieseke, B.G. and Baldwin, R.H. (1991) *Heat Resistant Materials*, Natesan, K. and Tillack, D.J., Eds., ASM, Materials Park, OH, p. 363.
681. Phillips, J., Eggeler, G., Ilschner, B. and Batawi, E. (1997) *Scripta Mater.*, **36**, 693.
682. Deevi, S.C. and Swindeman, R.W. (1998) *Mater. Sci. Eng.*, **258**, 203.
683. Voyzelle, B. and Boyd, J.D. (1998) *Mater. Sci. Eng. A*, **258**, 243.
684. Morris-Muñoz, M.A. (1999) *Intermetallics*, **7**, 653.
685. Sundar, R.S., Kutty, T.R.G. and Sastry, D.H. (2000) *Intermetallics*, **8**, 427.
686. Sastry, D.H., Prasad, Y.V.R.K. and Deevi, S.C. (2001) *Mater. Sci. Eng. A*, **299**, 157.
687. Málek, P., Kratochvíl, O., Pešička, J., Hanus, P. and Šedivá, I. (2002) *Intermetallics* **10**, 895.
688. Whittenberger, J.D. (1986) *Mater. Sci. Eng. A*, **77**, 103.
689. Whittenberger, J.D. (1983) *Mater. Sci. Eng. A*, **57**, 77.
690. Morris, D.G. (1998) *Intermetallics*, **6**, 753.
691. Lin, D., Shan, D. and Li, D. (1994) *Scripta Metall. Mater.*, **31**, 1455.

692. Lin, D., Li, D. and Liu, Y. (1998) *Intermetallics*, **6**, 243.
693. Chu, J.P., Wu, J.H., Yasuda, H.Y., Umakoshi, Y. and Inoue, K. (2000) *Intermetallics*, **8**, 39.
694. Chu, J.P., Kai, W., Yasuda, H.Y., Umakoshi, Y. and Inoue, K. (2002) *Mater. Sci. Eng. A*, **878**, 329–331.
695. Lin, D. and Liu, Y. (2002) *Mater. Sci. Eng. A*, **863**, 329–331.
696. García-Oca, C., PhD dissertation, Universidad Complutense de Madrid-CENIM, Madrid, Spain, 2003.
697. Muñoz-Morris, M.A., Garcia-Oca, C. and Morris, D.G. (2002) *Acta Mater.*, **50**, 2825.
698. Kolluru, D.V. and Baligidad, R.G. (2002) *Mater. Sci. Eng. A*, **328**, 58.
699. Baligidad, R.G., Radhakrishna, A. and Prakash, U. (1998) *Mater. Sci. Eng. A*, **257**, 235.
700. Morris-Muñoz, M.A. (1999) *Intermetallics*, **7**, 653.
701. Sundar, R.S. and Deevi, S.C. (2003) *Mater. Sci. Eng. A*, **357**, 124.
702. Aoki, K. and Izumi, O. (1979) *J. Japan Inst. Metals*, **43**, 1190.
703. Hemker, K.J., Mills, M.J. and Nix, W.D. (1991) *Acta Metall. Mater.*, **39**, 1901.
704. Zhu, W.H., Fort, D., Jones, I.P. and Smallman, R.E. (1998) *Acta Mater.*, **46**, 3873.
705. Knobloch, C., Glock, K. and Glatzel, U., *High-Temperature Ordered Intermetallic Alloys VIII*, 1999, MRS 556, p. 1.
706. Miura, S., Peng, L.-Z. and Mishima, Y. (1997) *High-Temperature Ordered Intermetallic Alloys VII*, MRS 460, p. 431.
707. Shah, D.H. (1983) *Scripta Metall.*, **17**, 997.
708. Hemker, K.J. and Nix, W.D. (1993) *Metall. Trans. A*, **24**, 335.
709. Nathal, M.V., Diaz, J.O. and Miner, R.V. (1989) *High Temperature Ordered Intermetallic Alloys III*, MRS 133, p. 269.
710. Wolfenstine, J., Kim, H.K. and Earthman, J.C., *Mater. Sci Eng. A* 192/193 (1994) 811.
711. Ham, R.K., Cook, R.H., Purdy, G.R. and Willoughby, G. (1972) *Metall. Sci. J.*, **6**, 205.
712. Anton, D.L., Pearson, D.D. and Snow, D.B., *High Temperature Ordered Intermetallic Alloys II*, 1987, MRS 81, p. 287.
713. Uchic, M.D., Chrzan, D.C. and Nix, W.D. (2001) *Intermetallics*, **9**, 963.
714. Rong, T.S., Jones, I.P. and Smallman, R.E. (1997) *Acta Mater.*, **45**, 2139.
715. Miura, S., Hayashi, T., Takekawa, M., Mishima, Y. and Suzuki, T. (1991) *High-Temperature Ordered Intermetallic Alloys IV*, MRS 213, p. 623.
716. Hsu, S.E., Lee, T.S., Yang, C.C., Wang, C.Y. and Hong, C.H. (1992) *Applied Sciences*, **213**, 597.
717. Hsu, S.E., Tong, C.H., Lee, T.S. and Liu, T.S. (1989) *High-Temperature Ordered Intermetallic Alloys III*, MRS 133, p. 275.
718. Glock, K., Knobloch, C. and Glatzel, U. (2000) *Metal. Mater. Trans. A*, **31**, 1733.
719. Schneibel, J.H. and Horton, J.A. (1988) *J. Mater. Res.*, **3**, 651.
720. Rong, T.S., Jones, I.P. and Smallman, R.R. (1995) *Acta Metall. Mater.*, **43**, 1385.
721. Smallman, R.E., Rong, T.S., Lee, S.C.D. and Jones, I.P., *Materials Science and Engineering A*, **852**, 329–331.
722. Wolfenstine, J., Kim, H.K. and Earthman, J.C. (1992) *Scripta Metall. Mater.*, **26**, 1823.
723. Flinn, P.A. (1960) *Trans. AIME*, **218**, 145.
724. Nicholls, J.R. and Rawlings, R.D. (1977) *J. Mater. Sci.*, **12**, 2456.
725. Schneibel, J.H., Petersen, G.F. and Liu, C.T. (1986) *J. Mater. Res.*, **1**, 68.
726. Rong, T.S., Jones, I.P. and Smallman, R.E. (1994) *Scripta Metall. Mater.*, **30**, 19.

727. Hayashi, T., Shinoda, T., Mishima, Y. and Suzuki, T., *High-Temperature Ordered Intermetallic Alloys IV*, 1991, MRS 213, p. 617.
728. Schneibel, J.H. and Porter, W.D. (1988) *J. Mater. Res.*, **3**, 403.
729. Yang, H.S., Jin, P. and Mukhejee, A.K. (1992) *Mat. Trans. JIM*, **33**, 38.
730. Nemoto, M., Takesue, H. and Horita, Z. (1997) *Mater. Sci. Eng. A*, **327**, 234–236.
731. Zhang, Y. and Lin, D.L. (1993) *High-Temperature Ordered Intermetallic Alloys V*, MRS 288, p. 611.
732. Rawlings, R.D. and Staton-Bevan, A.E. (1975) *J. Mater. Sci.*, **10**, 505.
733. Uchic, M.D. and Nix, W.D. (2001) *Intermetallics*, **9**, 1053.
734. Thornton, P.H., Davis, R.G. and Johnston, T.L. (1970) *Metall. Trans.*, **1**, 207.
735. Lunt, M.J. and Sun, Y.Q. (1997) *Mater. Sci. Eng. A*, **445**, 239–240.
736. Hazzdine, P.M. and Schneibel, J.H. (1989) *Scripta Metall.*, **23**, 1887.
737. Lapin, J. (1999) *Intermetallics*, **7**, 599.
738. Carreño, F., Jiménez, J.A. and Ruano, O.A. (2000) *Mater. Sci. Eng. A*, **278**, 272.
739. Klotz, U.E., Mason, R.P., Gohring, E. and Artz, E. (1997) *Mater. Sci. Eng. A*, **231**, 198.
740. Mason, R.P. and Grant, N.J. (1995) *Mater. Sci. Eng. A*, **192/193**, 741.
741. Mason, R.P. and Grant, N.J. (1995) *High-Temperature Ordered Intermetallic Alloys VI*, MRS 364, p. 861.
742. Veyssiere, P., Guan, D.L. and Rabier, J. (1984) *Phil. Mag. A*, **49**, 45.
743. Stoloff, N.S. (1989) *Int. Mater. Rev.*, **34**, 153.
744. Raj, S.V., *Mater. Sci. Eng. A*. (2003) in press.
745. Knobloch, C., Toloraia, V.N. and Glatzel, U. (1997) *Scripta Mater.*, **37**, 1491.
746. Link, T., Knobloch, C. and Glatzel, U. (1999) *Scripta Mater.*, **40**, 85.
747. Peng, Z.L., Miura, S. and Mishima, Y. (1997) *Mater. Trans. JIM*, **38**, 653.
748. Caron, P., Khan, T. and Veyssiere, P. (1989) *Phil. Mag. A*, **60**, 267.
749. Brehm, H. and Glatzel, U. (1998) *Int. J. Plasticity*, **15**, 285.
750. Hsu, S.E., Hsu, N.N., Tong, C.H., Ma, C.Y. and Lee, S.Y. (1989) *High Temperature Ordered Intermetallic Alloys II*, MRS 81, p. 507.
751. Fujita, A., Matsmoto, T., Nakamura, M. and Takeda, Y. *High-Temperature Ordered Intermetallic Alloys* III, 1989, MRS 133, p. 573.
752. Shah, D.M. and Duhl, D.N. (1987) *High Temperature Ordered Intermetallic Alloys II*, MRS 81, p. 411.
753. Nazmy, M. and Staubli, M. (1991) *Scripta Metall. Mater.*, **25**, 1305.
754. Khan, T., Caron, P. and Naka, S., *High Temperature Aluminides and Intermetallics*, Whang, S.H., Liu, C.T., Pope, D.P. and Stiegler, J.O., Eds., 1990, TMS, Warrendale PA, p. 219.
755. Pope, D.P. and Ezz, S.S. (1984) *Int. Metals Rev.*, **29**, 136.
756. Whittenberger, J.D., Nathal, M.V. and Book, P.O. (1993) *Scripta Metall.*, **28**, 53.
757. Timmins, R. and Artz, E. (1988) *Scripta Metall.*, **22**, 1353.
758. Miracle, D.B. (1993) *Acta Metall. Mater.*, **41**, 649.
759. Noebe, R.D., Bowman, R.R. and Nathal, M.V. (1993) *Int. Mater. Reviews*, **38**, 193.
760. Whittenberger, J.D., Mannan, S.K. and Kumar, K.S. (1989) *Scripta Metall.*, **23**, 2055.
761. Whittenberger, J.D., Westfall, L.J. and Nathal, M.V. (1989) *Scripta Metall.*, **23**, 2127.
762. Whittenberger, J.D., Reviere, R., Noebe, R.D. and Oliver, B.F. (1992) *Scripta Metall. Mater.*, **26**, 987.
763. Whittenberger, J.D., Ray, R., Jha, S.C. and Draper, S. (1991) *Mater. Sci. Eng. A*, **138**, 83.

764. Arzt, E. and Grahle, P. (1998) *Acta Mater.*, **46**, 2717.
765. Xu, K. and Arsenault, R.J. (1999) *Acta Mater.*, **47**, 3023.
766. Whittenberger, J.D., Ray, R. and Jha, S.C. (1992) *Mater. Sci. Eng. A*, **151**, 137.
767. Whittenberger, J.D., Artz, E. and Luton, M.J. (1992) *Scripta Metall. Mater.*, **26**, 1925.
768. Whittenberger, J.D., Noebe, R.D., Johnson, D.R. and Oliver, B.F. (1997) *Intermetallics*, **5**, 173.
769. Yang, W., Dodd, R.A. and Strutt, P.R. (1972) *Metall. Trans.*, **3**, 2049.
770. Yang, W.J. and Dodd, R.A. (1973) *Met. Sci. J.*, **7**, 41.
771. Strutt, P.R., Polvani, R.S. and Kear, B.H. (1973) *Scripta Metall.*, **7**, 949.
772. Prakash, A. and Dodd, R.A. (1981) *J. Mater. Sci.*, **16**, 2495.
773. Rudy, M. and Sauthoff, G. (1986) *Mater. Sci. Eng. A*, **81**, 525.
774. Whittenberger, J.D. (1988) *J. Mater. Sci.*, **23**, 235.
775. Whittenberger, J.D. (1987) *J. Mater. Sci.*, **22**, 394.
776. Whittenberger, J.D., Viswanadham, R.K., Mannan, S.K. and Sprissler, B. (1990) *J. Mater. Sci.*, **25**, 35.
777. Rudy, M. and Sauthoff, G. (1985) *High Temperature Ordered Intermetallic Alloys*, MRS 39, p. 327.
778. Vandervoort, R.R., Mukherjee, A.K. and Dorn, J.E. (1966) *Trans. ASM*, **59**, 930.
779. Hocking, L.A., Strutt, P.R. and Dodd, R.A. (1971) *J. Inst. Met.*, **99**, 98.
780. Bevk, J., Dodd, R.A. and Strutt, P.R. (1973) *Metall. Trans.*, **4**, 159.
781. Forbes, K.R., Glatzel, U., Darolia, R. and Nix, W.D. (1993) *High Temperature Ordered Intermetallic Alloys V*, MRS 288, 45.
782. Raj, S.V. and Farmer, S.C. (1993) *High Temperature Ordered Intermetallic Alloys V*, MRS 288, 647.
783. Rudy, M. and Sauthoff, G. (1985) *High Temperature Ordered Intermetallic Alloys*, MRS 39, p. 327.
784. Jung, I., Rudy, M. and Sauthoff, G. (1987) *High Temperature Ordered Intermetallic Alloys*, MRS 81, 263.
785. Whittenberger, J.D., Noebe, R.D., Cullers, C.L., Kumar, K.S. and Mannan, S.K. (1991) *Metall. Trans. A*, **22**, 1595.
787. Waddington, J.S. & Lofthouse, K.J. (1967) *J. Nuc. Mater.*, **22**, 205.
788. Stiegler, J.O., Farrell, K., Loh, B.T.M. & McCoy, H.E. (1967) *Trans. ASM Quart.*, **60**, 494.
789. Courtney, T.H. (1990) *Mechanical Behavior of Materials*, McGraw-Hill, New York.
790. Chen, I.W. & Argon, A.S. (1981) *Acta Metall.*, **29**, 1321.
791. Cocks, A.C.F. & Ashby, M.F. (1982) *Prog. Mater. Sci.*, **27**, 189.
792. Nix, W.D. (1988) *Mater. Sci. and Eng.*, **A103**, 103.
793. Needleman, A. & Rice, J.R. (1980) *Acta Metall.*, **28**, 1315.
794. Beere, W. (1983) *Scripta Metall.*, **17**, 13.
795. Chen, I.-W. (1983) *Scripta Metall.*, **17**, 17.
796. Argon, A.S. (1983) *Scripta Metall.*, **17**, 5.
797. Nix, W.D. (1983) *Scripta Metall.*, **17**, 1.
798. Goods, S.H. & Nieh, T.G. (1983) *Scripta Metall.*, **17**, 23.
799. Dyson, B.F. (1983) *Scripta Metall.*, **17**, 31.
800. Riedel, H. (1987) *Fracture at High Temperatures*, Springer-Verlag, Berlin.

References

801. Mackerle, J. (2000) *Int. J. of Pressure Vessels and Piping*, **77**, 53.
802. Feltham, P. & Meakin, J.D. (1959) *Acta Metall.*, **7**, 614.
803. Dunand, D.C., Han, B.Q. & Jansen, A.M. (1999) *Metall. Trans.*, **30A**, 829.
804. Dobes, F. & Miliska, K. (1976) *Met. Sci.*, **10**, 382.
805. Molinie, E., Piques, R. & Pineau, A. (1991) *Fat. Fract. Eng. Mater. Struct.*, **14**, 531.
806. Larson, F.R. & Miller, J. (1952) *Trans. ASME*, **74**, 765.
807. Murty, K.L., Zhou, Y. & Davarajan, B. (2002) *Creep Deformation: Fundamentals and Applications*, Mishra, R.S., Earthman, J.C. & Raj, S.V. Eds., TMS, Warrendale, PA. p. 3.
808. Ayensu, A. & Langdon, T.G. (1996) *Metall. Trans.*, **27A**, 901.
809. Chen, R.T. & Weertman, J.R. (1984) *Mater. Sci. and Eng.*, **64**, 15.
810. Arai, M., Ogata, T. & Nitta, A. (1996) *JSME Intl. J.*, **39**, 382.
811. Hosokawa, H., Iwasaki, H., Mori, T., Mabuchi, M., Tagata, T. & Higashi, K. (1999) *Acta Mater.*, **47**, 1859.
812. Lim, L.C. & Lu, H.H. (1994) *Scripta Metall. Mater.*, **31**, 723.
813. Yavari, P. & Langdon, T.G. (1983) *J. Mater. Sci. Lett.*, **2**, 522.
814. Weertman, J. (1986) *J. Appl. Phys.*, **60**, 1877.
815. Raj, R. & Ashby, M.F. (1975) *Acta Metall.*, **23**, 699.
816. Davanas, K. & Solomon, A.A. (1990) *Acta Metall. Mater.*, **38**, 1905.
817. Argon, A.S., Chen, I.-W. & Lau, C.W. (1980) *Proc. Symp. Creep Fatigue-Environment Interactions*, Pelloux, R.M. & Stoloff, N.S. Eds., TMS, Warrendale. p. 46.
818. Riedel, H. (1984) *Acta Metall.*, **32**, 313.
819. Chen, C.W. & Machlin, E.S. (1956) *Acta Metall.*, **4**, 655.
820. Fleck, R.G., Taplin, D.M.R. & Beevers, C.J. (1975) *Acta Metall.*, **23**, 415.
821. Gandhi, C. & Raj, R. (1982) *Acta Metall.*, **30**, 505.
822. Dewald, D.K., Lee, T.C., Robertson, I.M. & Birnbaum, H.K. (1990) *Metall. Trans.*, **21A**, 2411.
823. Chen, I.-W. (1983) *Metall. Trans.*, **14A**, 2289.
824. Yoo, M.H. & Trinkaus, H. (1986) *Acta Metall.*, **34**, 2381.
825. Trinkaus, H. & Yoo, M.H. (1987) *Philos. Mag.*, **55**, 269.
826. Nieh, T.G. & Nix, W.D. (1980) *Scripta Metall.*, **14**, 365.
827. Dyson, B.F., Loveday, M.S. & Rodgers, M.J. (1976) *Proc. Royal Soc. London*, **A349**, 245.
828. Watanabe, T. & Davies, P.W. (1978) *Phil. Mag.*, **37**, 649.
829. Greenwood, J.N., Miller, D.R. & Suiter, J.W. (1954) *Acta Metall.*, **2**, 250.
830. Goods, S.H. & Nix, W.D. (1978) *Acta Metall.*, **26**, 739.
831. Lombard, R. & Vehoff, H. (1990) *Scripta Metall. Mater.*, **24**, 581.
832. Cane, B.J. (1979) *Metal. Sci.*, **13**, 287.
833. McClintock, F.A. (1968) *J. App. Mech.*, **35**, 363.
834. Lee, Y.S. & Yu, J. (1999) *Metall. Trans. A*, **30A**, 2331.
835. Oh, Y.K., Kim, G.S. & Indacochea, J.E. (1999) *Scripta Mater.*, **41**, 7.
836. George, E.P., Kennedy, R.L. & Pope, D.P. (1998) *Phys. Stat. Sol. A*, **167**, 313.
837. Yousefani, A., Mohamed, F.A. & Earthman, J.C. (2000) *Metall. Trans.*, **31A**, 2807.
838. Wei, R.P., Liu, H. & Gao, M. (1997) *Acta Mater.*, **46**, 313.
839. Svoboda, J. & Sklenicka, V. (1990) *Acta Metall. Mater.*, **38**, 1141.
840. Kassner, M.E., Kennedy, T.C. & Schrems, K.K. (1998) *Acta Mater.*, **46**, 6445.
841. Randle, V. (1993) *Mater. Sci. Forum*, **113–115**, 189.

842. Yang, M.S., Weertman, J.R. & Roth, M. (1984) *Proc. Sec. Int. Conf. Creep and Fracture of Eng. Mater. and Structures*, Wilshire, B. & Owen, D.R.J. Eds., Pineridge, Swansea. p. 149.
843. Andersen, P.M. & Shewmen, R.G. (2000) *Mech. Mater.*, **32**, 175.
844. Hull, D. & Rimmer, D.E. (1959) *Phil. Mag.*, **4**, 673.
845. Speight, M.V. & Beere, W. (1975) *Met. Sci.*, **9**, 190.
846. Raj, R., Shih, H.M. & Johnson, H.H. (1977) *Scripta Met.*, **11**, 839.
847. Weertman, J. (1973) *Scripta Metall.*, **7**, 1129.
848. Nieh, T.G. & Nix, W.D. (1979) *Acta Metall.*, **27**, 1097.
849. Nieh, T.G. & Nix, W.D. (1980) *Acta Metall.*, **28**, 557.
850. Miller, D.A. & Langdon, T.G. (1980) *Metall. Trans.*, **11A**, 955.
851. Svensson, L.-E. & Dunlop, G.L. (1982) *Met. Sci.*, **16**, 57.
852. Hanna, M.D. & Greenwood, G.W. (1982) *Acta Metall.*, **30**, 719.
853. Cho, H.C., Yu, J. & Park, I.S. (1992) *Metall. Trans.*, **23A**, 201.
854. Broyles, S.E., Anderson, K.R., Groza, J. & Gibeling, J.C. (1996) *Metall. Trans.*, **27**, 1217.
855. Cane, B.J. (1981) *Met. Sci.*, **15**, 302.
856. Mintz, J.M. & Mukherjee, A.K. (1988) *Metall. Trans.*, **19A**, 821.
857. Raj, R. (1978) *Acta Metall.*, **26**, 341.
858. Needham, N.G. & Gladman, T. (1980) *Met. Sci.*, **14**, 64.
859. Needham, N.G. & Gladman, T. (1986) *Met. Sci. and Tech.*, **2**, 368.
860. Stanzl, S.E., Argon, A.S. & Tschegg, E.K. (1986) *Acta Metall.*, **34**, 2381.
861. Chuang, T.-J. & Rice, J.R. (1973) *Acta Metall.*, **21**, 1625.
862. Nix, W.D., Yu, K.S. & Wang, J.S. (1983) *Metall. Trans.*, **14A**, 563.
863. Evans, H.E. (1969) *Metal Sci. J.*, **3**, 33.
864. Chakraborty, A. & Earthman, J.C. (1997) *Acta Mater.*, **45**, 4615.
865. Adams, B.L. (1993) *Mater Sci. and Eng.*, **A166**, 59.
866. Watanabe, T. (1993) *Mater Sci. and Eng.*, **A166**, 11.
867. Dyson, B.F. (1976) *Met. Sci.*, **10**, 349.
868. Rice, J.R. (1981) *Acta Metall.*, **29**, 675.
869. Riedel, H. (1985) *Z. Metall.*, **76**, 669.
870. Cho, H.C., Jin Yu and Park, I.S. (1992) *Metall. Trans.*, **23A**, 201.
871. Anderson, P.M. & Rice, J.R. (1985) *Acta Metall.*, **33**, 409.
872. Nix, W.D., Earthman, J.C., Eggeler, G. & Ilschner, B. (1989) *Acta Metall.*, **37**, 1067.
873. Yousefiani, A., El-Nasr, A.A., Mohamed, F.A. & Earthman, J.C. (1997) *Creep and Fracture of Engineering Materials and Structures*, Earthman, J.C. & Mohamed, F.A. Eds., TMS, Warrendale, p. 439.
874. van der Giessen, E. and Tvergaard, V. (1991) *Int. Jour. Fract.*, **48**, 153.
875. Dyson, B.F. (2002) *Creep Deformation: Fundamentals and Applications*, Mishra, R., Earthman, J.C. & Raj, S.V. Eds., TMS, Warrendale, PA. p. 309.
876. Hancock, J.W. (1976) *Met. Sci.*, **10**, 319.
877. Beere, W. & Speight, M.V. (1978) *Metal Sci.*, **12**, 172.
878. Wang, J.S., Martinez, L. & Nix, W.D. (1983) *Acta Metall.*, **31**, 873.
879. Chen, I.-W. & Argon, A.S. (1981) *Acta Metall.*, **29**, 1759.
880. Lee, Y.S., Kozlosky, T.A. & Batt, T.J. (1993) *Acta Metall. Mater.*, **41**, 1841.
881. Schneibel, J.M. & Martinez, L. (1987) *Scripta Metall.*, **21**, 495.
882. Lu, M. & Delph, T.J. (1993) *Scripta Metall.*, **29**, 281.

883. Cadek, J. (1989) *Mater. Sci. and Eng.*, **A117**, L5.
884. van der Giessen, E. van der Burg, M.W.D. Needleman, A.I. & Tvergaard, V. (1995) *J. Mech. Phys. Solids*, **43**, 123.
885. Delph, T.J. (2002) *Metall. Mater. Trans.*, **33A**, 383.
886. Cocks, A.C.F. & Ashby, M.F. (1982) *Met. Sci.*, **16**, 465.
887. Forero, L.E. & Koss, D.A. (1994) *Scripta Met. et Mater.*, **31**, 419.
888. Han, B.Q. & Dunand, D.C. (2002) *Creep Deformation: Fundamentals and Applications*, Mishra, R.S., Earthman, J.C. & Raj, S.V. Eds., TMS, Warrendale, PA. p. 377.
889. Huang, Y., Hutchinson, J.W. & Tvergaard, V. (1991) *J. Mech. and Phys. Sol.*, **39**, 223.
890. Harlow, D.G. & Delph, T.G. (2000) *J. Eng. Mater. and Tech.*, Trans. ASME, **122**, 342.
891. Sherry, A.H. & Pilkington, R. (1993) *Mater. Sci. and Eng.*, **A172**, 51.
892. Ai, S.H., Lupinc, V. & Maldini, M. (1992) *Scripta Metall. Mater.*, **26**, 579.
893. Nix, W.D., Matlock, D.K. & Dimelfi, R.J. (1977) *Acta Metall.*, **25**, 495.
894. Landes, J.D. & Begley, J.A. (1976) *Mechanics of Crack Growth*, ASTM STP 590, ASTM, 128.
895. Wiesner, C., Earthman, J.C., Eggeler, G. & Ilschner, B. (1989) *Acta Metall.*, **37**, 2733.
896. Staley, J.T. & Saxena, A. (1990) *Acta Metall. et Mater.*, **38**, 897.
897. Tabuchi, M., Kubo, K. & Yogi, K. (1993) *Creep and Fracture of Engineering Materials and Structures*, Wilshire, B. & Evans, R.W. Eds., Inst. Metals, London, p. 449.
898. Raj, R. & Baik, S. (1980) *Metal. Sci.*, **14**, 385.
899. Churley, W.E. & Earthman, J.C. (1997) *Metall. Trans.*, **28A**, 763.
900. Wilkinson, D.S. & Vitek, V. (1982) *Acta Metall.*, **30**, 1723.
901. Wilkinson, D.S. (1981) *Mater. Sci. Eng.*, **49**, 31.
902. Miller, D.A. & Pilkington, R. (1980) *Metall. Trans.*, **11A**, 177.

Index

α, 36, 43, 49, 65, 71–74, 85, 87, 88, 92, 99, 101, 105, 184, 185, 189, 206, 233
α-Brass, 87, 88, 227
αTi$_3$Al, 177
304 Austenitic stainless steel, 34, 45, 49
Activation energy, 180, 182, 184, 185, 194, 203, 205–208
Activation energy for creep, 13, 15, 21–23, 164, 203, 206
Activation volume, 15, 60
Ag, 23, 73, 113, 223, 228
Al, 15, 22, 23, 26–27, 36, 42, 46–47, 49–51, 53, 55–56, 58, 60, 64, 67, 69–72, 81–88, 99, 106, 111–113, 117, 119, 128–131, 136, 137, 147, 153, 155, 161, 167, 169, 176–180, 184, 185, 189–202, 207–210, 218, 223
Al-5.8at%, 46
Al$_3$Ti, 175
Alloy additions, 183
Alloying elements, 206
Alpha, 73, 74
Aluminum, 7, 8, 14–16, 21–24, 26–27, 29, 33, 36, 42, 45, 49–51, 55, 69, 71–72, 74–76, 78–80, 82–85, 87, 99–101, 103, 113–115, 144, 216, 217, 234, 237
Anomalous yield, 191
Anomalous yield point phenomenon, 190
APB, 191, 192, 199, 203

B2, 173, 189–192, 208
Backstresses, 181
Bauschinger effect, 55–58
Boundary cohesion, 174
Brittleness, 174, 208

Carbides, 197
Cavity growth, 123, 227–237, 266
Cavity nucleation, 213, 217, 219, 221, 222, 224, 225, 227–230, 237, 240, 266
Class I (A) alloys, 113, 115, 117

Climb locking mechanism, 191
Climb-controlled creep, 183
Coarsening of γ-lamellae, 182
Coble, 8, 9, 20, 82, 91–92, 95, 99, 203, 206, 211, 217
Coble creep, 203, 206, 211
Coherent particles, 149, 151, 157, 158, 163, 165, 168, 169
Constant strain-rate, 3–5, 7, 14–15, 45, 47, 49–50, 52, 53, 74–77, 208
Constant stress, 3, 5, 7, 14, 31, 47, 49, 50, 74–76, 94, 114, 200, 209, 216
Constant structure stress-sensitivity exponent, N, 4
Constant-structure creep, 180, 197
Constitutional vacancies, 192
Constitutive equations, 185
Constrained diffusional cavity growth, 229, 230, 232, 234, 235
Continuous reactions, 81, 128, 147, 148, 150
Continuous recrystallization, 147, 148, 194
Copper, 27–28, 60, 82, 92, 95, 222
Crack formation, 188
Creep crack growth, 231, 237
Creep Curves, 199
Creep Fracture, 8, 215, 216, 223, 237
Creep life, 188
Creep mechanisms, 194
Creep transient, 40, 48, 52, 53, 51, 113, 167
Cross slip, 191
Cu, 23, 41, 55–56, 59–60, 71, 73, 82, 83, 88, 92, 119, 146, 164, 217–219, 227, 228
Cube cross-slip, 201, 202

Detachment model, 158, 207
Detachment-controlled dislocation creep, 211
Diffusion, 208
Diffusional creep, 194
Discontinuous dynamic, 42, 145
Discontinuous dynamic recrystallization, 194

Dislocation climb, 173, 179–184, 192, 194, 203, 207–209, 211
Dislocation density, 26, 30–33, 35–36, 41, 44–46, 49, 51, 53, 55–56, 66, 68, 71–77, 81–83, 88, 100–104, 106, 113, 115, 119, 134, 147, 167, 182, 202
Dislocation glide, 180, 181
Dislocation hardening equation, 44, 101
Dislocation networks, 64
Dislocation pile-ups, 181
Dislocation pinning, 191
Dislocation slip, 184
Dislocation-particle interactions, 197
Disordering temperature, 191
Dispersion-strengthened, 87, 118, 151, 161, 162, 170, 201, 206, 211, 215, 217, 222–223, 237, 244
Dispersions, 197
Dissolution of α_2 lamellae, 182
DO_{19}, 173
DO_{22}, 173
DO_{23}, 173
DO_3, 173, 189–191
Duplex, 177, 178, 186
Dynamic recrystallization, 180

ECAP, 129, 139
Elastic Modulus, 20–21, 177

Fe, 36, 39, 49, 51, 73, 85, 87, 92, 99, 101, 113, 119, 147, 155, 173, 188–197, 204, 206, 209
Fe_3Al, 188
FeAl, 119, 173, 188–194, 197
Fine structure superplasticity, 123, 124, 127
Five-Power-Law Creep, 8, 9, 15, 17, 21, 24, 26, 28, 30, 58, 60–61, 63, 65, 67, 69, 74, 77–79, 81–82, 231
Frank network, 31–32, 35–36, 64, 69–71, 75, 104, 143
Fully lamellar, 176–178, 186

γ-TiAl, 176–180, 184, 186
GB ledges, 218

Geometric dynamic recrystallization, 50, 80–81, 146
Geometrically necessary boundaries, 81
Glide-controlled creep, 181, 183
Grain growth, 194
Grain refinement, 194
Grain size, 8, 26, 33, 35, 45, 72, 79, 82, 83, 91, 94–96, 99–102, 119, 121, 124, 127–128, 130–134, 136, 139, 143, 146, 168, 179–181, 188, 205, 209, 211, 231
Grain-boundary, 43, 65, 91, 92–94, 99, 123, 124, 127, 129, 131–133, 135, 136, 139, 168, 215, 218, 221, 228, 231, 234, 237, 241
Grain boundary morphology, 188
Grain-boundary sliding, 94–95, 123–124, 127, 129, 131–133, 135, 139, 140, 165, 185, 194, 215, 218, 221, 229

Hall–Petch, 35, 43–45, 67, 82, 184, 206, 84, 85
Hard orientations, 184
Harper–Dorn Creep, 20, 26, 29–30, 95, 99–107, 194
Herring, 8, 20, 91–92, 95, 99–103, 207, 211
High strain, 128
High Stress-High Temperature Regime, 179
Hyperbolic sine, 13, 18, 26

Incoloy, 165–167
Instantaneous strain, 187
Interface boundary sliding, 187
Interfacial (Shockley) dislocations, 184, 187
Interfacial sliding, 184, 187
Interlocking grain boundaries, 184
Intermetallics, 173–175, 178, 189, 192, 200, 201, 203
Internal stress, 26, 41, 43, 56–59, 67–68, 70, 84, 95, 101–102, 115, 123
In situ, 47, 50, 58, 70, 131, 147, 161, 223
Inverse creep, 200
Iron aluminides, 174, 175, 188, 189, 192, 194–198

Jog height, 182
Jogger-screw creep model, 182
Jogs, 17, 61–62, 69, 105, 182

Kear-Wilsdorf (KW) locks, 201

L'1$_2$, 173
L10, 176
L12, 198
L1$_0$, 173, 175
L1$_2$, 173, 175, 199
L2$_1$, 173
Lamellar grains, 176, 177, 188
Lamellar interface spacing, 180, 181, 183–185
Lamellar interfaces, 180, 187
Lamellar orientation, 184
Larson–Miller, 218
Lattice diffusion, 180
Lattice order, 197
Ledges, 183
Ledge dislocations, 187
Ledge motion, 183
Long-range order, 173, 174
Low stress regime, 184

Mechanically alloyed, 124, 128, 134–136, 210
Metadynamic recrystallization, 144
Metal–matrix composites, 120, 123, 127–134, 136, 138
Minimum strain rate, 178, 188
Misfit dislocations, 183
Misorientation dislocations, 183
Mobile, 30, 36, 42, 49, 62–63, 74–76, 81, 104, 106, 115, 119, 128, 134, 192, 193, 202, 218
Monkman Grant, 13, 216, 218, 223, 224, 227–231
MoSi$_2$, 173

Nabarro–Herring creep, 9, 19, 95, 97, 100, 101, 203, 206

NaCl, 22, 23
Narrow lamellae, 181
Natural three-power-law, 29
nearly lamellar, 177, 178
Ni$_2$AlTi, 172
Ni$_3$Al, 119, 136, 137, 160, 171, 173, 198, 199–201, 203, 205, 208
Ni$_3$Ga, 201
NiAl, 173, 199, 203, 208, 211
Nickel, 5, 92, 174, 175, 189, 197, 199, 207, 208
Nickel aluminides, 174, 175, 198
Nickel superalloys, 174
Nimonic, 164, 166, 167, 223, 225
ñ-Model, 207

Octahedral planes, 198
Octahedral slip systems, 202
Off-stoichiometry, 206
Orientation, 204
Orowan bowing stress, 152, 154, 155, 158, 159, 169
Orowan stress, 181
Oxide dispersion strengthened (ODS), 151, 152, 155, 158, 159, 161, 164, 206, 207, 208, 210, 211
Oxide dispersions, 197
Oxide-dispersion strengthened, 206

Particle-dislocation, 197
Particle-strengthened alloys, 162, 166
Peierls stresses, 174
Phase transformation, 189
Power law models, 185
Precipitation, 187
Precipitation hardening, 197
Precipitation strengthening, 206
Primary creep, 3, 14, 24, 47–52, 74, 76, 113, 177, 186–188, 200, 201

Rate-controlling creep mechanisms, 174, 178, 184
Recovery, 180, 197
r-Type, 215, 216

Second phases, 197
Secondary creep, 3, 14, 52, 177, 182, 184, 188
Shockley dislocations, 199
Sigmoidal (or inverse) creep, 87, 88, 200–203
Sigmoidal creep, 200–202
Sn, 22, 24, 49, 123
Soft orientations, 184
Solid-solution strengthening, 197
Stacking fault energy, 24, 25, 33, 71, 81, 92, 169
Stage I, 3, 14, 73, 116
Stage II, 3, 14, 73, 219
Stage III, 3, 116, 145, 215, 218
Static recrystallization, 144–145
Steady-state creep, 3, 7, 13–16, 18, 20, 23–26, 30, 39, 43, 50, 71, 82, 87, 99, 102, 111, 114, 116, 154, 164, 165, 203, 216, 217, 228, 231, 238
Steady-state stress exponent, n, 7
Stiffness, 189
Strength, 189
Strengthening mechanisms, 197, 210
Stress anomaly, 193
Stress exponent, 178–180, 182, 184, 185, 194, 200, 203, 205–211
Stress reduction tests, 180
Stress-drop (or dip) tests, 40
Stress-induced phase transformations, 186
Subgrain formation, 182
Subgrain misorientation, 31
Subgrain size, 31, 33, 35, 39, 41, 44–46, 49–51, 55, 68, 71, 79–83, 85, 87, 104, 167, 169, 180
Super-partial dislocations, 199
Superalloy, 152, 159, 162, 163, 169, 174, 175, 198, 199, 206–208, 210
Superdislocations, 190, 191
Superlattice, 173
Superplasticity, 8, 15, 26, 117, 119, 121, 139, 194, 196

Taylor factor, 14, 24, 45, 78–79, 184
TD–Nichrome, 167
Tertiary creep, 3, 13, 92, 188, 200, 201, 218
Texture, 14, 77–79, 81, 126, 128, 145, 147, 148, 209
Thermal vacancies, 192, 193
Threading dislocation lines, 184
Threading dislocations, 184
Three-Power-Law, 29–30, 61, 111–113, 115–117
Threshold stress, 92, 125, 131–136, 138, 158, 160–162, 164, 167, 169, 180, 187, 197, 207, 221
Ti_3Al, 119, 173, 175, 176, 177
TiAl, 119, 173, 175, 182, 184–187, 201
Titanium aluminides, 174, 175
Torsion, 14, 24, 39, 49, 77–81, 129, 146, 149, 225
Transient strain, 187
Twinning, 187, 188

Vacancy hardening mechanism, 192
Viscous drag, 194
Viscous glide, 185, 200
Viscous glide creep, 8, 112, 199
Viscous glide of interfacial dislocations, 185

Wedge, 215
Wider lamellae, 181

X-ray peaks, 59

Yield stress anomaly, 201

Zirconium, 28, 93, 218

Printed in the United States
99585LV00004B/7/A